Sensors for Marine Biosciences

Next-generation sensing approaches

Online at: https://doi.org/10.1088/978-0-7503-5999-3

IOP Series in Sensors and Sensor Systems

The IOP Series in Sensors and Sensor Systems includes books on all aspects of the science and technology of sensors and sensor systems. Spanning fundamentals, fabrication, applications and processing, the series aims to provide a library for instrument and measurement scientists, engineers and technologists in universities and industry.

The series seeks (but is not restricted to) publications in the following topics:
- Advanced materials for sensing
- Biosensors
- Chemical sensors
- Industrial applications
- Internet of Things (IoT)
- Lab-on-a-chip
- Localization and object tracking
- Manufacturing and packaging
- Mechanisms, modelling and simulations
- Microelectromechanical systems/nanoelectromechanical systems
- Micro and nanosensors
- Non-destructive testing
- Optoelectronic and photonic sensors
- Optomechanical sensors
- Physical sensors
- Remote sensors
- Sensing for health, safety and security
- Sensing principles
- Sensing systems
- Sensor arrays
- Sensor devices
- Sensor networks
- Sensor technology and applications
- Signal processing and data analysis
- Smart sensors and monitoring
- Telemetry

Authors are encouraged to take advantage of electronic publication through the use of colour, animations, video and interactive elements to enhance the reader experience.

A full list of titles published in this series can be found here: https://iopscience.iop.org/bookListInfo/iop-series-in-sensors-and-sensor-systems.

Sensors for Marine Biosciences

Next-generation sensing approaches

Edited by
Shyam S Pandey
Graduate School of Life Science and Systems Engineering, Kyushu Institute of Technology, Wakamatsu, Kitakyushu, Japan

Rout George Kerry
Department of Biotechnology, Utkal University, Bhubaneswar, Odisha, India

Kshitij RB Singh
Graduate School of Life Science and Systems Engineering, Kyushu Institute of Technology, Wakamatsu, Kitakyushu, Japan

IOP Publishing, Bristol, UK

To those who care, conserve, protect but do not spoil/destroy the beauty and unique characteristics of Kshit (Earth), Jal (Water), Pawak (Fire), Gagan (Sky), and Sameera (Air).

Contents

7 Biosensor-based detection of major aquatic pathogens in the marine ecosystem

Pooja Singh and Ravindra Pratap Singh

8 Advancement of sensors in preservation and packaging of marine products

Ekta Poonia, Rohit Ranga, Heena Dahiya, Vinita Bhankar and Krishan Kumar

9   Nucleic acid-based biosensor for detection of infectious pathogens in marine products   9-1

Kanishk Singh, Parshant Kumar Sharma, Yi-Hsiang Huang, Sucharita Khuntia, Getaneh Berie Tarekegn, Wei-Chen Huag and Li-Chia Tai

Preface

Monitoring the emerging rate of pollution and transmission of multiple life-threatening pathogens via the aquatic ecosystem is critical for survival. As about 71% of Earth's surface is covered with water and 96.5% of it comes from the marine ecosystem, it is unequivocally safe to state that any threat to the marine ecosystem would directly and inherently influence the survivability of any species on Earth. Thus, sustainable development of the human species would be a myth, if the current and upcoming challenges and threats towards the marine ecosystem are not addressed in due time. Now at this point of time humanity is at the best of its ability in innovation and discoveries, where the applications of sensors have been proven to be instrumental. Their preciseness and consistency as well as the diversified customizable application in heterogeneous fields of science have revolutionized the perception of researchers in the present decade. The genesis of addressing any particular challenges in the present decade would not make sense without the foresight of a sensor. Therefore, to cordially and cautiously understand the application of sensors in safeguarding the marine ecosystem and its future prospects for the sustainability of mankind is described in eleven consecutive chapters in the book entitled *Sensors for Marine Biosciences: Next-generation sensing approaches.*

As an introduction to sensors, chapter 1 provides a basic and brief outline and application of sensors in monitoring the marine ecosystem. This is followed by chapter 2, where monitoring of major aquatic pollutants, their deposition and their impacts on aquatic ecosystem is covered. Further, chapter 3 provides the possible advancements in sensors for the detection of specific marine organic and inorganic pollutants that are critical for the sustenance of the marine ecosystem. Chapter 4 orchestrates the innovations and advancements of polymeric matrix-based biosensors for effective detection. Chapter 5 describes the possible roles of remote sensing and satellite biological sensors in monitoring and comprehending the diversified marine ecosystem and its deterioration by pollution. The advancement of computer science specifically in the section of artificial intelligence (AI), machine learning (ML) and Internet of Things (IoTs) is critical for the evolution of sensors, thus chapter 6 presents an interesting prospective of advanced cybertic sensor for the efficient detection of marine pollutants. Chapter 7 is an introduction of various biotic stress factors of aquatic ecosystems, which directly or indirectly affects the well-being of humanity. Chapter 8 provides the sensor-based innovations made for the viable preservation and packaging of marine products for commercialized or domestic purposes. Chapter 9 foresees the indisputable and profound application of biological macromolecule-based sensors for effective and precise detection of infectious pathogens in marine products such as fish, crab, prawn, etc. to shield from the possibility of zoonoses. Additionally, the horizon of sensors and their application in marine biology can be further broadened via the amalgamation of now emerging proteomics and genomics tools and evolving gene editing tools which is briefly presented in chapter 10. Finally, chapter 11 is dedicated towards commercialization and the present and future development of sensors in marine aquaculture. Hence, this book provides a brief outlook on how sensors could be beneficial in both shielding the marine ecosystem from pollutants and simultaneously safeguarding the interest and well-being of humans as a whole. From an educational and research point of view the book will be of great assistance for students, professionals/practitioners, scientists, researchers, and academicians in various research domains.

Shyam S Pandey
Rout George Kerry
Kshitij RB Singh

Acknowledgements

It gives us immense pleasure to acknowledge our affiliated institutions for providing constant assistance in all the possible ways. It is also our great pleasure to acknowledge and express our enormous debt to all the contributors who have provided their quality material to prepare this book. We are grateful to our beloved family members, who joyfully supported and stood with us in many hours of our absence to finish this book project. Thanks are also due to Dr John Navas, Phoebe Hooper, and the entire publishing team for their patience and extra care in publishing this book.

Shyam S Pandey
Rout George Kerry
Kshitij RB Singh

Editor biographies

Shyam S Pandey

Professor Pandey completed his PhD from, the National Physical Laboratory, New Delhi, India in 1997 in the area of synthesis, characterization and application of organic conjugated polymers. He came to Japan as a post-doctoral fellow in 1998. He worked as a Fukuoka IST-sponsored post-doctoral fellow at the Kyushu Institute of Technology from 1998 to 2001 in the area of photo-functional materials and devices. He was JSPS post-doctoral fellow from 2001 to 2003 (soft actuators and artificial muscles) and Knowledge Cluster invited researcher from 2003 to 2007 (Protein Biochips). He received the National Technology Award from the National Research Development Corporation, Government of India in 2005 for the development of glucose biosensors, which are currently being manufactured and marketed. He worked at Kyushu Institute of Technology as an Assistant Professor from 2009 to 2012 in the area of design and development of novel photofunctional materials for their application in the area of next-generation solar cells. As an Associate Professor, he started his Organic Photofunctional Materials Devices research group in the same institute from April 2012–March 2022 and worked in the area of research and development of next generational cells, organic electronic devices and bio-image sensors. Currently, he is a Professor in the Green Electronics department of the Graduate School of Life Science and Systems Engineering, Kyushu Institute of Technology. He has published more than 300 papers in international refereed journals, more than 500 papers in domestic and international conferences and about 30 patents in India, Japan, Europe and the USA. His research interests deal with dye-sensitized and organic solar cells, quantum chemical calculations, organic electronics and optoelectronics, organic conducting polymers, biosensors and protein biochips.

Rout George Kerry

Dr Kerry is an accomplished biotechnologist with a PhD from Utkal University, specializing in nanotechnology-based drug delivery systems for treating diseases such as type-II diabetes and neuro-degenerative disorders. Currently, he is a Research Associate at the Centre for Biotechnology, Siksha 'O' Anusandhan University, and has previously served as a Senior Research Fellow at Odisha University of Agriculture and Technology (OUAT). His academic journey includes a Master's degree in biotechnology from Berhampur University, where he graduated with distinction, and a Bachelor's degree from Utkal University. Dr Rout's professional career spans over a decade, during which he has contributed to both teaching and cutting-edge research. He has authored 15 first-authored international peer-review articles, and corresponded and co-authored numerous other international chapters, with his work appearing in high-impact journals such as *Phytomedicine* and *Environmental Research*. His research spans diverse areas, including nanotechnology, molecular biology, and bioinformatics, with a specific focus on using nanoparticles for therapeutic and diagnostic applications. In addition to his research publications, Dr Rout has contributed to 15 book chapters and edited several books, including *Biotechnological Advances for Microbiology*, *Molecular Biology*, and *Nanotechnology* (2021). His work has earned him several accolades, including being listed in the AD Scientific Index's World Scientist and University Rankings 2024. Mr Rout is also an invited Research Fellow at INTI International University, Malaysia, and holds a professional membership with the Institute de Diplomatie Publique, UK. His expertise extends to mentoring students and conducting workshops on biotechnology, making him a key figure in his field.

Kshitij RB Singh

Mr Singh is a postgraduate in biotechnology from Indira Gandhi National Tribal University, Amarkantak, Madhya Pradesh, India. He is currently working in the laboratory of Professor Shyam S Pandey, Graduate School of Life Science and Systems Engineering, Kyushu Institute of Technology, Fukuoka, Japan. He has more than 80 peer-reviewed publications to his credit, has edited 15 books, and has authored more than 60 book chapters published by internationally reputed publishers, namely American Chemical Society, Royal Society of Chemistry, Elsevier, IOP Publishing, Springer Nature, Wiley, and CRC Press. He has also been involved in editing books with international publishing houses, including IOP Publishing, Elsevier, Wiley, and Springer Nature. His research interests include biotechnology, biochemistry, nanotechnology, nanobiotechnology, biosensors, and materials sciences.

List of contributors

Charles O Adetunji
Applied Microbiology, Biotechnology and Nanotechnology Laboratory, Department of Microbiology, Edo State University Uzairue, Iyamho, Edo State, Nigeria

Tinku Basu
Amity Centre for Nanomedicine, Amity University, Noida 201313, Uttar Pradesh, India

Vinita Bhankar
Department of Biochemistry, Kurukshetra University Kurukshetra, Haryana 136119, India

Dhiraj Bhatia
Department of Biological Sciences and Engineering, Indian Institute of Technology Gandhinagar, Palaj, Gandhinagar, Gujarat 382355, India

Ravi Butola
University School of Automation and Robotics (USAR), GGSIP University, East Delhi Campus, Delhi 110092, India

Prakash Chandra
Department of Biotechnology, Delhi Technological University, New Delhi 110042, India

Chansi
Amity Centre for Nanomedicine, Amity University, Noida 201313, Uttar Pradesh, India

Heena Dahiya
Department of Chemistry, Baba Mastnath University, Asthal Bohar, Rohtak, Haryana 124021, India

Asmita Das
Department of Biotechnology, Delhi Technological University, New Delhi 110042, India

Adil Denizli
Department of Chemistry, Hacettepe University, Beytepe 06800, Ankara, Turkey

Nida E Falak
Department of Biotechnology, Delhi Technological University, New Delhi 110042, India

Shweta Gulia
Department of Biotechnology, Delhi Technological University, New Delhi 110042, India

Karan Hadwani
Amity Centre for Nanomedicine, Amity University, Noida 201313, Uttar Pradesh, India

Wei-Chen Huag
Department of Electrical and Computer Engineering, National Yang Ming Chiao Tung University, Hsinchu 30010, Taiwan

Yi-Hsiang Huang
Department of Electrical and Computer Engineering, National Yang Ming Chiao Tung University, Hsinchu 30010, Taiwan; Institute of Electrical and Control Engineering, National Yang Ming Chiao Tung University, Hsinchu 30010, Taiwan

Rout George Kerry
Department of Biotechnology, Utkal University, Vani Vihar, Bhubaneswar 751004, Odisha, India; Centre for Biotechnology, Siksha 'O' Anusandhan (Deemed to be University), Kalinga Nagar, Bhubaneswar 751003, Odisha, India

Sucharita Khuntia
Department of Electrical and Computer Engineering, National Yang Ming Chiao Tung University, Hsinchu 30010, Taiwan; Institute of Electrical Engineering and Computer Science, National Yang Ming Chiao Tung University, Hsinchu 30010, Taiwan

S. Jone Kirubavathy
Department of Chemistry, PSGR Krishnammal College for Women, Coimbatore 641 0014, Tamil Nadu, India

Krishan Kumar
Physical Chemistry Research Laboratory, Department of Chemistry, D. C. R. University of Science and Technology, Murthal, Sonipat, Haryana 131039, India

Cansu İlke Kuru-Sumer
Graduate School of Natural and Applied Sciences, Department of Biotechnology, Ege University, 35040, İzmir, Türkiye; Buca Municipality Buca Science and Art Center, 35390, Izmir, Türkiye

Arunadevi Natarajan
Department of Chemistry, PSGR Krishnammal College for Women, Coimbatore 641 0014, Tamil Nadu, India

Sanghamitra Nayak
Centre for Biotechnology, Siksha 'O' Anusandhan (Deemed to be University), Kalinga Nagar, Bhubaneswar 751003, Odisha, India

Chioma A Ohanenye
Department of Anatomy, Rhema University Aba, Abia State, Nigeria

Gloria E Okotie
Laboratory for Reproductive Biology and Developmental Programming, Department of Physiology, Rhema University Aba, Abia State, Nigeria

Olugbemi T Olaniyan
Laboratory for Reproductive Biology and Developmental Programming, Department of Physiology, Rhema University Aba, Abia State, Nigeria

Merve Asena Özbek
Department of Chemistry, Hacettepe University, Beytepe 06800, Ankara, Turkey

Shyam S Pandey
Graduate School of Life Science and Systems Engineering, Kyushu Institute of Technology, 2–4 Hibikino, Wakamatsu, Kitakyushu 808-0196, Japan

Ekta Poonia
Physical Chemistry Research Laboratory, Department of Chemistry, D. C. R. University of Science and Technology, Murthal, Sonipat, Haryana 131039, India

Rohit Ranga
Physical Chemistry Research Laboratory, Department of Chemistry, D. C. R. University of Science and Technology, Murthal, Sonipat, Haryana 131039, India

Parshant Kumar Sharma
RFIC Bio Centre, Department of Electronics Engineering, Kwangwoon University, 20 Kwangwoon-ro, Nowon-Gu, Seoul 01897, Republic of Korea; Department of Life Sciences Imperial College London, UK; NDAC Centre, Kwangwoon University, 20 Kwangwoon-ro, Nowon-Gu, Seoul 01897, Republic of Korea

Kanishk Singh
Department of Electrical and Computer Engineering, National Yang Ming Chiao Tung University, Hsinchu 30010, Taiwan; Institute of Electrical and Control Engineering, National Yang Ming Chiao Tung University, Hsinchu 30010, Taiwan

Pooja Singh
Department of Biotechnology, Indira Gandhi National Tribal University, Amarkantak, Madhya Pradesh 484887, India

Bhupendra Pratap Singh
Department of Environmental Studies, Central University of Haryana, Jant-Pali, Mahendergarh, Haryana 123031, India

Ravindra Pratap Singh
Department of Biotechnology, Indira Gandhi National Tribal University, Amarkantak, Madhya Pradesh 484887, India

Simran Singh
Department of Biotechnology, Delhi Technological University, New Delhi 110042, India

Kshitij RB Singh
Graduate School of Life Science and Systems Engineering, Kyushu Institute of Technology, 2–4 Hibikino, Wakamatsu, Kitakyushu 808-0196, Japan

Li-Chia Tai
Department of Electrical and Computer Engineering, National Yang Ming Chiao Tung University, Hsinchu 30010, Taiwan; Institute of Electrical and Control Engineering, National Yang Ming Chiao Tung University, Hsinchu 30010, Taiwan

Getaneh Berie Tarekegn
Department of Electrical and Computer Engineering, National Yang Ming Chiao Tung University, Hsinchu 30010, Taiwan; Institute of Electrical and Control Engineering, National Yang Ming Chiao Tung University, Hsinchu 30010, Taiwan

Aykut Arif Topçu
Medical Laboratory Program, Vocational School of Health Services, Aksaray University, Aksaray 68100, Turkey

Fulden Ulucan-Karnak
Institute of Health Sciences, Medical Biochemistry Department, Ege University, 35040, Bornova, Izmir, Türkiye

Damini Verma
Centre for Nanotechnology, Indian Institute of Technology Roorkee, Roorkee, Uttarakhand, 247667, India

Young N Wike
Laboratory for Reproductive Biology and Developmental Programming, Department of Physiology, Rhema University Aba, Abia State, Nigeria

Amit K Yadav
Department of Biological Sciences and Engineering, Indian Institute of Technology Gandhinagar, Palaj, Gandhinagar, Gujarat 382355, India

Sumit K Yadav
Department of Biotechnology, Vinoba Bhave University, Hazaribagh, Jharkhand 825301, India

Gaye Ezgi Yılmaz
Department of Chemistry, Hacettepe University, Beytepe 06800, Ankara, Turkey

IOP Publishing

Sensors for Marine Biosciences
Next-generation sensing approaches
Shyam S Pandey, Rout George Kerry and Kshitij RB Singh

Chapter 1

Introduction to sensor in marine bioscience

Rout George Kerry, Kshitij RB Singh, Sanghamitra Nayak and Shyam S Pandey

This chapter introduces the importance of marine biosensors in monitoring oceanic biodiversity and addressing challenges like pollution and pathogen transmission. It highlights the role of sensors in advancing our understanding of the marine ecosystem, essential for sustainable development. Various biosensor technologies, including microbial, optical, and electrochemical sensors, are discussed for their applications in detecting pollutants, toxins, and bioavailability in marine environments. The chapter emphasizes innovations in biosensor technology and their critical role in real-time monitoring, environmental protection, and responding to marine emergencies.

1.1 Introduction

Monitoring and safeguarding the biodiversity of the ocean is imperative due to its significant impact on our lives. Furthermore, the vast potential of the ocean for industrial resources like health-related products, oil and minerals underscores the importance of continuous, real-time monitoring [1]. Developing effective marine sensors is crucial for improving our understanding of the ocean environment and contributes towards sustainable development of the marine ecosystem. It is crucial to monitor the increasing pollution levels and the transmission of various pathogens through aquatic ecosystems for survival. With water constituting 71% of the total surface of the globe and marine accounting to 96.5% of it, any threat to the ocean directly affects the survival of all species [2]. Hence, achieving sustainable human development is impossible without addressing the current and future challenges to the marine ecosystem. Today, humanity's innovative prowess has reached new heights, with sensors playing a pivotal role. Their accuracy, consistency, and versatility across various scientific disciplines have transformed researchers' perspectives in recent years [3, 4]. Tackling present-day challenges without the aid of sensors would be futile, given their indispensable role in addressing various issues.

The focus of the chapter commences with the advancement of various sensors and devices designed for use in underwater ocean applications. Faria *et al* [5] introduced a energy harvesting linear electromagnetic device specifically designed for underwater use, capable of operating within frequencies ranging between 0.1 and 0.4 Hz. Their research showcased the device's ability to generate adequate energy to replenish the sustained power supply (battery or capacitor) to the sensors. Dyomin *et al* [6] designed an underwater holographic sensor specifically for studying oceanic particles, while Martins *et al* [7] focused on developing and characterizing a polyvinylidene difluoride-based ultrasound transducer designed for use as a wireless emitter in underwater communications. In two other articles addressing this topic, Nguyen *et al* [8] employed convolutional neural networks to classify underwater sonar images for detecting submerged human bodies, while Sheng *et al* [9] proposed a bioinspired twin-inverted multiscale matched filtering technique to identify underwater moving targets [4].

Biosensors are analytical tools designed to convert biological responses into electrical signals. They must possess high specificity, be unaffected by physical factors like temperature and pH, and ideally be recyclable [10]. Cammann coined the term 'biosensor' which was defined by IUPAC. Their fabrication involves interdisciplinary research spanning chemistry, biology, and engineering, with materials falling into three categories: biocatalytic i.e., enzyme-based, bioaffinity i.e., antibodies-based, and microbe-based [11]. Pioneered by Clark and Lyons in the 1960s, biosensors encompass varied types such as DNA biosensors, immunosensors enzyme-based, tissue-based and thermal and piezoelectric biosensors [12]. Despite over 35 years of development, only glucose sensors have achieved widespread commercial success. Initially, biosensors were perceived as probes, often associated with pH or ion selective electrodes [13]. However, modern definitions include sensors integrated into automated instruments, with some advocating for mass spectrometric, chromatographic, or electrophoretic sensor components [14]. Notably, sensors for pH, temperature, pressure, electrocardiograms and similar parameters are not considered biosensors, while surface plasmon resonance (SPR) devices and labelled nanoparticles within cells are recognized [15].

A physical transducer in a biosensor combines with an element of biological recognition to generate a quantifiable signal proportional to the concentration of the target analyte. In this setup, the biological material detects the targeted compound, and the biological response is converted into a detectable signal by the transducer, which can be measured using a number of methods i.e., optical, electrochemical, acoustic, mechanical, calorimetric, or electronic techniques. Since the development of the first glucose detecting biosensor by Clark and Lyons in 1962, biosensors have been widely explored and used in various applications, including environmental monitoring, food safety, homeland security, and public health [11]. Different recognition elements of biological entity, such as microorganisms, cells from higher organisms, tissues, organelles, enzymes, cofactors, and antibodies, have been employed in biosensor fabrication. Enzymes are particularly popular due to their high specificity and sensitivity, although their purification can be expensive and time-consuming. Microbes, such as bacteria, algae and yeast, offer an alternative as they can be mass-produced and are easier to manipulate, providing simplification in biosensor fabrication and enhancing performance. In the case of microalgae-based biosensors for water toxicity testing, the colonial nature of the microalgae used eliminates the need for immobilization procedures, making the preparation of each analysis quicker and more straightforward [12]. These biosensors exhibit sensitivity to specific pollutants, but variations in sensitivity may occur due to biological factors, such as cell wall adsorption characteristics and metabolic pathways, or chemical factors, such as the existence of competing agents and pH of the experiment's medium [13].

The biochemical and biological processes within the ocean are crucial for the global economy as well as climate regulation. Human actions have significantly altered ocean conditions, leading to phenomena such as acidification, warming, deoxygenation, pollution, eutrophication, reduced nutrient flux, habitat loss, decline in fisheries and biodiversity. Recent observational data indicate accelerated ocean warming, contributing to glacier and ice cap melting, rising sea levels, decreased primary production, and alteration in ocean stratification in polar regions [14]. To understand these complex ocean dynamics, regional and global *in situ* observing networks have been established, including initiatives like the Biogeochemical-Argo program (BGC-Argo), the Global Ocean Observing System (GOOS), the US Integrated Ocean Observing System (IOOS), the Australian Integrated Marine Observing System (IMOS), the Ocean Observatories Initiative (OOI) and the EXPORTS project [15]. In coastal regions, legislative measures are in place to reduce the transport of pollutants into the marine ecosystem, such as the Marine Strategy Framework Directive (MSFD) aiming for the 'good environmental status of the EU's marine waters' and the Water Framework Directive (WFD) aiming for 'good ecological status.' *In situ* sensing technology has played a vital role in coastal monitoring efforts and oceanography [15, 16].

Currently, sensor networks claim to be the most promising technique for acquiring vertically and temporally detailed information of biogeochemical processes across the oceanic ecosystem [22], real-time information for decision-making in the coastal regions [23, 24], validating airborne and environmental observations [25], generating data of high-frequency and big data for predictive models [26]. Sensors of the marine ecosystem are subjected to adverse conditions and extended deployments must be economically feasible. The sensors must be resilient, energy-conserving, and fitted with effective antifouling guards to endure the hostile marine environment [27]. There are plenty of *in situ* sensors, ranging from commercial to research-based, which possess varying levels of technological competence. The focus of this review is investigating current information on the prevailing *in situ* technologies, whether at the prototype level, laboratory or commercial stages, and suited for implementation in the marine environment. Therefore, (a) this article reviews current sensors used for coastal and marine monitoring, emphasising those mostly accessible with a comprehensive list of sensors and manufacturers. Modern advancements have been discussed, including novel breakthroughs in hybrid technology, wet chemistry–based, and optical-based detectors. (b) It presents an overview of technologies and modern resources utilized in the manufacturing of sensors and vital approaches used for attaining marine rating, robustness and cost effectiveness.

A microbe-based biosensor contains an immobilized transducer with microbial cells either live or non-living. Non-living cells, typically achieved after permeabilization, or whole cells containing periplasmic enzymes, are commonly used as a cost-effective alternative to isolated enzymes [17]. The metabolic activity and respiratory functions of the cell are exploited by viable cells, and the target analyte acts as an inhibitor or a substrate of such processes. Bioluminescence microbial biosensors use genetically modified microorganisms, where the lux gene is fused with an inducible gene to serve as a promoter for toxicological and bioavailability testing [18].

Recent microbial biosensors can be classified into two main categories on the basis of their sensing techniques: optical and electrochemical. Amperometry is extensively used in electrochemical microbial biosensors, particularly in environmental applications. However, it is tedious and not ideal for online monitoring. Meanwhile, potentiometric, conductometric, voltametric transducers, and microbial fuel cell (MFC)-based systems are employed in other electrochemical microbial biosensors. Conductometric microbial biosensors offer rapid and sensitive responses to analytes, with examples including biosensors employing *Chlorella vulgaris* microalgae for detecting ionized heavy metal as well as pesticides in water samples [17].

The use of electrochemical microbial biosensors for detecting marine biotoxicity, particularly with a focus on *Staphylococcus aureus*, is now attracting the attention of researchers worldwide (figure 1.1) [19]. It highlights the advantages of integrated microbial sensors (IMS) over traditional dispersed systems, emphasizing their user-friendliness, efficiency, and ability to provide real-time monitoring (figure 1.2). Challenges such as high salinity and complex marine environments can affect electrochemical readings. However, the biosensors demonstrated a robust capability for detecting toxic contaminants, and effective techniques for evaluating biotoxicity in

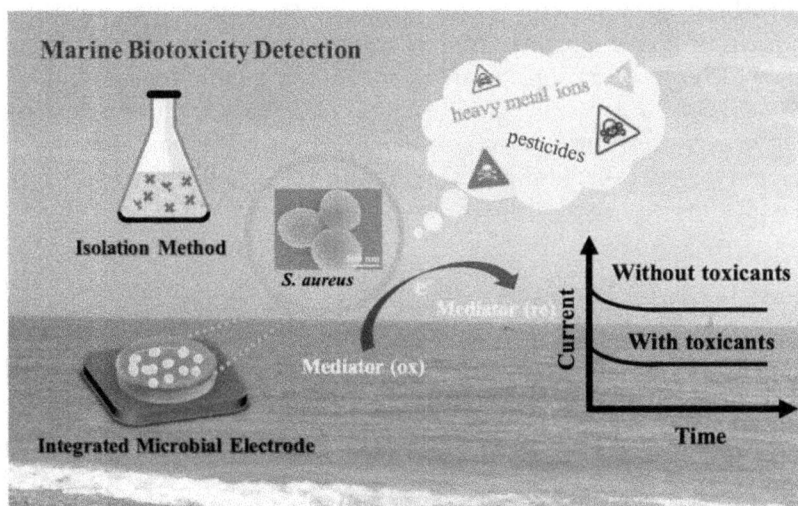

Figure 1.1. Microbial electrochemical biosensors for marine detection of toxicants and pollutants in marine environment. Reprinted from Liu *et al* [20] with permission from Elsevier.

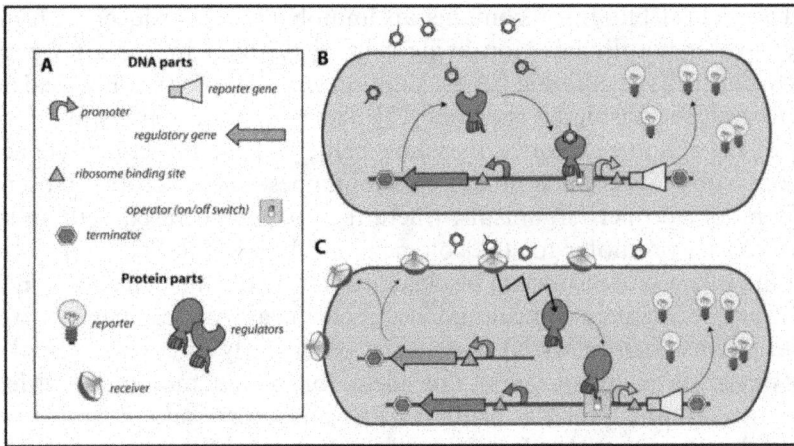

Figure 1.2. A bacterial sensor-reporter cell operates by detecting specific environmental signals through a sensor component and then producing a measurable output with a reporter component. (a) A sensor-reporter circuit's DNA components are made up of different portions. Both the reporter gene and the regulatory gene have distinct roles; the reporter gene serves as the system's output, while the regulatory gene detects the target. The regulation of gene expression involves additional factors. (b) The sensor operates through the binding of a regulatory protein to a target molecule in a single-regulatory protein mechanism. A reporter gene's transcription factor is activated by this binding, which amplifies the signal. (c) The receptor protein identifies the target molecule in a method that combines sensor and regulatory functions. It then initiates a signaling cascade to activate regulatory proteins. The target gene is expressed as a result of this activation. Reprinted form Tecon and van der Meer [26] CC BY 4.0.

real seawater samples were established. Overall, IMS could significantly enhance marine pollution monitoring and provide timely responses to environmental emergencies [20].

The microbial biosensors involving potentiometric principles measure analyte levels by measuring the potential difference between the working and reference electrodes, which are often separated by a selective membrane. Examples include biosensors developed on the basis of pH electrodes altered by permeable *Pseudomonas aeruginosa* for the quick identification of cephalosporin antibiotics and for identifying Beta-lactam residues in milk [21, 22]. Further, these biosensors can be classified as discussed in the below sub-sections.

1.1.1 Fluorescent-based biosensors

Fluorescent-based biosensors can be grouped into two types on the basis their sensing mode: *in vitro* and *in vivo*. In the case of *in vivo* fluorescent-based microbial biosensors, there is use of bioengineered microorganisms involving a transcriptional fusion between an inducible promoter and reporter gene encoding for a fluorescent protein [22]. Among these, the gfp gene is coded for green fluorescent protein (GFP) and widely favoured due to its stability, sensitivity, and the ease with which its fluorescence detection can be done using modern optical equipment without posing any significant threat to the host system [17, 21, 23].

1.1.2 Bioluminescent biosensor

Bioluminescent biosensors detect changes in luminescence emitted by living micro-organisms, typically in response to target analytes. This change is mediated by the lux gene-encoded luciferase enzyme, which responds dose-dependently to the analyte. The lux gene expression can either be inducible or constitutive, with constitutive expression allowing for direct bioluminescence changes upon exposure to specific chemicals. For instance, there was development of a microbial biosensor with *Vibrio fischeri*-based bioluminescent to rapidly assess the toxicity of pollutants on the environment [18, 21]. Additionally, a whole-cell biosensor with tetracycline (TC) luminescent was created for swift detection of specific muscle tissue TC residue in poultry, by using EDTA and polymyxin B [22].

1.1.3 Sodium channel-based biosensors

Sodium channel-based biosensors exploit inhibitory action of PSP toxins (such as tetrodotoxin, STX, and gonyautoxin) on sodium (Na) channels for identification. For instance, a tissue biosensor used a sodium (Na) electrode wrapped in bladder membrane of frog enriched with sodium channels, detecting toxin presence by monitoring Na^+ ion transport. The biosensor's sensitivity correlated with mouse bioassay results, detecting tetrodotoxin concentrations over ten times lower than the bioassay's limit [17, 21]. Another biosensor utilized cultured mammalian neurons on a microelectrode array to detect STX and PbTx-3 toxins. Despite the several effects of toxins on nervous tissue, they lowered the average spike rate of neuronal networks. sensing limits were remarkably low, far below regulatory and bioassay limits. The biosensor also responded specifically to toxin-producing algae. While not capable of identifying individual toxins, these biosensors serve as effective screening tools [24].

1.1.4 Enzyme inhibition-based biosensors

Enzyme inhibition-based biosensors have been designed for detecting okadaic acid (OA). A biosensor utilizing PP2A inhibition and phosphate ion consumption by pyruvate oxidase (PyOx) in a system with flow injection analysis (FIA) was developed. Despite immobilizing the second enzyme, they achieved an identification limit of 0.1 ng ml^{-1} for OA, making their biosensor fifty times more susceptible than ELISA. Extensive studies are being carried out to construct a biosensor for OA sensing based on an electrochemical PP2A inhibition. Approaches include enzyme immobilization by using water-soluble photopolymer polyvinyl alcohol azide-unit-pendant (PVA-AWP) on screen-printed electrodes of carbon [25]. This strategy, simpler than previous methods, directly detects enzyme inhibition using electrochemically active PP2A substrates only after enzyme-mediated de-phosphorylation [22].

1.1.5 Aptamer-based biosensors

There is an urgent need for effective methods to detect marine toxins because of their detrimental impact on human health through seafood consumption.

Aptasensors, which use aptamers, have become advanced alternatives to traditional detection techniques, providing high sensitivity, specificity, and low detection limits. Recent progress in molecular biology, materials science, and electronics has greatly enhanced the performance of these biosensors (figure 1.3) [27]. The systematic evolution of ligands by exponential enrichment (SELEX) and the use of aptasensors for detecting various marine toxins highlight their potential to improve food safety and public health. First, aptamers with high specificity and affinity can be selected *in vitro* for any target, including tiny compounds, big proteins, and cells, allowing for the development of diverse aptamer-based biosensors. Second, once selected, aptamers can be produced with great reproducibility and purity using commercial sources. Unlike protein-based antibodies or enzymes, DNA aptamers are often chemically stable. Third, aptamers frequently undergo considerable conformational changes upon target binding. This provides significant flexibility in the design of innovative biosensors with high detection sensitivity and selectivity. Recently, a range of aptamer-based biosensors have been developed to detect pollutants. These include AuPs aptasensors, gap-based electrochemical aptasensors, photochemical aptasensors, fluorescence-based aptasensors, single-walled carbon nanotube fluorescence-based aptasensors, graphene quantum dot aptasensors, SPR-based aptasensors, microfluidic disc-based aptasensors, among others (figure 1.3) [27, 28].

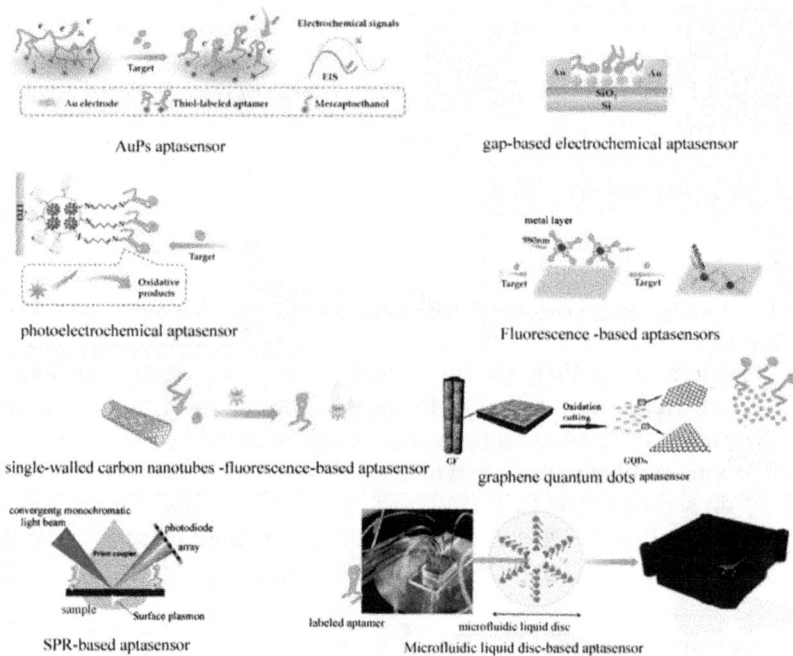

Figure 1.3. Recently developed Aptamer-based biosensor. Reprinted from Ye *et al* [27] CC BY 4.0.

1.2 Benefits of microbial biosensors in the marine environments

Recent incidents of shellfish and fish poisonings caused by various phycotoxins highlight the urgent need for effective detection technologies. While several analysis methods such as bioassays, chromatography, enzyme inhibition-based and immune-based assays have been developed, there is still a demand for easy-to-handle, fast working, and cost-effective devices capable of handling complex matrices. Biosensors emerge as promising tools, proving fast, simple, inexpensive, and as an alternative for toxicity screening or complement to conventional techniques [21]. For instance, a whole-cell sensor system using *Spirulina subsalsa* algae coupled with a Clark-type oxygen electrode has been employed to analyze seawater pollution by monitoring changes in photosynthetic activity. Similarly, a system utilizing *C. vulgaris* algae and fiber optic signal has been developed to detect atrazine, simazine, isoproturon, and diuron, achieving detection limits in the sub-parts per billion (ppb) range for photosystem II inhibitors [17]. Whole-cell sensors offer advantages in marine monitoring by providing information on bioavailability and potential physiological responses relevant to marine processes. However, their signals are usually less specific compared to enzymatic or affinity sensors, often requiring additional chemical analysis to establish the causative relationships between contaminants [22]. A significant issue in monitoring marine environments is the introduction of pollutants from drilling fluids and water produced in oil exploration platforms, which can contain various chemicals with acute or chronic toxic effects. Utilizing genetically modified microorganisms to counter to specific toxicity or stresses has been explored, but coupling these organisms to suitable transducers is necessary for them to function as biosensors [29]. Another prevalent marine issue is the sediments with antifouling agents and contamination of water, particularly tributyltin (TBT).

Despite the International Maritime Organization's (IMO) ban on tributyltin (TBT), its presence remains widespread in sediments, presenting a challenge for marine scientists and regulators for the foreseeable future. The documented effects of TBT on organisms include immediate toxicity and shell thickening in oysters, reduced endocrine disruption and recruitment of early developmental stages. Detection methods for TBT must be highly sensitive, as it can yield effects at very low concentrations [22, 30]. A bacterial bioluminescence-based assay has been designed for the identification of specific organotin compounds, with detection limits of 0.0001 μM for DBT and 0.08 μM for TBT. The application of the assay to environmental samples is still in progress and will be contingent on contamination levels relative to the bioassay's detection limit. A flow-through sensor utilizing bacteria immobilized on a chip for luminescent detection was presented at the Eighth World Congress on Biosensors [31]. Enhanced detection limits (1 nM TBT) are attributed to organotin accumulation within the immobilization matrices [22].

The biosensor's capability to identify heavy metals (Cu^{2+}, Zn^{2+}, $Cr_2O_7^{2-}$, Ni^{2+}) and pesticides demonstrates its effectiveness in evaluating biotoxicity in seawater. It is noted that salinity impacts the biosensor's performance, with IC_{50} values indicating varying levels of toxicity [26]. This method presents benefits compared to traditional analytical techniques, such as shorter preparation times and lower costs, making it well-suited for ongoing monitoring of marine environments [19, 20]. In summary, this approach enhances the detection and management of marine pollution.

1.2.1 Trace metals

In marine environments trace metals may be serve for twofold roles: they can be essential limiting trace elements, like iron for algae growth, but they also pose pollution risks [32]. Various biosensors have been developed for metal detection, including zinc and chromate luminescent bacterial sensors, a cellular biosensor utilizing algae for alkaline phosphatase activity to detect cadmium, and ampere-metric systems for heavy metal screening in soil leachates [22]. Bioavailability is essential for determining the fate and effects of metals in the environment and biosensors can address such concerns effectively compared to chemical analysis [33]. Various sensing systems, including protein-based and whole-cell approaches, have been developed for assessing metal bioavailability. Considering the vast field of trace metal analysis and speciation in marine chemistry, integrating biosensors for additional information seems like a practical approach.

1.2.2 Food safety

Algal toxin pollution in seafood is a significant concern necessitating comprehensive monitoring programs. Tissue biosensors utilizing frog-bladder membranes and sodium electrodes have been designed for precise detection of paralytic shellfish toxins [22]. Additionally, electrochemical immune sensors have been described for saxitoxin, brevetoxin, okadaic acid, and tetrodotoxin detection, utilizing various enzyme labels and electrode systems [34]. These biosensors offer diverse applications in marine science, complementing chemical methods by providing insights into interactions with living matter and offering benefits in terms of practical application, automation, or cost efficiency [35].

1.3 Recent advances in biosensor

Nernstian biosensor response enables the measurement of high concentrations of glucose without sample dilution, while potentiometric CO_2 electrodes prove effective in calibrating biochemical oxygen demand (BOD)-like biosensors with standard artificial wastewaters [36]. Fluorescence-based microbial biosensors, categorized as *in vivo* and *in vitro* types, demonstrate versatility in analytical chemistry. *In vivo* biosensors utilize microorganisms to produce fluorescent substances, whereas *in vitro* biosensors detect changes in light emission due to microbial metabolic activity [37]. Examples include *Pseudomonas putida* for detecting pollutants like BTE and 2,4-dichlorophenoxyacetic acid, *Candida tropicalis* for ethanol biosensors, and a *Gluconobacter oxydans*-based amperometric biosensor for ethanol measure-ment in an FIA system. These advancements underscore the potential of biosensors across diverse analytical applications, benefiting from their straightforward con-struction using standard molecular biology techniques.

1.4 Conclusion

In conclusion, monitoring pollution and the transmission of life-threatening pathogens through aquatic ecosystems is crucial for survival. Approximately 71% of Earth's surface is surrounded by water and 96.5% of that water is marine ecosystems; any threat to these systems has a direct impact on the survival of all species on the planet. Sustainable development of humanity hinges on addressing current and emerging challenges facing marine ecosystems. Fortunately, advancements in innovation and discoveries have equipped humanity with powerful tools such as sensors. Their precision, consistency, and customizable applications across various fields of science have revolutionized researchers' perspectives in the present decade. Indeed, addressing challenges without the foresight of sensor technology would be incomplete in today's era of innovation and discovery.

References

[1] Benedetti-Cecchi L *et al* 2018 Strengthening Europe's capability in biological ocean observations http://marineboard.eu/sites/marineboard.eu/files/public/publication/EMB_FSB3_Biological_Ocean_Observation.pdf (accessed 16 June 2024)

[2] Mishra R K 2023 Fresh water availability and its global challenge *Br. J. Multidiscip. Adv. Stud.* **4** 1–78

[3] Bhatia D, Paul S, Acharjee T and Ramachairy S S 2024 Biosensors and their widespread impact on human health *Sensors Int.* **5** 100257

[4] Gonçalves L, Martins M S, Lima R A and Minas G 2023 Marine sensors: recent advances and challenges *Sensors* **23** 2203

[5] Faria C L, Martins M S, Matos T, Lima R, Miranda J M and Gonçalves L M 2022 Underwater energy harvesting to extend operation time of submersible sensors *Sensors* **22** 1341

[6] Dyomin V, Davydova A, Polovtsev I, Olshukov A, Kirillov N and Davydov S 2021 Underwater holographic sensor for plankton studies *in situ* including accompanying measurements *Sensors* **21** 4863

[7] Martins *et al* 3991 Wideband and wide beam polyvinylidene difluoride (PVDF) acoustic transducer for broadband underwater communications *Sensors* **19** 3991

[8] Nguyen H-T, Lee E-H and Lee S 2019 Study on the classification performance of underwater sonar image classification based on convolutional neural networks for detecting a submerged human body *Sensors* **20** 94

[9] Sheng, Dong C, Guo L and Li L 2019 A bioinspired twin inverted multiscale matched filtering method for detecting an underwater moving target in a reverberant environment *Sensors* **19** 5305

[10] Joshi R 2006 *Biosensors* (Gyan Publishing House)

[11] Koyun A, Koca A E and Kara S 2012 *A Roadmap of Biomedical Engineers and Milestones* (Books on Demand)

[12] Sharma S K, Sehgal N and Kumar A 2003 Biomolecules for development of biosensors and their applications *Curr. Appl. Phys.* **3** 307–16

[13] Ma Z, Meliana C, Munawaroh H S H, Karaman C, Karimi-Maleh H, Low S S and Show P L 2022 Recent advances in the analytical strategies of microbial biosensor for detection of pollutants *Chemosphere* **306** 135515

[14] Garcia-Soto C *et al* 2008 An overview of ocean climate change indicators: sea surface temperature, ocean heat content, ocean pH, dissolved oxygen concentration, arctic sea ice extent, thickness and volume, sea level and strength of the AMOC (Atlantic Meridional Overturning Circulation) *Front. Mar. Sci.* **8** 642372

[15] Wang Z A *et al* 2019 Advancing observation of ocean biogeochemistry, biology, and ecosystems with cost-effective *in situ* sensing technologies *Front. Mar. Sci.* **6** 519

[16] Cinnirella S *et al* 2014 Steps toward a shared governance response for achieving good environmental status in the Mediterranean Sea *Ecol. Soc.* **19** 47

[17] D'Souza S F 2001 Microbial biosensors *Biosens. Bioelectron.* **16** 337–53

[18] Luo J, Liu X, Tian Q, Yue W, Zeng J, Chen G and Cai X 2009 Disposable bioluminescence-based biosensor for detection of bacterial count in food *Anal. Biochem.* **394** 1–6

[19] Hernandez-Vargas G, Sosa-Hernández J E, Saldarriaga-Hernandez S, Villalba-Rodríguez A M, Parra-Saldivar R and Iqbal H M N 2018 Electrochemical biosensors: a solution to pollution detection with reference to environmental contaminants *Biosensors (Basel)* **8** 29

[20] Liu Y, Yang Y, Fan Y, Zhao Q, Gao G and Zhi J 2023 Feasibility investigation and development of microbial electrochemical biosensors for marine pollution monitoring *Talanta* **255** 124204

[21] Su L, Jia W, Hou C and Lei Y 2011 Microbial biosensors: a review *Biosens. Bioelectron.* **26** 1788–99

[22] Balootaki P A and Hassanshahian M 2014 Microbial biosensor for marine environments *Bull. Environ. Pharmacol. Life Sci.* **3** 1–13

[23] VanEngelenburg S B and Palmer A E 2008 Fluorescent biosensors of protein function *Curr. Opin. Chem. Biol.* **12** 60–5

[24] Valina-Saba W-W, Smetazko M C, Anrather M M, Schmutzer D, Bauer S E, Schalkhammer G D and T G 1998 New highly sensitive gated ion channel biosensor for antibody detection: setup, ion channel design, and supported membranes *Proc. Volume 3199, Biomedical Systems and Technologies II BiOS Europe '97, 1997 (San Remo)* https://doi.org/10.1117/12.301103

[25] Bachan Upadhyay L S and Verma N 2013 Enzyme inhibition based biosensors: a review *Ana. Lett.* **46** 225–41

[26] Tecon R and van der Meer J R 2008 Bacterial biosensors for measuring availability of environmental pollutants *Sensors (Basel)* **8** 4062–80

[27] Ye W, Liu T, Zhang W, Zhu M, Liu Z, Kong Y and Liu S 2020 Marine toxins detection by biosensors based on aptamers *Toxins* **12** 1

[28] Song S, Wang L, Li J, Fan C and Zhao J 2008 Aptamer-based biosensors *TrAC, Trends Anal. Chem.* **27** 108–17

[29] Gray J S 1992 Biological and ecological effects of marine pollutants and their detection *Mar. Pollut. Bull.* **25** 48–50

[30] Antizar-Ladislao B 2008 Environmental levels, toxicity and human exposure to tributyltin (TBT)-contaminated marine environment: a review *Environ. Int.* **34** 292–308

[31] Bryan G W and Gibbs P E 1991 Impact of low concentrations of Tributyltin (TBT) on marine organisms *Metal Ecotoxicology Concepts and Applications* ed M C Newman and A W McIntosh (CRC Press)

[32] Twining B S and Baines S B 2013 The trace metal composition of marine phytoplankton *Ann. Rev. Mar. Sci.* **5** 191–215

[33] Patrick R 1978 Effects of trace metals in the aquatic ecosystem: the diatom community, base of the aquatic food chain, undergoes significant changes in the presence of trace metals and other alterations in water chemistry *Am. Sci.* **66** 185–91

[34] Cheun B, Endo H, Hayashi T, Nagashima Y and Watanabe E 1996 Development of an ultra high sensitive tissue biosensor for determination of swellfish poisoning, tetrodotoxin *Biosens. Bioelectron.* **11** 1185–91

[35] Mishra P and Sahu P P 2022 *Biosensors in Food Safety and Quality: Fundamentals and Applications* (Boca Raton, FL: CRC Press)

[36] Kröger S, Piletsky S and Turner A P F 2002 Biosensors for marine pollution research, monitoring and control *Mar. Pollut. Bull.* **45** 24–34

[37] Camarca A, Varriale A, Capo A, Pennacchio A, Calabrese A, Giannattasio C, Murillo Almuzara C, D'Auria S and Staiano M 2021 Emergent biosensing technologies based on fluorescence spectroscopy and surface plasmon resonance *Sensors* **21** 906

IOP Publishing

Sensors for Marine Biosciences
Next-generation sensing approaches
Shyam S Pandey, Rout George Kerry and Kshitij RB Singh

Chapter 2

Sensors for monitoring global deposition of pollutants in aquatic ecosystems

Shweta Gulia, Simran Singh, Nida E Falak, Bhupendra Pratap Singh, Ravi Butola, Asmita Das and Prakash Chandra

This chapter delves into the critical role of sensors in monitoring global deposition of pollutants in aquatic ecosystems. The chapter begins by highlighting the significance of monitoring these ecosystems in light of the pervasive impact of pollutant deposition. With an emphasis on real-time data collection, the limitations of conventional sampling methods are discussed, underscoring the importance of timely data for effective environmental management. A diversity of tools are available for pollutant monitoring. Chemical sensors, including electrochemical and optical variants, offer precise detection of heavy metals and organic pollutants. Physical sensors, for example those measuring pH, temperature, dissolved oxygen (DO), and turbidity, provide understanding of water quality dynamics. Biological sensors extend the scope by using algal bioindicators, microbial markers, and behavioral responses of aquatic organisms to measure pollution levels. The chapter also discusses the design and operation of monitoring networks, including considerations for sensor placement and the establishment of reliable communication networks. Furthermore, the integration of remote sensing (RS) and geographic information systems (GIS) amplifies the capacity to map pollutant distributions over expansive aquatic areas. Real-time data analysis and interpretation methodologies, encompassing visualization tools and statistical techniques, form a crucial segment of this discussion. This chapter underscores the pivotal role of sensors in acquiring real-time, accurate, and comprehensive data about pollutant deposition in aquatic ecosystems. By emphasizing collaborative efforts, innovative technologies, and informed decision-making, the chapter advocates for a holistic approach towards safeguarding our aquatic environments.

2.1 Introduction

Marine ecosystems (oceans, coral reefs, estuaries and coastal ecosystems) and the freshwater ecosystems (lotic, lentic and wetland ecosystems) collectively constitute the aquatic ecosystem [1, 2], forming an intricate web of life that sustains biodiversity, provides essential ecosystem services, and supports human well-being. However, these vital ecosystems face an ever-growing challenge in the form of pollutant deposition. Pollutants, ranging from industrial chemicals to airborne particulate matter, traverse continents and settle into aquatic environments, posing significant threats to their health and functionality (figure 2.1) [1].

2.1.1 Importance of monitoring aquatic ecosystems

Aquatic habitats support a huge variety of flora and fauna species, many of which are rare, and monitoring not only aids in preserving this biodiversity that supports the stability of ecosystems but also provides information related to the general conditions of aquatic habitats. Early warning signs of ecological stressors or disturbances might include changes in water quality, habitat conditions, and species numbers. The health of every living thing on the planet depends on the quality of the water [3]. For determining if water is fit for use in agriculture, recreation, and human consumption, it is essential to monitor water quality indicators including pH, DO, and nutrient levels. These ecosystems act as natural laboratories for scientific research studies contributing to our understanding of ecological processes, species interactions, and ecosystem dynamics that in turn aids in locating the sources of

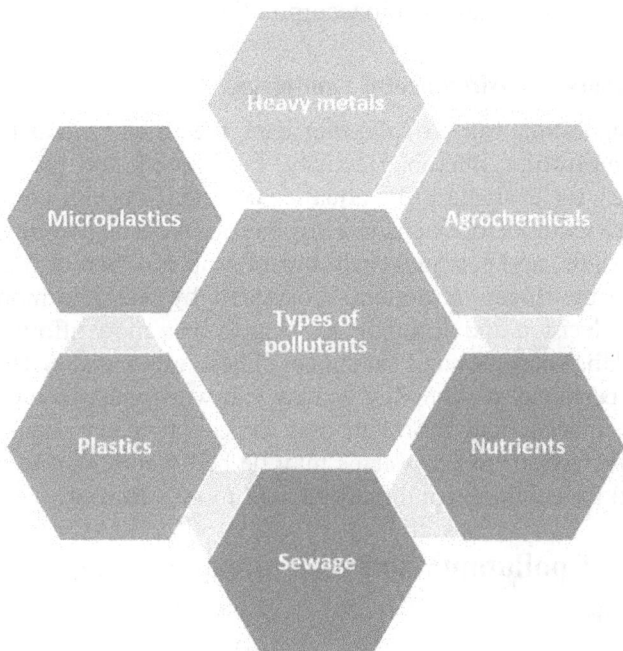

Figure 2.1. Types of pollutants in aquatic ecosystems.

pollution and devising corrective measures. It plays a pivotal role in informed decision-making and proactive environmental stewardship [2].

2.1.2 Global significance of pollutant deposition

Pollutant deposition can introduce a variety of chemical compounds, including heavy metals, pesticides, and organic pollutants, into aquatic ecosystems, altering the physicochemical properties of water. According to Evans *et al*, industries are responsible for dumping tonnes of pollutants into water bodies each year. Marine pollution was not given much of a thought until a certain threshold was achieved, as most aquatic systems naturally possess a tendency to dilute pollution to a certain degree, but once the contamination crosses this threshold, it has adverse effects on living organisms [3].

Additionally, 80% of the wastewater is discharged into freshwater ecosystems without any prior treatment. Agricultural fields also contribute in deterioration of water quality by releasing massive amounts of agrochemicals from fertilizers and pesticides, drug residues, sediments and saline drainage. Such changes in water's pH, temperature and oxygen levels can stress the aquatic biodiversity and disturb habitats. These pollutants can take a toll on human health if ingested through contaminated water causing gastrointestinal problems and even major issues like cancer and organ damage [1]. Pollutants may have an immediate and observable negative influence on the environment or they may have a more subtle negative impact on the delicate balance of the biological food web that is only evident over extended time periods as many pollutants settle in aquatic environments and can enter the food chain and keep on getting concentrated as they move up in the food chain to the higher trophic levels (bioaccumulation) [1, 3].

2.1.3 Role of sensors in environmental monitoring

Sensors play a significant role in collecting data regarding several factors including turbidity, pH, temperature, microbial activity, biodiversity and presence of chemical elements, necessary for evaluating the aquatic ecosystem's health and condition. This data serves as a pre-requisite for extensive ecosystem assessment, assisting in the detection of trends, abnormalities and potential environment problems. Sensors not only provide a quick response to the threats in aquatic ecosystems by real-time monitoring but the results guarantee precision and accuracy, proving to be a more affordable option than the cumbersome lab procedures and techniques. These automated systems don't require regular qualified personnel or sampling as they can work independently for long time periods [4]. With quicker and thorough data collection made possible by this cost-effectiveness, the behaviour of ecosystems may be better comprehended and resources can be channelled for environmental management more effectively.

2.2 Overview of pollutants and their impacts on aquatic ecosystems

2.2.1 Heavy metals

Due to extensive water use in industrial operations and the discharge of effluents from industrial and urban regions, aquatic habitats, in particular freshwater ecosystems, present a higher sensitivity to pollution than other environments (figure 2.1). One class of pollutants with significant ecological implications is heavy metals. In contrast to certain other contaminants that can be naturally degraded in water, heavy metals tend to persist and build up inside aquatic ecosystems. Heavy metals gradually settle and accumulate in the sediment over time. The sediment layer's metal contents may significantly rise as a result of this accumulating process. These accumulating metals become available for aquatic organisms to absorb over time. This bioaccumulation of heavy metals can result in elevated concentrations of these toxic substances in organisms at the top of the food chain, including humans. Consequently, heavy metal contamination poses risks not only to aquatic ecosystems but also to human health. Heavy metal contaminants like mercury, chromium, lead, cadmium, copper, zinc, and nickel have increased quickly and alarmingly as a result of the expanding industrial and agricultural sectors, stormwater runoffs, and urban wastewater discharges. The exceptionally high toxicity degree of arsenic, cadmium, lead, mercury and chromium has made them the priority metals that are of great concern for public health [5]. Prior studies have shown that heavy metals can have a significant impact on a variety of cell organelles in biological systems, including the cell membrane, ER, mitochondria, lysosomes, nuclei, and an array of enzymes that regulate metabolism, detoxification, and damage repair. These metal ions are known for their interactions with critical cellular constituents such as DNA and nuclear proteins, resulting in DNA damage and conformational alterations. These molecular changes could alter the cell cycle, promote cancerous processes, or cause apoptosis. Many of the high degree toxic heavy metals cause oxidative stress by aiding in reactive oxygen species (ROS) production, the major driving force towards cancer and toxicity [5].

When these metals enter into the nucleus, the production of RNA coding for metallothioneins, a type of low molecular weight peptides having high cysteine content. These metallothioneins can bind to heavy metals by their thiol groups. Once induced they can alter a variety of biochemical processes in aquatic organisms [5].

2.2.2 Agrochemicals

The ever-growing demand for food due to a surge in the global population growth has played a pivotal role in driving the demand for food resulting in expansion of agriculture, leading to increased land clearance. Moreover, in response to this demand, the agricultural industry has ramped up production, which, in turn, has resulted in the amplified use of agrochemicals. Numerous substances, including fertilizers, pesticides insecticides, herbicides, trace elements and plant hormones, are included in this category of agricultural pollutants. Pesticides rank as the most prevalent organic contaminants on a global scale. Their remarkable persistence raises concerns, primarily because they are ubiquitous in a wide range of ecosystems. Pesticide residues, despite being subjected to various environmental processes such as chemical, physical, and biological break-down, often exhibit remarkable resistance [1, 3]. This resilience is attributed to their high stability and, in certain cases, their ability to dissolve in water. These characteristics enable these contaminants to endure in the environment, posing challenges to the ecosystem's equilibrium and raising environmental concerns [6].

The constant and frequently unsustainable use of these agrochemicals, with the goal of maximising agricultural productivity, has increased environmental contamination. It has an impact on a variety of ecosystems, including those in rivers, lakes, aquifers, and coastal waters, all of which are essential to preserving the health of our world. Significantly, agricultural regions serve as primary receptors of agrochemicals, accumulating a diverse range of these substances from adjacent fields. Surface runoffs, direct dispersion, and leaching contribute to the gathering of agrochemicals in these areas [1, 3, 6, 7].

2.2.3 Nutrients

When fertilizers are applied in quantities exceeding the capacity of the soil to absorb or crops to utilize, they are carried away by surface runoff. These fertilizers being a rich source of nitrogen, potassium and phosphate can percolate into groundwater or find their way into surface water bodies through runoff. Mostly manures are stored in an open environment, so they can also be washed off by heavy rains and become a matter of concern, polluting water bodies by nutrient enrichment [7]. The elevated nutrient levels, along with the presence of additional substances, lead to the phenomenon known as nutrient enrichment or eutrophication within water bodies. This process can be observed in numerous aquatic environments, including lakes, ponds, coastal regions and reservoirs. Eutrophication results in the excessive proliferation of aquatic vegetation, notably including the formation of algae blooms. These overgrowths have the potential to detrimentally impact the surrounding aquatic ecosystems by disrupting the balance of other aquatic plants and animals, ultimately posing a threat to their overall health and biodiversity [1, 3].

When dense algal blooms ultimately decompose, microbial breakdown begins and significantly lower quantities of DO are produced. As a result, an area that lacks enough oxygen to support the majority of aquatic life is known as a hypoxic dead zone. Many freshwater lakes contain these dead zones. Eutrophication-related hypoxia and anoxia continue to be a global hazard to both lucrative commercial and recreational fishing. Some of these algal blooms also provide an extra risk by producing toxic byproducts [8]. In particular, cyanobacteria are in charge of producing chemicals like geosmin and methylisoborneal that emit disagreeable odours. These substances have the potential to harm aqua-cultured fish as well as municipal drinking water systems, which would have a significant negative financial impact on local and regional economies. This highlights the extensive economic and environmental effects of algal blooms as well as the requirement for proactive management techniques [5]. Nitrate infiltration from agricultural activities into groundwater stands as the most prevalent chemical contaminant affecting global groundwater. This excessive accumulation of nitrate leads to concerning health consequences including the blue-baby syndrome [5].

2.2.4 Sewage

Sewage is a complex mixture encompassing industrial, municipal, and domestic waste, including pollutants from various sources. Unfortunately, our freshwater resources function as primary receptacles for such sewage pollutants. This has had far-reaching consequences, leading to extensive ecological degradation. This degradation involves a decline in water quality and availability, heightened instances of flooding, the loss of aquatic species, and significant alterations in the composition and structure of aquatic ecosystems. The severity of the sewage's negative impact depends on the type and concentration of contaminants present, as well as the volume and frequency of wastewater discharges into water bodies. Sewage is a complex cocktail of constituents, including microorganisms, heavy metals, nutrients like nitrites, nitrates, and phosphorus, along with pharmaceutical and personal care products. Sewage has a high biological oxygen demand (BOD) on receiving waters due to its organic nature. This may cause the oxygen levels in aquatic habitats to drop, which is harmful to the species that depend on those ecosystems. This, in turn, raises sedimentation rates, increases biomass of plants and animals, reduces species diversity, and may even result in anoxic (oxygen-depleted) conditions [1, 3]. Efforts to eliminate pharmaceuticals and personal care products from wastewater treatment plants are often inadequate. Consequently, a significant portion of these chemicals is released into the environment and accumulates in plants. This occurs as a result of practices like irrigating with treated wastewater and applying biosolids and fertilizers to agricultural land. As a consequence, plants can uptake pharmaceuticals and personal care products, and these substances are found in plant tissues at concentrations that typically range from nanograms per kilogram (ng kg^{-1}) to micrograms per kilogram (μg kg^{-1}) [6]. Sewage effluent entering surface waters may carry various pathogenic microorganisms, potentially leading to the transmission of waterborne diseases if people come into contact with contaminated water for domestic or recreational purposes. Figure 2.2 demonstrates various sources of aquatic pollutants and the passage of these pollutants into the aquatic ecosystems and how they contaminate them. Proper treatment and management of sewage are vital to mitigate these ecological and public health risks [1, 6, 9].

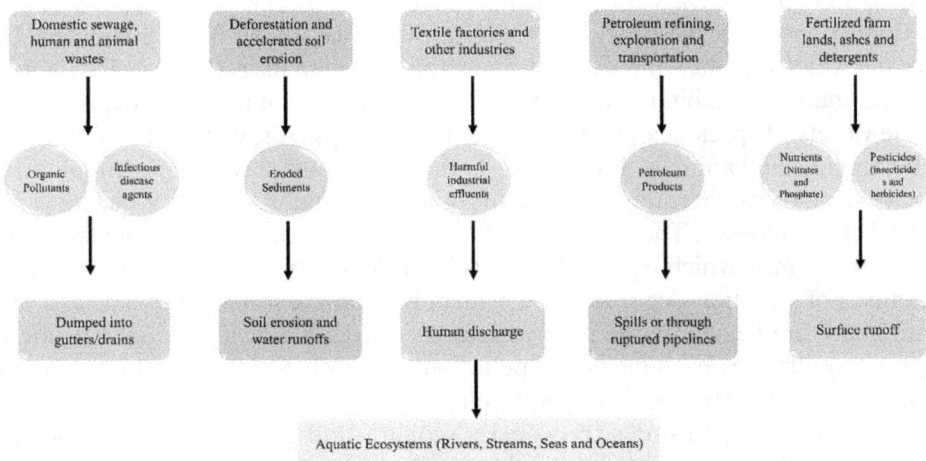

Figure 2.2. The sources of various aquatic pollutants and the route by which these pollutants reach and contaminate the aquatic ecosystems.

2.2.4.1 Plastics and microplastics

In regions with inadequate waste removal infrastructure, coupled with logistical challenges in informal settlements and remote communities, a substantial problem arises where plastic waste accumulates in natural habitats. The widespread mobility of plastic debris has practically infiltrated the entire global marine environment. Rough estimates suggest that a significant portion of marine litter, with plastics being the predominant component, originates from inland sources, often making their way into the oceans through rivers. Plastic waste is discarded in large quantities on beaches, in lakes, and other aquatic ecosystems. Within the marine environment, plastics of varying sizes and origins are widely prevalent, impacting several aquatic species through entanglement or ingestion [9]. Large plastic items, known as macroplastics (exceeding 5 mm in their longest dimension), include items like polybags, packing foam, and stems of earbuds. Over time, these large debris items disintegrate into smaller pieces known as microplastics (measuring less than 5 mm in their longest dimension). Small fibres released from fabrics, frequently while washing, are also considered microplastics. Through processes like abrasion, chemical interactions, biological breakdown, and exposure to ultraviolet (UV) radiation, rivers play a critical role in the fragmentation of plastic debris into tiny bits. Both macroplastics and microplastics are direct dangers to wildlife, endangering it by causing entanglement, asphyxia, and ingestion. These plastics can also take up other substances from the environment, such as mercury and persistent organic pollutants. Plastic waste is stored in freshwater sediments as well, and when floods or heavy flows occur, these secondary sources can spread plastic pollution across the environment [1, 6, 9].

2.2.5 Environmental impact of pollutant deposition

Pollutant deposition leads to water pollution, with contaminants ranging from heavy metals to chemicals and particulate matter infiltrating the aquatic system. In addition to upsetting the ecosystems' delicate balance, this pollution can seriously harm the water's quality and make it dangerous for human usage as well as aquatic life. Degradation of habitat is yet another important effect of pollution deposition [1]. A wide variety of species can be found in aquatic ecosystems, which are vital habitats for both plants and animals. Pollutants, on the other hand, have the potential to ruin these environments, harming aquatic life and changing the water's chemical and physical characteristics. The general health of these ecosystems is seriously threatened by this disturbance, which frequently results in the loss of breeding or feeding grounds. The effects of pollution deposition are evident economically. Governments and local communities are heavily burdened by the costs of treating water to make it safe for use, treating illnesses brought on by pollution, and losing money in the fishing and tourism sectors. International cooperation is required due to the worldwide nature of pollution deposition. In order to properly monitor and manage the transboundary migration of pollutants, which frequently crosses national lines, cooperative efforts are needed to address this complex and linked issue [1, 6].

2.3 Need for real-time monitoring

2.3.1 Limitations of conventional sampling methods

Traditional approaches to assess water quality consist of collecting water samples at designated locations and times, followed by laboratory analysis. These methods encompass grab sampling, depth profiling, bailer sampling, composite sampling, passive sampling, water quality testing kits, bioassays, chemical analysis, and microbiological testing. Although these techniques have proven valuable, they tend to be the following.

2.3.1.1 Time-consuming and labor-intensive
Traditional methods involve manual collection of water samples at specific locations and subsequent analysis in a laboratory. This process is not only time-consuming but also demands significant labor and expertise [4, 10].

2.3.1.2 Limited temporal resolution
The periodicity of data collection, such as weekly or monthly, provides only a snapshot of water quality. This may not capture sudden changes or transient events that could have significant environmental implications [4, 10].

2.3.1.3 Transportation issues causing sample contamination
Transporting water samples introduces the risk of contamination or alteration. Chemical reactions can occur in the bottles during transport, potentially skewing the results and misrepresenting the actual water quality [4, 10].

2.3.1.4 First flush phenomenon
The 'first flush' effect refers to the initial runoff during a rain event, which often carries a more concentrated load of pollutants. In traditional sampling methods, by the time samples are collected, this critical phase may have passed, and important data could be missed [4].

2.3.2 Advantages of real-time monitoring

Real-time water quality monitoring offers a multitude of benefits compared to traditional approaches. It excels in swiftly detecting shifts in water quality, ensuring the rapid response to any emerging issues or unexpected deviations in parameters. By providing a continuous and up-to-the-minute data stream, it paints a dynamic picture of water quality trends and variations, a stark contrast to sporadic data collection from conventional methods. This immediacy can also serve as an early warning system, allowing authorities to take swift action to safeguard public health and the environment. Moreover, real-time sensors and monitoring devices tend to be more accurate, minimizing the potential for human error in manual sampling and analyses. Over time, these systems can prove cost-efficient, reducing the need for frequent site visits and lab analyses, particularly in remote or challenging-to-reach locations [10]. They enable trend analysis over time, shedding light on patterns, seasonal fluctuations, and long-term water quality changes that may elude traditional periodic sampling. With the ability to send alerts when predefined thresholds are crossed, real-time systems ensure immediate intervention when water quality parameters deviate from acceptable levels. This streamlined resource allocation, especially beneficial in expansive water bodies and watersheds [4]. Real-time monitoring supports scientific research efforts, offering a rich data source for ecosystem studies, pollutant transport investigations, and climate change impact assessments on water quality. Moreover, these systems often enable the public to access data, fostering greater community awareness and engagement in water quality issues, ultimately contributing to more responsible water resource management. To account for the first flush phenomenon, it is crucial to manage the sampling time intervals smartly. Instead of collecting samples at fixed intervals, the focus should be on the peak loading periods. This means deploying automated devices to capture the most critical data during the initial stages of flooding when pollutant concentrations are typically higher [4, 10].

2.3.3 Case studies highlighting the importance of timely data

2.3.3.1 Harmful algal blooms (HABs)
A serious hazard to aquatic ecosystems and public health, harmful algal blooms occur in many water bodies. Early warning signs of HAB growth can be obtained through timely data collection and monitoring, allowing authorities to respond appropriately. For example, in Lake Erie, prompt identification of a toxic algal bloom made it possible for authorities to declare water advisories, shielding the public from exposure to pollutants and directing the distribution of resources for water treatment [11].

2.3.3.2 Coral bleaching events
Increases in water temperature have the potential to cause coral bleaching episodes, as coral reefs are extremely vulnerable to environmental changes. Marine biologists and conservationists can respond quickly to temperature increases by using timely data from satellite-based temperature sensors and underwater monitoring devices. For instance, the Australian Great Barrier Reef uses real-time data to initiate interventions like lowering human activity in the vicinity of impacted areas and starting coral regeneration projects [12].

2.3.3.3 Fisheries management and invasive species control
Timely data is pivotal in aquatic ecosystems, exemplified in fisheries management and invasive species control. Sustainable fisheries rely on real-time data about fish populations, enabling managers in the Gulf of Mexico to adjust quotas and seasons promptly. Real-time data collected through monitoring systems proves indispensable in detecting the presence and movement of these intruders. In the Great Lakes region, for example, monitoring systems equipped to track zebra mussels utilize timely data to formulate and guide management strategies [13].

2.3.3.4 Response to oil spills

Responding quickly to marine oil spills is crucial to minimizing the harm to the environment and the economy. Sensor data from remotely operated vehicles (ROVs) in real time is essential. In order to reduce the damage on the environment, ROVs monitoring the Deepwater Horizon leak in real time helped to deploy booms and skimmers, among other cleanup tools. They also provided chemical dispersants. ROVs equipped with sensors evaluated the effects of the Exxon Valdez oil leak and guided response operations, underscoring the crucial role that real-time data plays in oil spill response [11].

2.4 Sensor technologies for aquatic pollutant monitoring

2.4.1 Overview of sensor technologies

Over time, as delays occur between the collection of water samples and their subsequent analysis, there's a risk of compromising the integrity of the samples. The evaluation of water quality encompasses the assessment of physical, chemical, and biological indicators like turbidity, temperature, pH, total organic content, total suspended solids (TSS), electrical conductivity (EC), DO, and N, P and nutrient concentrations. Essentially, sensors are responsible for detecting environmental stimuli, converting them into signals and then storing this data in a dedicated platform for future use [10]. Data collection, signal processing to limit noise, amplifying that data and then transmitting it, followed by data handling, including computational processes, are crucial aspects in the context of a wireless water quality monitoring system. Sensors positioned in the water body or at designated points continuously measure and record various parameters. These sensors can detect changes in pH, EC, DO, turbidity, and more, ensuring real-time data acquisition. Subsequently, the collected data undergoes signal processing to eliminate background noise and ensure the accuracy of the measurements. This procedure is essential to getting reliable, high-quality data. After signal processing, more amplification is applied to the data, which is subsequently transmitted wirelessly to a central repository or data platform. This immediate data transfer guarantees quick access to data from remote locations, providing a fast and thorough summary of the state of the water quality [14]. When the data gets to the central repository, it goes through several data management processes. These processes make it possible to quickly make decisions and respond to pollution incidents or changes in water quality by highlighting trends, anomalies, and patterns in the data on water quality [4].

2.4.1.1 Physical sensors

Using *in situ* approaches, environmental parameters can be directly assessed in their natural habitat, usually at regular or semi-regular intervals. After collection, the data is transported to facilities onshore for analysis. These *in situ* sensors have shown to be quite helpful in assessing various physical characteristics in their natural environments. For instance, they have a long history of detecting critical components in saltwater, such as pH, oxygen concentrations, and CO_2 concentrations, in addition to characteristics like temperature and the quantification of nephelometric turbidity units (NTU), conductivity, and depth [8]. In recent times, the availability of low-cost and commercially accessible sensors has expanded the range of applications for *in situ* monitoring. Furthermore, advanced tools such as spectrometers and water quality probes equipped with fluorescent detectors can be deployed to assess downwelling spectral irradiance, Chl-a, turbidity and dissolved nutrient levels, along designated transects. Meanwhile, for underway sampling, a towed body probe is a practical choice. Surface light sensors also come into play for monitoring variations in the ambient environment's light profiles. In sum, *in situ* monitoring plays a pivotal role in the efficient and comprehensive assessment of various environmental parameters [4].

2.4.1.2 Chemical monitoring sensors

Nitrate and pH are the most often used and easily accessible solutes for chemical sensors. Other solutes continue to be studied using conventional wet analytical chemistry techniques, despite improvements in 'lab-on-a-chip' sensor technologies that have lowered power requirements and addressed problems with optical absorbance measurements. Optical sensors including fluorimeters, have also proven to be capable of measuring organic compounds like chlorophyll a and other pigments involved in photosynthesis [7]. By calculating the potential difference between a working pH probe and a reference electrode, pH is traditionally determined. The pH level of the water sample demonstrates a direct link with this voltage output (mV) from the electrode. Even with limited sample volumes, the exact *in situ* measurement of phosphate and multi-analyte detection is now possible because to the development of micro-electro mechanical systems (MEMS) in conjunction with microelectrode array sensors. Additionally, electrochemical and optical sensors can be combined with microfluidic devices to provide water quality monitoring for heavy metals, nutrients, and pathogens in a microchannel system [4, 10].

2.4.1.3 Biological monitoring sensors

Based on the type of the recognition unit and the characteristics of the transducer, biosensors are majorly categorized. Receptor proteins, DNA and enzymes of immune systems are just a few examples of these recognition units, which can all be combined with electrochemical, chemical, or piezoelectric mechanisms for transduction. Biosensors are remarkably sensitive and specific, which allows them to adapt to various environmental matrices. However, the difficulties associated with calibration and validation frequently prevent their widespread usage in industrial production and long-term application. Nevertheless, they do show promise in specific applications, such as drinking water purification and water treatment plants monitoring. Biosensors can be useful in these situations for the detection of live organisms, such as parasites and pathogens.

A few types of sensors and their selection criteria are enlisted in figure 2.3. The rapid and precise identification of such contaminants by them can improve the security and effectiveness of water treatment procedures, protecting public health [4, 8, 10].

2.4.2 Selection criteria for sensor types

Selecting the appropriate sensors for aquatic ecosystem monitoring is a critical decision that hinges on various factors. The sensor's capabilities must match the

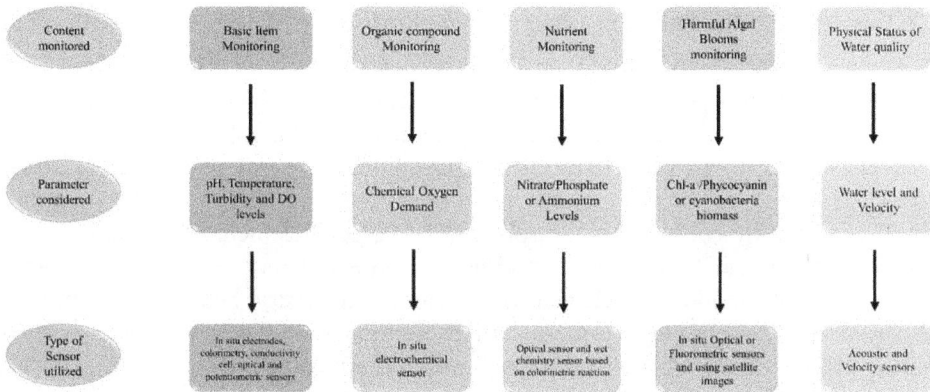

Figure 2.3. Selection criteria for sensor types.

Figure 2.4. Illustration of sensor network in aquatic environment. The nodes (sensors) are dispersed throughout the water body making a mesh form-like network and communicate acoustically (represented as lines between nodes). The host node is connected to the computers on land. The Uplink node communicating with both sensors and host node, consists of transducer and RF transceiver for communicating with the sensors submerged in water body and host node, respectively.

characteristics of interest, such as temperature, DO, pH, and nutrient levels (figure 2.4). Key factors to take into account include accuracy, location compatibility, calibration, response time, and data system integration [6]. Additionally, it is important to consider the decision-making process's power needs, data transmission, cost, regulatory compliance, and long-term durability. The use of redundant sensor deployments can improve the accuracy of the data. The chosen sensors can provide accurate, timely, and reliable data for efficient aquatic ecosystem monitoring by carefully evaluating these characteristics [8].

2.4.3 Emerging technologies

The monitoring of aquatic ecosystems has been completely transformed by modern technologies such as RS, biosensors, and nanotechnology, which provide creative and incredibly effective methods for gathering and analysing data. Water quality monitoring has never been more sensitive thanks to the development of tiny sensors made possible by nanotechnology. These sensors can identify even the smallest amounts of contaminants in water. In contrast, biosensors use biological elements to identify certain substances or biological reactions, providing real-time information about the condition of aquatic habitats and species. These biosensors are very helpful in determining whether toxic algal blooms are present or evaluating the effects of contaminants on aquatic life [8]. The use of satellite imagery and unmanned aerial vehicles (UAVs) is an example of RS technologies that allow for thorough evaluations of large water areas. These cutting-edge instruments provide an aerial perspective of variables such as temperature, water quality, and even the distribution of aquatic life. We can collect more accurate and timely data by incorporating this cutting-edge technology into the monitoring of aquatic ecosystems. This will enable us to take proactive conservation measures and make well-informed judgments in these crucial areas [6].

2.5 Design and operation of monitoring networks

2.5.1 Considerations for sensor placement

Accuracy and usefulness of the gathered data depend on where the sensors are positioned precisely. Important considerations for aquatic ecosystems include water depth, flow velocity, and the particular characteristics of the water body. To record changes in temperature, DO, nutrient content, and other characteristics, sensors must be strategically placed. Sensors should be placed in stratified water bodies at different depths to keep an eye on vertical profiles. Sensor location may also be influenced by ecological characteristics, such as the existence of underwater vegetation and wildlife, in order to record the effects of these variables on the environment.

The placement of sensors close to probable pollution sources, as illustrated in figure 2.4, should also be taken into account because it will enable early identification and action [10]. The operation and design of an aqueous sensor network on the bank of a water body, where a pollution source is also present is demonstrated in figure 2.4. In the aquatic ecosystem, there is rapid attenuation of electromagnetic waves which necessitate usage of acoustic communication between the nodes, under the water. The nodes are represented by multicoloured spheres in figure 2.4 and are, in reality, data routers with user-defined environmental sensors. To achieve efficient wide-area coverage with a minimal consumption of power, these nodes communicate acoustically (shown by connecting grey lines), enabling long-term monitoring. Acoustic communication enables widespread low-power implementations to achieve long-term monitoring. All the nodes in the system are similar, except for the following two special nodes: the host node and the uplink node. The onshore host node interfaces to the computer through an RF transceiver to communicate wirelessly. The uplink node interfaces with submerged sensors and transmits data across the water—air boundary. This arrangement ensures efficient and continuous monitoring of the aquatic environment for pollutants. Bayesian models are used in more sophisticated monitoring systems to aid in the optimization of sensor location. These algorithms evaluate the most likely sites for sensors to successfully gather important environmental characteristics using probabilistic reasoning. For instance, they can recommend where sensors should be placed for the best data gathering by taking into account stochastic fluctuations in nutrient concentrations and stratification patterns [15]. Optimizing sensor placement in aquatic ecosystem monitoring can be augmented by machine learning models. These models can examine past data and find spatial patterns that guide sensor positioning. They are frequently based on algorithms like k-means clustering or random forests. Decisions about where to locate sensors are more accurate when using machine learning models, which take into account complex interactions among environmental characteristics [14].

2.5.2 Establishing reliable communication networks

In order to monitor aquatic ecosystems, it is essential to establish efficient communication networks. Rapid accessibility to vital information is ensured via real-time data transmission from sensors to central data repositories or laboratories. Data recorders with remote telemetry capabilities, satellite systems, and wireless communication technologies are frequently used. The preferred means of communication should, however, be in line with the infrastructure and accessibility requirements of the monitoring site. In situations where data confidentiality and authenticity are crucial, such as in the monitoring of water quality for public health, it is also crucial to ensure data integrity and security throughout transmission [14].

2.5.3 Data management and quality control

Analysing and comprehending an enormous amount of data gathered by aquatic ecosystem sensors requires effective data management and quality control. Different characteristics and frequency of data generation may be included in the data collection. These data are organized, stored, and analysed using cutting-edge software and data management systems. To achieve accurate measurements, quality control includes frequent sensor calibration, validation, and maintenance. For the purpose of identifying abnormal readings and assuring data dependability, it is essential to create data validation and verification procedures. To verify the correctness and consistency of the data, it must be combined and cross-validated with data from other sensors. Additionally, it is important to keep past statistics around in order to spot ongoing patterns and aberrations that can provide information on the health and evolution of the aquatic ecosystem [15]. Bayesian networks can also be used to create probabilistic models for data validation, where they estimate the likelihood of a particular measurement based on the relationship between different parameters. This method makes it possible to spot abnormalities and outliers in the data, which eventually improves the reliability of the data. Furthermore, Bayesian models can be used in data fusion to help with the fusion and validation of data from diverse devices and sources, permitting a more precise depiction of the state of the aquatic ecosystem. Monitoring systems benefit from a potent combination of probabilistic reasoning and data-driven insights when machine learning models are used alongside Bayesian models. This improves the accuracy, dependability, and overall efficacy of efforts to monitor aquatic ecosystems [14, 15].

2.6 Chemical biosensor

Electrochemical biosensors are useful analytical tools that turn a bio-signal into a substantial electrical response using biorecognition components. These electronic devices benefit from simple operation, portability, and quick reaction. They are also capable of being downsized, have a long lifespan, respond quickly, and have good sensitivity and selectivity, making them suitable for use as portable biosensing assays. In order to generate an electrical signal proportional to the analyte concentration, electrochemical biosensors interact with the target analyte or molecules. A reference electrode and a sensing electrode (working electrode), which are separated by an electrolyte, make up a traditional electrochemical biosensor. Most electrochemical biosensors are three-electrode systems with a reference electrode connected to a potentiostat. A counter electrode can be added to complete the circuit so that current can flow.

2.6.1 Electrochemical sensors for metal ions detection

Due to their noxious, non-biodegradable nature, and bioaccumulation in ecological systems (aquatic system), heavy metals (lead, cadmium, mercury, and arsenic) are among the most difficult contaminants to remove [16]. Electrochemical sensors because of their sensitivity, selectivity, and affordability, are frequently employed to detect metal ions. They work by turning a chemical signal (concentration of a metal ion) into an electrical signal that can be detected.

There are different types of electrochemical sensors, i.e. potentiometric, voltametric, ion-selective electrodes (ISEs). ISEs are the specialized sensor which has specialized electrodes that react only to a certain ion. They are made of an ion-selective substance and a sensor membrane. A voltage potential proportionate to the concentration of the target metal ion is created by the membrane's selective interaction with it. It works on the Nernst equation which connects the electrode's measured potential to the logarithm of the ion activity in the solution.

There are three components: an ion-selective membrane (generally made from a polymer matrix), ion-filling solution (also called ion reference solution, which has fixed concentration), and reference electrodes.

ISEs are used for a variety of operations, including as pH measurement, ion measurement (e.g., Na^+, K^+, Cl^-, etc), water quality analysis, environmental monitoring, clinical diagnosis (e.g., blood electrolyte analysis), and industrial process control [17].

2.6.2 Optical sensors for organic pollutants

In order to obtain analyte information, optical devices convert optical phenomena changes that come from the interaction between the analyte and the receptor portion into digital or electrical signals using various transduction techniques [18, 19].

Optical sensors for organic pollutants are instruments that make use of interactions between light and matter to identify and measure the presence of particular organic chemicals in a sample. High sensitivity, selectivity, and the ability for real-time monitoring are just a few benefits they provide. optical chemical sensors can be categorized: reflectance is measured in non-transparent media, usually using an immobilized indicator. Luminescence is estimated by the intensity of light emitted by a chemical reaction in the receptor system. Fluorescence and phosphorescence are both measured as the positive emission effect caused by irradiation. Absorbance is measured in a transparent medium and caused by the absorptivity of the analyte itself or by a reaction with some suitable indicator. The presence of particles of various sizes in the examined sample causes scattering, which may be described as a variation in the refraction index due to the composition of the analyzed sample, and refraction, which can be expressed as a variation in the wavelength of the emitted irradiation [20]. Various kinds of electrochemical and optical sensors are listed in table 2.1.

2.7 Physical sensors

2.7.1 pH sensors

To develop a fundamental understanding of surface processes, it is crucial to monitor pH changes at the micro- and nanoscale. Micro-/nano-sized pH sensors can be used to detect local pH changes at the electrode/electrolyte contact [21]. By calculating the hydrogen ion (H^+) concentration in a solution, pH sensors determine whether a solution is acidic or alkaline. They use the Nernst equation, which connects pH to an electrode's electrochemical potential. Water treatment, agriculture, food and beverage production, medicines, and environmental monitoring are just a few of the industries that use pH sensors [22].

2.7.2 Temperature sensors

Temperature sensors are instruments or transducers that change temperature into an electrical signal. They are employed to gauge a system's, process's, or environment's thermal conditions. They rely on a material's temperature-dependent physical characteristics, such as a material's resistance or the voltage applied across a diode. In general, there are four types of temperature sensor thermocouples, resistance temperature detectors (RTDs), thermistors, infrared sensors.

These temperature sensors are widely used in HVAC (heating, ventilation, and air conditioning), manufacturing, the automotive industry etc [23].

2.7.3 Dissolved oxygen sensors

The concentration of oxygen molecules dissolved in a liquid is measured via DO sensors. They may run on polarography, galvanic cells, or optical sensors, among other operating systems. The major component involved is a sensing element that is made from a fluorescent dye and an electrode. The sensor interacts with oxygen molecules to provide a signal that is proportional to the oxygen concentration. This signal is often electrical or optical. DO sensors are essential for monitoring the

Table 2.1. Examples and types of electrochemical and optical sensors in environmental monitoring.

S. No.	Type of sensor	Method of detection	Target analyte	Component	Advantages	Applications	References
1	Scattering sensors	Disparities in refraction index	Particles present in sample	Refractive index measurement system	Measures composition and size of the particle	Monitoring of environment, monitoring of water quality	[20]
2	Flourescence sensors	Positive emission	Organic pollutants present in water	Irradiation source, fluorescent molecules	Highly selective and sensitive	Clinical diagnosis, environmental monitoring	[18, 19]
3	Absorbance sensors	Absorptivity	Organic pollutants present in water	Suitable indicator, transparent medium	Various analytes can be measured	Water quality monitoring, environmental monitoring	[18, 19]
4	Luminescence sensors	Light emission	Organic pollutants	Luminescent chemical reaction system	Real-time monitoring, highly sensitive	Chemical analysis and environmental monitoring	[18, 19]
5	Electrochemical sensors	Potentiometric	Numerous metal ions (e.g., Cd^{2+}, Pb^{2+}, As^{3+}, Hg^{2+})	Reference electrode, reference solution, ion-selective membrane	Affordability, highly selective and sensitive	Clinical diagnosis, analysis of water quality, environmental monitoring	[16, 17]
6	Ion-selective electrodes (ISEs)	Voltage potential	Particular ions such as Cl^-, K^+, Na^+	Sensor membrane, ion-filling solution, ion-selective substance	Works on Nernst equation, particular to target ion	Industrial process control, blood electrolyte analysis, pH measurement	[16, 17]
7	Optical sensors	Reflectance	Organic pollutants present in water	Immobilized indicator	Highly selective and sensitive	Real-time analysis, environmental monitoring	[18, 19]

environment, especially when analyzing the water quality for uses including, wastewater treatment, aquaculture, and ecological study.

2.7.4 Turbidity and suspended solid measurement

Measurements of turbidity and suspended particles are essential components of water treatment, industrial operations, and environmental monitoring. They serve as markers for a liquid sample's quality and clarity. The term 'turbidity' describes a fluid's cloudiness or haziness, which is brought on by a vast number of individual particles that are typically undetectable to the unaided sight. NTU or Formazin Turbidity Units (FTU) are commonly used measurements for turbidity [50]. Applications of turbidity and suspened solid measurements are demonstrated in figure 2.5 and is discussed further.

2.7.4.1 Applications

2.7.4.1.1 Water quality indicator
An essential factor in determining the overall water quality of aquatic environments is turbidity. Elevated levels of turbidity may suggest the existence of impurities, sediments, or other types of contaminants inside the water [24].

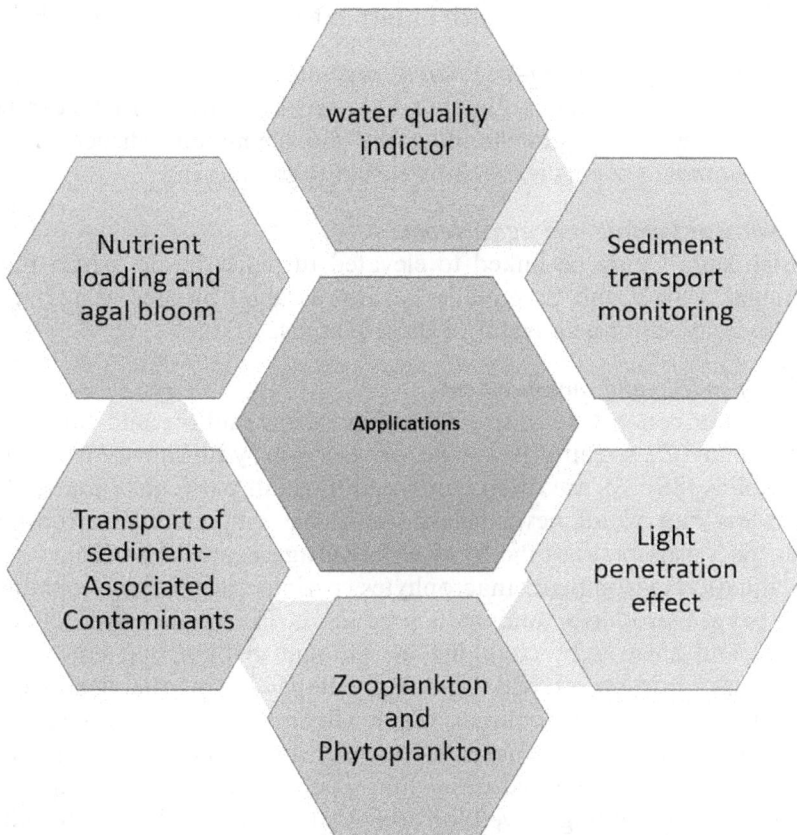

Figure 2.5. Application of turbidity and suspended solid measurement.

2-18

2.7.4.1.2 Sediment transport monitoring

Sediment transfer in rivers, lakes, and coastal regions is observed using turbidity. Elevated levels of turbidity may suggest the presence of erosion and sedimentation processes, thereby affecting the aquatic ecosystem's ecology.

2.7.4.1.3 Light penetration effect

Light ray penetration into the water column is decreased by turbidity. This has an effect on the maximum depth at which photosynthesis may take place, which has an effect on the distribution and development of aquatic algae and plants.

2.7.4.1.4 Effects on watery plant life

As less light is available for photosynthesis when there is high turbidity, submerged aquatic vegetation (SAV) development may be restricted. Given that SAV serves as both a habitat and a source of food for a variety of aquatic creatures, this might have a domino effect on the entire food web.

2.7.4.1.5 Impact on zooplankton and phytoplankton

The distribution and productivity of phytoplankton, the main producers in aquatic environments, are impacted by turbidity. Turbidity variations can affect the makeup of phytoplankton communities, which in turn affects zooplankton numbers.

2.7.4.1.6 Transport of sediment-associated contaminants

Contaminants such as organic pollutants, heavy metals, and nutrients can be found in suspended sediments. Understanding the movement and dispersion of these pollutants in aquatic systems is aided by turbidity monitoring.

2.7.4.1.7 Nutrient loading and agal bloom

Nutrient-rich runoff may be linked to elevated turbidity levels, which may cause excessive algal growth and potentially hazardous algal blooms. Aquatic life and water quality may suffer as a result of these blooms.

2.7.4.1.8 Suspended solid measurement

The mass (mg) or concentration (mg l^{-1}) of inorganic and organic particles held in the water column of a stream, river, lake, or reservoir by turbulence is referred to as suspended solids (SS). SS are often composed of small particulate matter having a diameter of less than 62 μm; nevertheless, studies have shown that for most cohesive solids, transport usually takes the form of larger aggregated flocs [25].

In the aquatic environment, macrophytes, periphyton, and phytoplankton are significant oxygen producers and food producers. The main way that SS affects macrophytes and algae is by changing the amount of light that enters the water column. Primary consumers will be directly impacted by the decrease in light penetration through the water column, which will limit the pace at which periphyton and emergent and submerged macrophytes may ingest energy through photosynthesis. It is noteworthy, nevertheless, that planktonic species, such as surface phytoplankton and floating-leaved or free-floating macrophytes, do not really depend on this mechanism.

2.8 Biological sensors

Biosensors are instruments that identify and measure particular contaminants using biological elements (enzymes, antibodies, or entire cells). They can be programmed to identify a variety of pollutants, such as pesticides, heavy metals, and organic chemicals. In aquatic ecosystems, biological sensors are crucial for tracking the global deposition of contaminants. They offer important information about the existence, concentration, and possible effects of different pollutants on aquatic life. They are used in bioindicators, i.e. algae, diatoms, and macroinvertebrates are a few examples of aquatic species that are extremely sensitive to variations in water quality [26]. The amount of pollution in the water can be estimated by scientists by examining the species' makeup and abundance. Apart from the bioindicators they are used to check the BOD of water as well as its toxicity. Scientists and environmental organizations can monitor the global deposition of contaminants in aquatic ecosystems by using these biological sensors and monitoring approaches. Understanding the effects of pollution, putting management plans into action, and preserving the well-being of aquatic habitats and the communities that depend on them all depend on this knowledge.

2.8.1 Algal bioindicators for water quality assessment

Algal bioindicators are certain kinds of algae that are employed in aquatic habitat to measure the water quality (table 2.2). They are susceptible to variations in

Table 2.2. Algal bioindicators for measuring water quality.

Algal bioindicator	Characteristics	Applications	References
Green algae (chlorophytes)	Photosynthetic microorganism Found in freshwater	Nutrient pollution indicators e.g., *Spirogyra, Cladophora*	[27]
Blue-green algae (cyanobacteria)	Photosynthetic bacteria, produce toxin in certain conditions nutrient-rich, found in eutrophic waters	Monitor harmful agal blooms (HAB) and associated toxins	[28]
Red algae (Rhodophytes)	Found in marine water, sometimes freshwater	Good water quality indicator	[29]
Euglenoids	Singled celled, flagellated organism	Indicator of organic pollution; presence in high numbers may indicate elevated levels of organic matter.	[30]
Dinoflagellates	Single-celled organisms, some photosynthetic, found in marine and freshwater	Indicate nutrient enrichment and potential harmful algal blooms	[30]

temperature, pH, nutrient levels, pollution levels, and other environmental factors. Scientists may learn a great deal about the health and condition of water bodies by examining the makeup and quantity of algae colonies.

2.8.2 Microbial sensors for detecting pollution indicators

Microbial sensors, sometimes referred to as biosensors, are analytical tools that identify and measure particular substances—often pollutants—in a sample by utilizing microorganisms or their constituent parts (figure 2.6). These sensors rely on how target substances affect microorganisms biologically. They can be quite specific and sensitive, which makes them useful instruments for monitoring the environment. Microbial sensors are used to detect and quantify contaminants in different contexts when it comes to identifying pollution indicators. The principle behind microbial sensors for pollution indicators is that they use microorganisms or parts of them to react to particular contaminants and provide a detectable signal.

2.8.2.1 Biological sensors
These sensors make use of particular bacterial strains that have undergone genetic modification to react differently to different contaminants. The response of the bacteria, such as fluorescence or bioluminescence, can be monitored to determine the amounts of pollutants present [31].

2.8.2.2 Enzymatic sensors
Biological catalysts known as enzymes have the ability to react uniquely with specific contaminants. Enzymatic sensors generate a detectable signal by catalyzing reactions between target chemicals and immobilized enzymes [32].

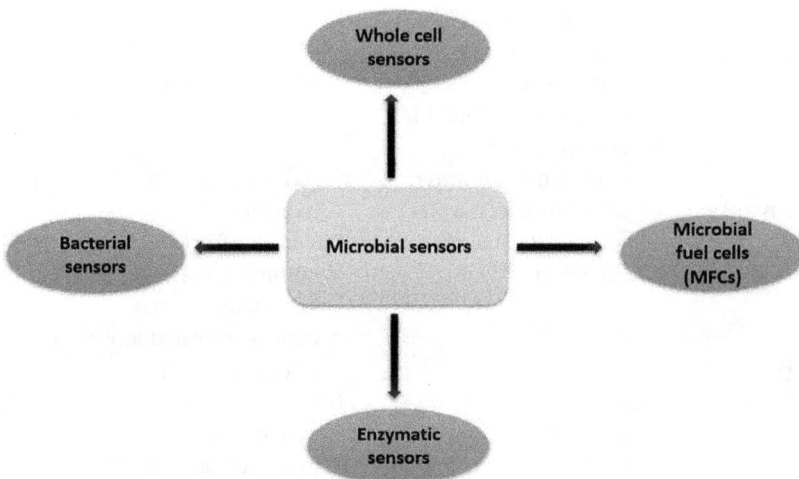

Figure 2.6. Types of microbial biosensors.

2.8.2.3 Whole cell sensors

These sensors react to contaminants by using intact microorganisms, most commonly yeast or bacteria. When particular contaminants are present, the microbes may emit a signal, such as a change in electrical conductivity or fluorescence [33].

2.8.2.4 Microbial fuel cells (MCFs)

MFCs produce electricity by using the metabolic processes of microbes, usually bacteria. Changes in electrical output can be used to detect the presence of pollutants, and they can be made to react to particular pollutants [34].

2.8.3 Fish and invertebrate behavior as a pollution indicators

Many pollutants have direct and indirect effects on the behavior of terrestrial and aquatic organisms, especially in fish. Keeping an eye on fish activity and health can reveal important information about the toxins present in the water and its quality. Fish tissues can get polluted with chemicals, herbicides, and heavy metals over time. Scientist can determine the extent of contamination in the aquatic environment by examining the concentrations of these pollutants in fish. In histopathological examinations, the tissues and organs of fish are examined for indications of damage or anomalies brought on by pollution exposure. Changes in fish behaviour can serve as important indicators of contamination [35]. Modifications in swimming, avoidance, and feeding behaviours can indicate the presence of contaminants or environmental stressors. Because of their sensitivity to particular contaminants, several fish species are referred to as 'sentinel species'. Fish are frequently employed as 'sentinel' animals in ecotoxicology studies and for behavioural and cognitive tests. Many different behaviors, including activity, exploration, avoidance, sociability, aggression, sexual behavior, and eating habits, are impacted by both inorganic and organic pollution. Contaminants frequently have a significant negative influence on spatial cognitive functions including spatial recall and spatial learning capacity. For example, aluminum pollution reduced the learning performance of Atlantic salmon (Salmo salar) in a labyrinth challenge, which may have negative effects on the fish's capacity to assimilate knowledge and adapt to new surroundings [36].

The unusual minnow Gobiocypris rarus and the zebrafish Danio rerio both showed altered activity and spatial memory as a result of organic contaminants like pesticides [37]. It is anticipated that such detrimental cognitive effects will have a significant impact on fish's capacity to acquire and retain knowledge in order to avoid contaminated regions and food sources, find food and mates, and elude predators [35, 38]. Because of this, contaminated fish may find it more difficult to gather, analyze, and retain information regarding the quality of their food and habitat. This might increase the amount of pollution that they are exposed to and create positive feedback loops. Furthermore, a lot of contaminants have an impact on migration and dispersal, which may have an impact on how exposed animals are to pollution.

Aquatic environments employ invertebrates, including crustaceans, molluscs, aquatic insects, and other non-vertebrate creatures, as indicators of pollution. Because of their position in the food chain, sensitivity to environmental changes and toxins, and other factors, they are excellent species for evaluating the quality of water.

Aquatic ecosystems can be exposed to the effects of pollution by tracking alterations in the invertebrate groups' composition. Water quality changes may be indicated by changes in the dominance of tolerant or sensitive species. Pollution can cause changes in the behaviours of invertebrates. Changes in burrowing behaviour, eating habits, or mobility, for instance, may be related to exposure to pollutants or environmental stresses. Changes in the quality of the sediment can affect benthic creatures, which are invertebrates that live in or on the sediment. Invertebrates that live in sediment can indicate levels of contamination and serve as markers of sediment pollution. In underwater environments, invertebrates serve as vital indicators of spoilage. A great deal may be learned about aquatic ecosystem health, pollution sources, and water quality from their behavioural responses, population dynamics, and community composition.

2.9 Remote sensing and GIS applications

Monitoring shoreline alterations and coastal activity is made possible by the beneficial resources provided by geospatial technology. Tracking temporal alterations in coastal regions is a particularly valuable application of RS. Even slight changes to the coast may be precisely analysed thanks to RS imagery's great spatial resolution and distinct spectral bands [39].

2.9.1 Satellite-based monitoring of aquatic ecosystems

An innovative method for keeping tabs on aquatic ecosystems is satellite-based monitoring, that utilizes the use of the power of orbiting satellites configured with specific sensors. These sensors are an effective and non-invasive way to monitor ecosystems since they are made to gather a plethora of data regarding expansive bodies of water. The temperature of the water is one of the main variables that is tracked since it is essential to comprehending the thermal dynamics in aquatic habitats. Turbidity is another important indicator that is tracked and evaluates the water's purity. High turbidity levels provide important information about water quality since they frequently point to problems like pollution or sediment discharge [40]. Monitoring sea surface heights is another way to keep tabs on variations brought about by elements like temperature and ocean currents. The dynamic oceanic phenomena and their effects on coastal ecosystems are shown by these data. The data obtained from satellite-based monitoring is a fundamental source of information for the creation of efficient conservation and management plans for these invaluable natural resources [39, 40].

2.9.2 Geographic information systems (GIS) in pollutant mapping

GIS serves as an essential asset in managing the dispersion of pollutants within water bodies in the overall context of monitoring aquatic ecosystems. The capacity of GIS to combine data from multiple sources, creating a central repository for a variety of environmental information, is what makes it unique. Data from satellite imaging, in-depth water quality tests, and on-site sensors are commonly used in this integration process. A thorough and graphical depiction of the distribution of pollutants in water bodies is created when several data sources merge within the GIS platform [39]. Making maps of pollutant concentrations is one of the main uses of GIS in the monitoring of aquatic ecosystems. These maps offer a clear understanding of the spatial distribution of contaminants in a particular body of water. GIS users can easily detect polluted hotspots—areas where contaminants are highly concentrated—by emphasising fluctuations in pollutant concentration. Finding pollution sources and locations where aquatic life may be most vulnerable is made much easier with the use of this data. GIS makes it possible to follow the advancement of contaminants over time in addition to identifying hotspots. The movement and distribution of pollutants within aquatic ecosystems can be visualized by examining both historical data and current readings [39, 40]. GIS is an essential tool for data visualization, but it also plays a big part in decision-making. This involves choosing the best locations and methods for putting pollution-reduction and aquatic ecosystem protection measures into action. Decision-making is further enhanced by the capacity to overlay and analyse different geographical data layers, guaranteeing that conservation activities are well-informed and successful [39].

2.9.3 Combining remote sensing data with sensor networks

Sensor networks—which are frequently installed in particular places—offer localized, real-time data on a range of characteristics, including temperature, nutrient levels, and water quality. However, using satellite-based or aerial data collecting, RS provides a wider view [39]. The integration of these two technologies facilitates a thorough comprehension of aquatic habitats. For example, sensors can provide high-resolution data at particular locations, but RS offers a wider coverage. Researchers and resource managers are able to make more informed judgements about the conservation and sustainable management of aquatic ecosystems because to the integration of various databases, which aids in the correlation of local changes with broader-scale environmental patterns [15, 39].

2.10 Advanced data tools and techniques for monitoring aquatic ecosystems

2.10.1 Real-time data visualization and analysis tools

Current aquatic ecosystem monitoring systems should incorporate real-time data visualization and analysis tools. Researchers and regulatory agencies can visualize data as it is acquired and do real-time analysis with the help of these advanced tools. Through instantaneous insights, they enable prompt reactions to alterations or new problems in aquatic environments. These technologies frequently have intuitive user interfaces that make it possible to visualize trends in data, deviations, and patterns, improving decision-making. Web-based data dashboards are useful for quick and simple accessibility to information; they offer interactive maps, charts, and alerts from any place. By taking data integration to the spatial level, GIS software makes it possible to create dynamic maps and visualizations that may be used to evaluate the spatial distribution of important characteristics [41]. Using both real-time data and historical information, advanced analytics powered by machine learning and artificial intelligence predict events like dangerous algal blooms or pollution crises [42].

2.10.2 Statistical approaches for identifying trends and anomalies

The foundation of data analysis in aquatic environment monitoring is provided by statistical techniques, which provide a methodical and quantitative way to extract significant insights from large and complicated datasets. Applying these methods is essential to comprehending the temporal and spatial variation of aquatic habitats. In order to make complex datasets easier to handle and understand, multifaceted statistical methods such as principal component analysis (PCA), hierarchical agglomerative cluster analysis (HACA), and others are based on the fundamental statistical principle of simultaneously monitoring and analysing multiple variables. By lowering the dimensionality of the data while maintaining critical information, PCA can assist in identifying the underlying, most significant variance in the set [43]. HACA is useful in classifying environmental circumstances or sources of variability by enabling the identification of unique groups or patterns within the data. Moreover, the Water Quality Index (WQI) analysis frequently uses these multifaceted statistical approaches. These techniques help determine the main variables impacting water quality as well as the connections between different metrics. As a result, scientists and ecological groups are able to evaluate the general condition of aquatic ecosystems, identify problem regions, and create focused management plans for water quality [43].

2.10.3 Integration of multiple sensor data streams

Multiple sensor data streams combined improve data interpretation reliability while also facilitating more efficient ecological management. It makes it possible to address environmental issues pro-actively, facilitating prompt interventions and well-informed decision-making. This comprehensive knowledge of the dynamics of aquatic ecosystems is essential for protecting these invaluable natural resources from pollution and environmental deterioration. In order to record changes in temperature, oxygen saturation, and nutrient concentrations, sensors may be positioned at different depths within an aquatic habitat [44]. By integrating data from these sensors, it becomes possible to visualize how changes in one parameter, such as temperature, correlate with variations in DO or nutrient levels. This integrated perspective reveals complex interactions within the aquatic ecosystem. Moreover, numerous variables, such as water temperature, nutrient content, and light availability, can affect harmful algal blooms [44]. It is possible to create predictive algorithms that can notify authorities of the possibility of a dangerous algal bloom based on the current conditions of these variables by combining data from sensors monitoring these parameters [5, 44].

2.11 Case studies and application

2.11.1 Global initiatives for aquatic pollutant monitoring

The objectives of international programs for aquatic pollution monitoring are to evaluate, control, and lessen the effects of pollutants on aquatic ecosystems and public health. Global governments, organizations, and scientific communities are working together on these projects.

A few well-known international programs for monitoring aquatic pollutants

1. *United Nations Environment Programme (UNEP)*

   The monitoring of aquatic contaminants is one of the many environmental protection-related projects and initiatives that UNEP is in charge of. It offers member nations information and advice on how to set up and carry out monitoring programs.

2. *Global Environment Monitoring System (GEMS)/Water*

   The UNEP initiative GEMS/Water is dedicated to the assessment of freshwater quality worldwide. It makes data on water quality, including pollution monitoring, easier to gather, analyze, and disseminate.

3. *Global Monitoring Plan for Persistent Organic Pollutants (POPs)*

   This strategy was created in accordance with the Stockhlm Convention on Persistent Organic Pollutants with the intention of tracking and minimizing POP releases into the environment. It entails member nations reporting on a regular basis the amounts of POPs in their surroundings.

4. *World Health Organization (WHO)—Guidelines for Drinking-Water Quality*

   WHO establishes global norms and regulations for the quality of safe drinking water. To protect the safety of drinking water, these recommendations specify recommended limits for various pollutants. The guidelines include microbiological, chemical, radiological, and physical characteristics, among other elements of drinking-water quality. They offer suggestions for the highest amounts of particular pollutants that are permitted.

2.11.2 Regional studies highlighting sensor success

Regional studies on sensor success demonstrate how different sensors are implemented and impactful in particular geographic areas. These studies demonstrate the ways in which sensor technologies have been applied to local problems, enhance monitoring capacities, and further scientific investigations. Some instances of local research highlighting the success of sensors are as follows.

1. *Urban air quality monitoring in Beijing, China*

   This study shows how a network of air quality monitors was successfully installed in Beijing to track pollution levels. Real-time data on pollutants such as PM2.5, PM10, sulfur dioxide (SO_2), and nitrogen dioxide (NO_2) was provided by the sensors. Pollution control efforts and public health recommendations were guided by the data [45].

2. *Water quality monitoring in the Great Lakes, USA and Canada*

   A variety of water quality sensors have been used by researchers in the Great Lakes region to track variables like temperature, DO, turbidity, and nutrient levels. These sensors have proved crucial in determining the origins of pollution, monitoring the ecosystem's health, and directing conservation activities [46].

3. *Volcano monitoring in Iceland*

   The active volcanoes of Iceland are well-known. Seismometers, gas analyzers, and thermal cameras are part of the sensor networks that are placed around volcanic sites to keep an eye on surface temperature fluctuations, gas emissions, and seismic activity. Authorities can anticipate and respond to possible eruptions with the use of this knowledge [47].

2.11.3 Industry and regulatory applications of monitoring data

Data monitoring is essential to different sectors and regulatory systems. It offers vital information for judgment calls, compliance checks, and performance reviews. Certain examples are listed below:

1. *Environmental regulation and protection*
   Compliance monitoring: Environmental rules frequently mandate that industries keep an eye on their emissions and discharges to make sure they adhere to set criteria for noise levels, air and water quality, and other factors.

2. *Environmental impact studies*
   Monitoring data is utilized to evaluate the possible environmental effects of initiatives that are being proposed. Regulators can use this information to make well-informed choices about project approval and mitigation strategies.

3. *Pharmaceuticals and healthcare*
   Drug development and clinical trials: Analyzing data from clinical trials is essential to determining the efficacy and safety of novel medications and medical procedures.
   Manufacturing quality control: To guarantee the uniformity and quality of goods, from raw materials to completed pharmaceuticals, data monitoring is employed throughout the pharmaceutical manufacturing process.

4. *Agriculture and food safety*
   Precision agriculture: Farmers can adjust irrigation and fertilization techniques to increase crop yields and lower resource inputs by monitoring data, such as soil moisture, temperature, and nutrient levels.
   Food protection and quality assurance: The use of monitoring data allows for the tracking and validation of adherence to food safety regulations, as well as the identification of allergens, pathogens, and pollutants.

5. *Water resource management*
   Monitoring of water quality: Sensor data is used to evaluate and control the quality of water in rivers, lakes, and reservoirs for a range of uses, such as recreation, drinking water supply, and ecosystem health.
   Flood forecasting and management: Water resources are managed during extreme weather events and early warning systems for flood events are provided by monitoring data.

2.12 Challenges and future directions

2.12.1 Calibration and maintenance of sensor networks

The process of calibrating sensors involves making necessary adjustments and fine-tuning them so that they continue to deliver reliable and accurate readings over time.
Procedure for calibration

1. Create a reference standard: Compare sensor data using a recognized and trustworthy reference standard.
2. Conduct calibration checks: To find any discrepancies, periodically compare the sensor's readings to the reference standard.
3. Modify or calibrate: In the event that differences are discovered, modify the sensor's parameters as needed to bring its readings into compliance with the reference standard.
4. Note calibration data: Keep a record of the calibration procedure, including reference values, modifications, and any other pertinent details.
5. Preserve calibration records: Record the dates, outcomes, and modifications made for each calibration that is carried out in a log.

Sensor type, ambient conditions, and manufacturer recommendations are some of the variables that affect how frequently a sensor has to be calibrated. Some sensors can only need to be calibrated once a year or even less often, while critical sensors could need to be done more often.

2.12.2 Maintenance

Routine tasks are needed for maintenance in order to maintain good functioning of the sensors and the sensor network as a whole. Steps involves regularly inspections, cleaning of sensors, monitoring the environmental conditions as well as record maintenance.

2.12.3 Addressing sensor drift and cross-sensitivity

In order to provide reliable and precise information from sensor networks, sensor drift and cross-sensitivity must be addressed. Use sensor structures and shields, isolate sensors, apply filtration techniques, calibrate for cross-sensitivity, and choose sensors carefully. It is possible to effectively handle sensor drift and cross-sensitivity by putting these ideas into practice, which will guarantee that your sensor network delivers precise and trustworthy data for your particular application. Quality assurance, regular inspection and calibration are required for management of a good sensor plan.

1. Frequent calibration: To account for any drift in sensor data, calibrate sensors using recognized reference standards on a regular basis.
2. Adopt stable environments: Reduce exposure to harsh or erratic environments, as these can exacerbate sensor drift.
3. Quality assurance procedures: To detect and address drift concerns during routine sensor inspections, implement quality assurance and quality control procedures.

2.13 Conclusion

2.13.1 Significance of sensor-based monitoring for environment protection

1. Real-time data: Sensor-provided real-time that enables prompt decision-making and action in response to emergencies or changes in the environment. This skill makes prompt interventions easier in urgent circumstances.

2. Early warning detection systems: By sensing changes in environmental parameters like humidity, temperature, or pollutant levels, sensors can issue early detection about possible dangers or pollution incidents.
3. Accuracy and precision: High degrees of precision and accuracy are delivered by well-calibrated sensors, lowering the possibility of mistakes and guaranteeing accurate information. By carefully calibrating the data, the likelihood of errors in the gathered information is reduced.
4. Cost-effective: By eliminating the need for labor-intensive and expensive manual sampling and laboratory analysis, automated sensor networks can cover large areas or numerous parameters at once.
5. Comprehensive coverage: By measuring a large number of variables, including as turbidity, pH, temperature, humidity, and contaminants, sensors can give a complete picture of the state of the environment.

2.13.2 Collaborative efforts in global aquatic ecosystem monitoring

In order to collect and analyse information about the state of aquatic environments across the globe, governments, organizations, researchers, and communities must coordinate and cooperate in collaborative efforts in global aquatic ecosystem monitoring. Understanding, controlling, and safeguarding the world's water supplies depends on these activities.

2.13.2.1 United Nations Sustainable Development Goals (SDGs)
Initiative: The topic of SDG 14, 'Life Below Water,' is the preservation and wise use of marine resources, including oceans and seas. It highlights how crucial international cooperation is to the observation and preservation of aquatic ecosystems. Research institutes, civic society, NGOs, Member States, and the United Nations are among the collaborators.

2.13.2.2 Global Ocean Observing System (GOOS)
Initiative: To offer real-time data on important ocean variables, GOOS is an international organization that organizes efforts to monitor the world's oceans. It seeks to deepen our knowledge of the oceans' involvement in the planet's climate system. Partners include the World Meteorological Organization (WMO), a number of national oceanographic agencies, research institutions, and the Intergovernmental Oceanographic Commission (IOC) of UNESCO [48].

2.13.2.3 International Joint Commission (IJC)
Initiative: International Joint Commission (IJC) In order to address shared water management and environmental challenges, especially those pertaining to the Great Lakes and other transboundary waterways, the United States and Canada founded the IJC, a binational organization. Federal agencies in the United States and Canada, state and local governments, indigenous groups, and stakeholders are collaborators.

2.13.2.4 Global Lake Ecological Observatory Network (GLEON)
GLEON is an international network of practitioners, scientists, and academics whose goal is to comprehend and forecast the behavior of reservoirs and lakes. In order to address ecological and environmental concerns, it makes joint monitoring and data exchange easier. Global government agencies, academic institutions, and research centre's are among the Collaborators [49].

References

[1] Bashir I, Lone F A, Bhat R A, Mir S A, Dar Z A and Dar S A 2020 Concerns and threats of contamination on aquatic ecosystems *Bioremediation and Biotechnology: Sustainable Approaches to Pollution Degradation* (Berlin: Springer International Publishing) pp 1–26

[2] Rose K A *et al* 2010 End-to-end models for the analysis of marine ecosystems: challenges, issues, and next steps *Mar. Coast. Fish.* **2** 115–30

[3] Evans A E, Mateo-Sagasta J, Qadir M, Boelee E and Ippolito A 2019 Agricultural water pollution: key knowledge gaps and research needs *Curr. Opin. Environ. Sustain.* **36** 20–7

[4] Yaroshenko I *et al* 2020 Real-time water quality monitoring with chemical sensors *Sensors (Switzerland)* **20** 1–22

[5] Ghorab M A 2018 Environmental pollution by heavy metals in the aquatic ecosystems of Egypt *Open Access J. Toxicol.* **3** 555603

[6] Rathi B S, Kumar P S and Vo D V N 2021 Critical review on hazardous pollutants in water environment: occurrence, monitoring, fate, removal technologies and risk assessment *Sci. Total Environ.* **797** 149134

[7] Gama A F, Cavalcante R M, Duaví W C, Silva V P A and Nascimento R F 2017 Occurrence, distribution, and fate of pesticides in an intensive farming region in the Brazilian semi-arid tropics (Jaguaribe River, Ceará) *J. Soils Sediments* **17** 1160–9

[8] Sumudumali R G I and Jayawardana J M C K 2021 A review of biological monitoring of aquatic ecosystems approaches: with special reference to macroinvertebrates and pesticide pollution *Environ. Manage.* **67** 263–76

[9] Verster C and Bouwman H 2020 Land-based sources and pathways of marine plastics in a South African context *S. Afr. J. Sci.* **116** 1–9

[10] Park J, Kim K T and Lee W H 2020 Recent advances in information and communications technology (ICT) and sensor technology for monitoring water quality *Water (Switzerland)* **12** 510

[11] World Health Organization *Guidelines on Recreational Water Quality* **1** (Coastal and Fresh Waters)

[12] McLeod I M *et al* 2022 Coral restoration and adaptation in Australia: the first five years *PLoS One* **17** e0273325

[13] Cooke S J *et al* 2023 Towards vibrant fish populations and sustainable fisheries that benefit all: learning from the last 30 years to inform the next 30 years *Rev. Fish. Biol. Fish.* **33** 317–47

[14] Demetillo A T, Japitana M V and Taboada E B 2019 A system for monitoring water quality in a large aquatic area using wireless sensor network technology *Sustain. Environ. Res.* **29** 12

[15] Zhu M *et al* 2022 A review of the application of machine learning in water quality evaluation *Eco-Environ. Health* **1** 107–16

[16] Lin Y, Gritsenko D, Feng S, Teh Y C, Lu X and Xu J 2016 Detection of heavy metal by paper-based microfluidics *Biosens. Bioelectron.* **83** 256–66

[17] Dimeski G, Badrick T and John A S 2010 Ion selective electrodes (ISEs) and interferences—a review *Clin. Chim. Acta* **411** 309–17

[18] Hulanicki A, Glab S and Ingman F 1991 Chemical sensors: definitions and classification *Pure Appl. Chem.* **63** 1247–50

[19] McDonagh C, Burke C S and MacCraith B D 2008 Optical chemical sensors *Chem. Rev.* **108** 400–22

[20] Caroleo F *et al* 2022 Advances in optical sensors for persistent organic pollutant environmental monitoring *Sensors* **22** 2649

[21] Al-Jeda M, Mena-Morcillo E and Chen A 2022 Micro-sized pH sensors based on scanning electrochemical probe microscopy *Micromachines (Basel)* **13** 2143

[22] Skoog D A, West D M and Holler F 2013 *Fundamentals of Analytical Chemistry* (Cengage Learning)

[23] Shukla R, Prakash Yadav S, Verma S, Kushwaha S and Kumar Verma V 2021 Temperature and pollution monitoring using sensor node in *2021 International Conference on Advance Computing and Innovative Technologies in Engineering (ICACITE)* (Piscataway, NJ: IEEE) pp 1030–4

[24] Hou Y *et al* 2022 A study on water quality parameters estimation for urban rivers based on ground hyperspectral remote sensing technology *Environ. Sci. Pollut. Res.* **29** 63640–54

[25] Bilotta G S and Brazier R E 2008 Understanding the influence of suspended solids on water quality and aquatic biota *Water Res.* **42** 2849–61

[26] Singh S, Kumar V, Dhanjal D S, Datta S, Prasad R and Singh J 2020 Biological biosensors for monitoring and diagnosis *Microbial Biotechnology: Basic Research and Applications* (Springer) pp 317–35

[27] Finger D, Schmid M and Wüest A 2007 Comparing effects of oligotrophication and upstream hydropower dams on plankton and productivity in perialpine lakes *Water Resour. Res.* **43** W12404

[28] Paerl H W and Paul V J 2012 Climate change: links to global expansion of harmful cyanobacteria *Water Res.* **46** 1349–63

[29] McCormick A R, Phillips J S and Ives A R 2019 Responses of benthic algae to nutrient enrichment in a shallow lake: linking community production, biomass, and composition *Freshw Biol.* **64** 1833–47

[30] Wasmund N 2017 The diatom/dinoflagellate index as an indicator of ecosystem changes in the Baltic sea. 2. Historical data for use in determination of good environmental status *Front Mar Sci.* **4** 153

[31] Bounegru A V and Apetrei C 2021 Development of a novel electrochemical biosensor based on carbon nanofibers–cobalt phthalocyanine–laccase for the detection of p-coumaric acid in phytoproducts *Int. J. Mol. Sci.* **22** 9302

[32] Nunes E W, Silva M K L, Rascón J, Leiva-Tafur D, Lapa R M L and Cesarino I 2022 Acetylcholinesterase biosensor based on functionalized renewable carbon platform for detection of carbaryl in food *Biosensors (Basel)* **12** 486

[33] Zhu Y, Elcin E, Jiang M, Li B, Wang H, Zhang X and Wang Z 2022 Use of whole-cell bioreporters to assess bioavailability of contaminants in aquatic systems *Front Chem.* **10** 1018124

[34] Zhou T, Han H, Liu P, Xiong J, Tian F and Li X 2017 Microbial Fuels Cell-Based Biosensor for Toxicity Detection: *A Review Sensors* **17** 2230

[35] Yancheva V 2016 Histological biomarkers in fish as a tool in ecological risk assessment and monitoring programs: a review *Appl. Ecol. Environ. Res.* **14** 47–75

[36] Grassie C *et al* 2013 Aluminum exposure impacts brain plasticity and behavior in Atlantic salmon (*Salmo salar*) *J. Exp. Biol.* **216** 3148–55

[37] Jacquin L, Petitjean Q, Côte J, Laffaille P and Jean S 2020 Effects of pollution on fish behavior, personality, and cognition: some research perspectives *Front Ecol Evol* **8** 86

[38] Grunst A S, Grunst M L, Daem N, Pinxten R, Bervoets L and Eens M 2019 An important personality trait varies with blood and plumage metal concentrations in a free-living songbird *Environ. Sci. Technol.* **53** 10487–96

[39] Parthasarathy K S S and Deka P C 2021 Remote sensing and GIS application in assessment of coastal vulnerability and shoreline changes: a review *ISH J. Hydraul. Eng.* **27** 588–600

[40] Singh A 2019 Remote sensing and GIS applications for municipal waste management *J. Environ. Manage.* **243** 22–9

[41] Parthasarathy K S S and Deka P C 2021 Remote sensing and GIS application in assessment of coastal vulnerability and shoreline changes: a review *ISH J. Hydraul. Eng.* **27** 588–600

[42] Zhu M *et al* 2022 A review of the application of machine learning in water quality evaluation *Eco-Environ. Health* **1** 107–16

[43] Isiyaka H A, Mustapha A, Juahir H and Phil-Eze P 2019 Water quality modelling using artificial neural network and multivariate statistical techniques *Model Earth Syst. Environ.* **5** 583–93

[44] Yaroshenko I *et al* 2020 Real-time water quality monitoring with chemical sensors *Sensors* **20** 3432

[45] Wang L *et al* 2013 Assessment of urban air quality in China using air pollution indices (APIs) *J. Air Waste Manage. Assoc.* **63** 170–8

[46] Burlakova L E, Hinchey E K, Karatayev A Y and Rudstam L G 2018 U.S. EPA Great Lakes National Program Office monitoring of the Laurentian Great Lakes: insights from 40 years of data collection *J. Great Lakes Res.* **44** 535–8

[47] Magnússon E, Gudmundsson M T, Roberts M J, Sigurðsson G, Höskuldsson F and Oddsson B 2012 Ice–volcano interactions during the 2010 Eyjafjallajökull eruption, as revealed by airborne imaging radar *J. Geophys. Res. Solid Earth* **117** B07405

[48] Miloslavich P *et al* 2018 Essential ocean variables for global sustained observations of biodiversity and ecosystem changes *Glob Chang Biol.* **24** 2416–33

[49] Leach T H *et al* 2018 Patterns and drivers of deep chlorophyll maxima structure in 100 lakes: the relative importance of light and thermal stratification *Limnol. Oceanogr.* **63** 628–46

[50] Wagner E D 2006 Turbidity and total suspended solids *Handbook of Water Analysis* (Boca Raton, FL: CRC Press) pp 127–46

IOP Publishing

Sensors for Marine Biosciences
Next-generation sensing approaches
Shyam S Pandey, Rout George Kerry and Kshitij RB Singh

Chapter 3

Sensors for detection of marine organic and inorganic pollutants

S Jone Kirubavathy and Arunadevi Natarajan

The contamination of marine and ocean water by oil, chemicals, industrial and agricultural wastes are a major threat to aquatic organisms and the issue will be solved only by studying the biochemical mechanism. Owing to the rapid enhancement in the population during the last few decades, constant monitoring is challenging. Advancements in sensor technology aided the determination of various pollutants in marine water and provided real-time information about the type and concentration of pollutants. Sensor technology possesses many advantages, like low sample requirement, accurate results, less time, more sensitivity, and low detection limits. Varied sensors are in use based on their accuracy in detecting different kinds of contaminants and their concentration in water bodies. This chapter aims to illustrate the remote sensing monitoring mechanism involved in marine water.

3.1 Introduction

The marine environment plays a crucial part in universal climatic regulation and serves as a significant reservoir of biodiversity. Leveraging marine resources can lead to the development of novel products and services, offering potential solutions to various challenges impacting our planet. These solutions encompass sustainable food and energy sources, innovative industrial developments, the discovery of bioactive materials, and advancements in healthcare treatments [1]. The ocean serves as a natural repository for carbon dioxide and other greenhouse gases. Nevertheless, human engagements have significantly contaminated the marine environment in recent years. Contaminants such as plastic, harmful chemicals, radioactive waste, and industrial sewage are prevalent in pollution. Additionally, marine pollution is exacerbated by sewage discharge into rivers and the influx of disproportionate nutrients from fertilizers/pesticides into the marine environment [2].

doi:10.1088/978-0-7503-5999-3ch3
3-1

The adverse effects of these pollutants extend to various components of the marine ecosystem, including sensitive habitats like coral reefs, mangroves, and aquaculture sites. Hence, besides mitigating the flow of pollutants into oceans, it is imperative to conduct mapping and monitoring of marine toxicants to safeguard the sustainability of the marine ecosystem [2]. The fundamental surveillance of marine ecosystems plays a pivotal part in the timely detection and evaluation of marine hydrometeorological conditions, climate shifts, and ecological crises. Over the past few years, numerous marine regulation activities have been deployed, including offshore platforms, sensors, and ships situated on custom-designed buoys/sub-merged marine infrastructures. They commonly utilize diverse sensors to deliver precise observations, yet they face constraints due to limited spatial resolution and high expenses associated with data acquisition [3].

In current years, cost-effective and efficient systems for monitoring surface water quality have been developed through the utilization of various computer programs and technologies [4]. While sensors have traditionally been employed in marine studies to assess physical parameters, there is a growing need for real-time data concerning chemical and biological parameters [5]. Besides the unsustainable harvesting of biological resources, human actions have introduced numerous contaminants into the ocean over recent years. Typically, these contaminants originate from terrestrial sources via industrial or urban activities and river runoff. Additionally, chemical pollutants can be directly discharged into the marine environment through oil and gas extraction as well as marine operations [6]. Prioritizing selective and sensitive management of these pollutants is an urgent concern in marine research and is essential for safeguarding the marine ecosystem [1].

The majority of remote and automated examinations rely on fluctuations in seawater specifications like temperature, conductivity, depth, and turbidity. Sensors deployed in marine settings face severe conditions and must endure extended deployments to ensure cost-effectiveness. To withstand the rigors of the marine environment, sensors need to be durable, energy-efficient, and equipped with effective antifouling mechanisms. A wide array of *in situ* sensors is available, ranging from commercially available options to those still in the research phase, each at different technology readiness levels. Certain traditional analytical techniques have been utilized in the identification of effluents. For instance, the detection of organic impurities in seawater predominantly relies on analytical methods, such as liquid and gas chromatography. Mass spectrometry is also extensively employed for environmental screening objectives [7].

Water resources have long captivated human interest and have emerged as a critical concern across various facets of human existence in recent times. These resources encompass four primary categories: drinking water, groundwater, surface water (including lakes, rivers, rainfall, and springs), and seawater. With the surge in global population, industrialization, and urbanization, there has been a substantial rise in the demand for drinking water, as well as for agricultural, medical, and industrial purposes. Consequently, this heightened demand has resulted in the discharge of diverse hazardous contaminants, like heavy metals, chemical contaminants, and hospital waste, among others, into water reservoirs, leading to severe water pollution [8].

Biosensors have undergone rapid and diverse advancements over the past few decades. Biosensors are classified based on the primary elements engaged in their detection mechanism: the bioreceptor and the transducer. Concerning the bioreceptors, biosensors are categorized into the following groups: enzymatic biosensors (biocatalytic group), immunosensors, nanosensors, and microbial biosensors (microbial group). On the other hand, biosensors are classified depending on the physicochemical properties and working principles of their transducers as follows: electrochemical biosensors, optical biosensors, and mass-based biosensors [9].

The primary harmful contaminants found in water bodies encompass physical contaminants (such as nanoplastics and microplastics), chemical pollutants (including heavy metals, aromatic compounds, and organic compounds), biological agents, and radiological substances. Various analytical techniques, such as liquid and gas chromatography, mass spectrometry, capillary electrophoresis, and spectrophotometry, have been employed to assess water quality parameters, including dissolved oxygen levels, pH, turbidity, and the presence of chemical contaminants [10].

Despite their excellent sensitivity and repeatability, traditional analytical techniques are often costly, time-consuming, and reliant on sophisticated instrumentation, skilled operators, and sometimes intricate pre-treatment procedures, which can increase the risk of sample loss and hinder real-time, *in situ* measurements [11]. Recognizing these challenges, online sensors, particularly electrochemical and optical sensors, have seen rapid development for water quality detection due to their inherent benefits, including rapid response times, affordability, portability, and the capability to facilitate real-time, *in situ* detection [12].

Chemical, electrochemical, and biosensors have garnered significant interest for their potential in early pollution detection, contributing to environmental preservation. These sensors are categorized based on their applications and transducers. Here is a comprehensive overview of the various chemical, electrochemical, and biosensors reported in the literature, utilized for the identification of both organic and inorganic chemicals [13]. Adsorption technology proves to be an economical, swift, and effective physicochemical approach for the removal of both organic and inorganic contaminants [14].

The high concentration of organic contaminants in water bodies has prompted significant research attention. Identifying the sources of these pollutants in both ground and surface water can be attributed to various human activities, including agriculture and industrial processes. Organic pollutants are not confined solely to water; they are also found in soil and sediments. These contaminants' bioaccumulation, toxicity, carcinogenicity, and mutagenicity collectively impact the ecosystem [15].

Inorganic pollutants, particularly heavy metals, have become one of the most pressing environmental concerns today. Their persistence and durability in the environment demand special attention [16]. Both natural processes and human activities have resulted in the accumulation of inorganic pollutants in groundwater [17]. A range of sensor technologies has been utilized for detecting marine pollutants, encompassing gas sensors, liquid sensors, optical sensors, biosensors, and mass spectrometry. Optical methods, such as fluorescence, absorption spectroscopy, and laser-induced breakdown spectroscopy (LIBS), are also viable for detecting

pollutants, particularly in aqueous settings. Apart from environmental considerations, sensors face constraints related to power supply, communication capabilities,
and the need for periodic recalibration. To achieve sustained monitoring of extensive
deep-sea regions over extended periods, fixed-platform sensors can be installed on
the ocean floor for multiple years [18].

3.2 Role of chemical/biosensors

To safeguard the living systems, it is essential to perform a clear thorough analysis of
toxic substances and consistently monitor their presence in environmental samples
such as soil, water, and air. As a result, biosensors have emerged as an efficient
analytical instrument for assessing environmental pollutants [19].

A biosensor can be defined as a transducer that possesses a biological recognition
component as its main functional element. It comprises major primary parts like the
bio-recognition element, the transducer, and the signal display [20]. The biological
components include tissues, acids, microorganisms, cells, enzymes, and more. The
type of enzyme and materials utilized in these biological elements determine whether
the transducer's output will be in the form of current or voltage [21] (electrically).
Aptamers, antibodies, and cells are commonly used as recognition parts in
biosensors for marine toxins. These bio-materials are known for their strong ability
to select and identify specific target materials accurately [22]. Over the past 15 years,
biosensor technologies have seen extensive use in various fields such as pharmacology, environmental regulation, and ensuring food safety [23]. There are many ways
to make a biosensor, but they can generally be grouped into types like electrochemical, piezoelectric, optical, mechanical, and thermal-based on the kind of
transducer they use [24]. A chemical sensor uses chemical substances as functional
elements. Chemical sensors are also getting a lot of attention because they are used
in many different areas like industry, monitoring the environment, space research,
medicine, and making drugs [25]. Various approaches, including platforms and
tools, have been devised for conducting automated marine measurements, such as
satellites, towed bodies, submersibles, remotely operated vehicles, drifters, and
'SmartBuoys.' These approaches depend on direct sampling, aerial and satellite
imaging, hydrological measurements using probes to measure conductivity, temperature, and depth, as well as remote sensing utilizing electromagnetic waves and
acoustic techniques [1]. However, the majority of self-directed measurements rely on
sensors to investigate factors like temperature, depth, conductivity, and turbidity in
oceanographic research [26]. The applications of chemical/biosensors for detecting
marine pollution in the past few decades are discussed below.

One promising option is the chemiresistor, a type of chemical sensor that detects
analytes by measuring the change in resistance. Chemiresistors have benefits like
using minimal power and being able to measure resistance very precisely [27].
Campanella *et al* proposed an algal-based biosensor for testing estuarine water
safety involving *Spirulina subsalsa* algae with a gas electrode. This setup measures
oxygen changes during photosynthesis and detects any harmful effects caused by
pollutants in the water [28]. A novel optical biosensor was developed to detect water

impurities using immobilized living algae (*Scenedesmus subspicatus*) cells. It measures chlorophyll fluorescence, which varies based on the presence of toxic compounds in water samples, using fiber-optic electronic equipment. This biosensor works fast and is a simple environmental monitoring tool [29]. A biosensor developed by combining *paramecium* and *chlorella* allows for real-time monitoring of marine water and assessment of toxicity levels [30]. Tsopela *et al* developed a lab-on-chip biosensor designed specifically for analyzing water toxicity, especially for detecting herbicides in water [31]. By combining the ability to make things smaller using digital microfluidics (DMF) with the high sensitivity of a fluorescence probe, a new sensor device that can detect mercury in coastal waters was developed [32]. Out of the wide range of marine toxins, there are over 1000 different types, and dozens of them have been successfully found. The entry of marine toxins into the food chain can result in toxicosis in humans, including fatal outcomes. For instance, shellfish poisoning leads to an estimated annual death toll of 750–7500 individuals world-wide [33]. Micro/nano biosensors for shellfish toxin detection were studied using optical biosensors (utilizing optical phenomena), electrochemical biosensors, electrogenerated chemiluminescent (ECL) biosensors, field-effect transistor (FET) biosensors, and surface acoustic wave (SAW) biosensors. These biosensors offered rapid, sensitive, and real-time detection capabilities, ensuring the safety of shellfish products by detecting harmful toxins accurately [34]. An advanced biosensor system was created by merging the miniaturization and automation capabilities of DMF with the broad-spectrum sensitivity and robustness of algae motion signals. This biosensor demonstrated significant promise for marine routine and early-warning monitoring, both in laboratory settings and on-site applications [35]. As per Albaladejo *et al*, wireless sensor networks stand out among small-scale sensor networks as the best solutions for monitoring the marine environment. They are favored for their ease of deployment, operation, and dismantling, as well as their cost-effectiveness. These networks typically include a processor, a radio module, a power supply, and multiple sensors interconnected [36]. Shin'ichiro Kako and their coworkers established a low-altitude remote sensing approach employing a balloon fitted with a digital camera to monitor marine and beach litter of different colors [37]. Miren *et al* carried out a practical strategy for using biomarkers to evaluate the result of pollution in the coastal environments of the Iberian Peninsula [38]. A fluorescence nanosensor capable of dual emission for detecting okadaic acid (OA) was developed by Lian *et al* and optimization was carried out for the ratiometric nanosensor using molecularly imprinted polymers. The nanosensor demonstrated self-referencing capability and enabled visual detection of OA. The nanosensor was used for the analysis of seawater and sediment samples [39]. Sarkar *et al* discussed the importance of utilizing molecular biomarkers as diagnostic and prognostic tools in monitoring marine pollution and reported that cytochrome P4501A induction, acetylcholinesterase activity, DNA integrity, and metallothionein induction were the important biomarkers gaining attention [40]. Fallati *et al* have shown that using unmanned aerial vehicle drones (UAVs) to monitor anthropogenic marine debris (AMD) works well. Also, they discovered that using deep learning can help detect and measure AMD automatically, which is a new finding [41].

The research on the use of biosensors especially for marine pollution detection is limited in number. Thus developing a chemical/biosensor needs more attention, particularly for detecting pollution in a marine environment.

3.3 Types of sensors used for water monitoring

One of the significant challenges humanity is expected to confront in this century is the deterioration of water quality. The availability of clean drinking water is diminishing steadily due to pollution caused by various sources. Two primary factors contribute to the worsening effects of water pollution: firstly, the increasing levels of contaminants in water without effective means to monitor natural changes; secondly, human population growth necessitates the exploration of new water sources whose quality is uncertain. Therefore, regular monitoring of water quality is crucial. Traditional monitoring methods, which are sensitive and reliable, are commonly used for this purpose [42]. There are different types of sensors that are used for water monitoring. These include spaceborne sensors, optical sensors, magnetic sensors, electrochemical sensors, chip sensors, remote sensing, and geographic information systems (GIS).

3.3.1 Spaceborne sensors

Lack of water resources is currently one of the most important global problems of human society. Over the past few decades, remote sensing techniques have seen extensive application in assessing the qualitative characteristics of water bodies [43]. Sensors utilized in remote sensing applications are categorized into two primary types based on the platforms they operate on spaceborne and airborne sensors. Spaceborne sensors are brought by satellites/spacecraft and operate in regions beyond the Earth's atmosphere [44] (figures 3.1(a)–(c)). Spaceborne hyperspectral coverage extends over larger areas, and the spatial/ spectral resolution is comparatively rough, rendering it unsuitable for small inland water bodies [45]. These sensors typically generate images at minimal to no cost and are applicable for large-scale monitoring of water quality [44]. Various spaceborne or satellite sensors are employed in remote sensing applications. Images captured by sensors on satellite platforms funded by the US government are typically accessible to users at no charge. Each sensor possesses distinct spatial, spectral, radiometric, and temporal resolutions [44]. In the last three decades, satellite remote sensing (SRS) has become increasingly valuable in providing insights into the conditions and stresses affecting biodiversity across various spatial scales, ranging from landscapes to global levels [46]. Spaceborne platforms encompass a range of systems such as the space station, along with satellites positioned at both low-level orbits (ranging from 700 to 1500 km) and high-level orbits (around 36 000 km). These platforms can acquire large extensive areas of data, facilitating the monitoring of Earth's resources, atmospheric dynamics, weather forecasting, and various other applications [46]. Examples of spaceborne sensors include Landsat satellite, Advanced Spaceborne Thermal Emission and Reflection Radiation (ASTER), Moderate Resolution Imaging Spectroradiometer (MODIS) Sensor, GeoEye, and IKONOS.

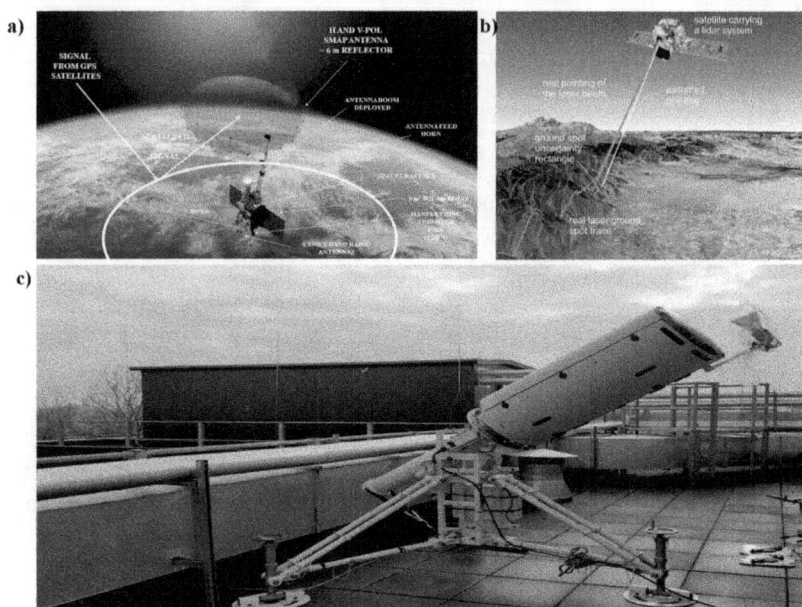

Figure 3.1. (a) Global Navigation Satellite Systems (reproduced with permission from [47] CC BY 4.0), (b) Spaceborne Lidars (reproduced with permission from [48] CC BY 4.0) and (c) Ground-Based Passive Surveillance System (reproduced with permission from [49] CC BY 4.0).

3.3.2 Optical sensor

In the past few decades, optical biosensors have gained a significant role in analyzing biomolecular interactions, including enzyme-substrate reactions, antigen–antibody binding, DNA–DNA interactions, and peptide nucleic acid (PNA)–DNA hybrid-izations. These sensors offer intricate insights into binding affinity and interaction kinetics. However, it is only with recent advancements in optical biosensors and nanotechnology that their application has expanded to include the sensing of living cells—a far more intricate and dynamic effort. Fluorescence serves as a prevalent optical technique in biosensing, valued for its selectivity and sensitivity. In fluorescence-based monitors, the alteration in the frequency of electromagnetic radiation emission is observed. This change is prompted by the absorption of radiation, followed by the generation of an ion in an excited state, which is transient and lasts for only a short period [50]. Typically, two types of sensors are utilized for sensing surface water: optical sensors and microwave sensors. Optical sensors have been extensively used in this field owing to their abundant data availability, as well as their appropriate spatial and temporal resolutions [51]. Optical biosensing is commonly based on detecting a fluorescence signal, typically imposing the utiliza-tion of a fluorescence moiety [52]. A schematic representation of optical sensors is shown in figure 3.2. Advancements in optical sensor technology offer enhanced analytical capabilities for continuously evaluating pollution levels both in the liquid and gas phases [53]. Optical chemical sensors (OCSs) hold a significant role and offer extensive versatility for the on-site detection of heavy metals. Belonging to a

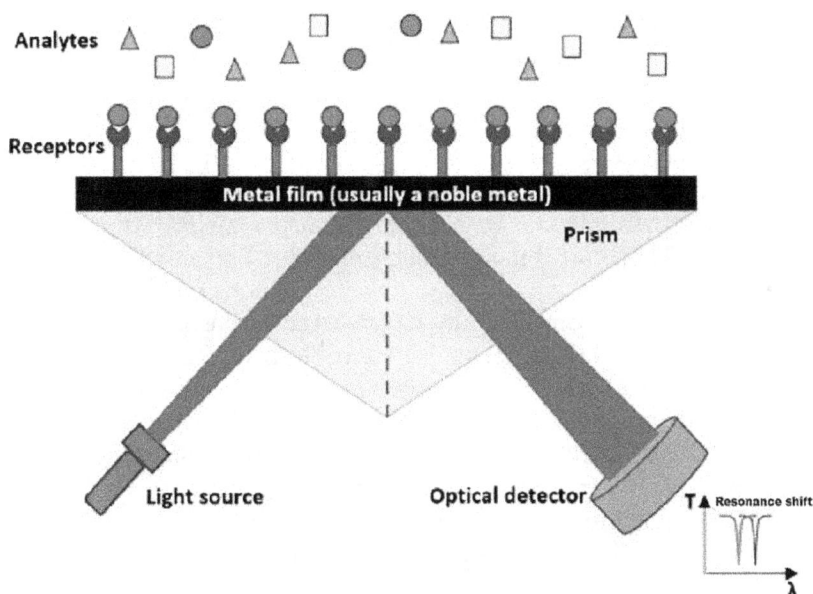

Figure 3.2. Schematic representation of plasmonic optical sensors. Adapted from [56] CC BY 4.0.

category of chemical sensors, OCSs utilize electromagnetic radiation to produce an analytical signal within a transduction element [54]. A reported optical fiber sensor demonstrates the ability to detect ethanol in water. This sensor comprises a single optical fiber incorporated into a 1 mm length of polymer-clad silica optical fiber with a core diameter of 62.5 μm. To enhance sensitivity, the sensor adopts a U-bend configuration, exposing the core directly to the fluid under test by removing the cladding [55].

3.3.3 Magnetic sensors

The disposal of hazardous pollutants in water resources stands as a paramount global concern, posing significant risks to both human health and aquatic ecosystems. Increasing the pollutants in water resources and groundwater represents a major global challenge, particularly in nations abundant in water reserves. Pharmaceutical and pesticide contaminants, among various water pollutants, increase the risk of cancer. The discharge of pollutants into water bodies presents numerous hazards to both human well-being and aquatic ecosystems. Hospital waste, industrial factory-derived phenolic compounds, and agricultural pesticides in groundwater foster the emergence of novel diseases [8]. A significant breakthrough in magnetic sensor technology occurred with the adoption of anisotropic magnetoresistance (AMR) sensors over Hall sensors in numerous applications, leveraging AMRs' superior sensitivity to their advantage [57]. The properties of magnetic nanoparticles lend themselves to various applications such as catalysis, photocatalysis, degradation of organic and mineral pollutants, and serving as magnetic sensors for monitoring and measuring pollutants in aquatic environments. Magnetic position sensors offer affordability and robustness, functioning effectively even in harsh environments such as those contaminated with oil, where optical sensors may fail. The magnetic proximity switch is probably the most used sensor type [58]. Crucial parameters for magnetic sensors include the temperature coefficient of offset and sensitivity, noise levels, linearity, and hysteresis. Multiple magnetic sensors hold promise for application in downhole environments, with magnetoresistive (MR) sensors emerging as leading candidates. This is attributed to their exceptional sensitivity, high tolerance to high temperatures, temperature stability, low power consumption, and established fabrication methods on silicon substrates [59]. Magnetic sensors have been created by harnessing diverse physical phenomena, including electromagnetic induction, the Hall effect, tunnel magnetoresistance (TMR), giant magnetoresistance (GMR), AMR, and giant magnetoimpedance (GMI). MR sensors are anticipated to see widespread adoption across various fields such as biomedical applications, flexible electronics, position sensing/human–computer interaction, non-destructive evaluation, and monitoring, as well as navigation and transportation. This trend is attributed to enhancements in sensing capabilities and operational performance [60].

3.3.4 Electrochemical sensors

Electrochemical sensors play a vital role in identifying chemical ions, molecules, and pathogens in various applications in water quality monitoring. These sensors offer sensitivity, portability, rapid response times, cost-effectiveness, and suitability for online and *in situ* measurements, distinguishing them from alternative methods. They possess the capability to detect compounds undergoing specific transformations within a defined potential range, facilitating their application across diverse scenarios. Their non-specific nature particularly enables efficient detection across multiple ions [61]. An electrochemical sensor is a device designed to detect electron exchange between the sensor and analyte. Typically, it consists of two fundamental parts: a chemical recognition layer and a physicochemical transducer. The latter includes various metal electrodes such as the working electrode, reference electrode, and often a counter electrode. The performance of a chemical sensor is predominantly defined by its response curve, which describes the correlation between the sensor signal and the concentration of the analyte. The primary objective behind employing nanomaterials in chemical sensors is to enhance sensitivity and selectivity. Due to their elevated surface-to-volume ratio, nanomaterials are anticipated to exhibit increased sensitivity [62]. Electrochemical detection systems are potentially ideal for determination due to their ease of scaling down apparatus size without significant performance differences. This monitoring system is both time and cost-effective, playing a significant role in addressing water contamination issues. In these sensors, the analyte's interaction with the electrode surface produces an electrical signal [63].

3.3.5 Chip sensors

Miniaturized electrochemical laboratory chip sensors hold significant promise in environmental monitoring, offering numerous advantages including markedly lower sensing expenses and enhanced portability of sensing systems [64]. Chip calorimetry presents a powerful tool for rapid and high-throughput analysis of biochemical processes. However, it is challenging to realize an inexpensive, easy-to-fabricate microfluidic chip-based calorimeter with high sensitivity [65]. The sensor chip was manufactured employing micro-electro-mechanical system (MEMS) techniques. The challenges posed by the large size, high cost, and intricate structure of conventional water determination sensors and devices make achieving real-time water monitoring on a large scale difficult. There is a need for development of a multi-parameter sensor chip, compact, cost-effective, and resilient, capable of simultaneously detecting water pH, conductivity, and temperature in water [66].

3.3.6 Remote sensing and GSI

Remote sensing provides detailed spatial and temporal information not only from accessible regions but also from areas that are otherwise difficult to reach, while GIS facilitates the storage, interpretation, and retrieval of spatial data. Various types of remotely sensed images, including both airborne and spaceborne, as well as active and passive microwave images, have become crucial for monitoring both surface and groundwater quality. Satellite sensors with different spatiotemporal and spectral resolutions play a significant role in water quality monitoring [67]. Recent progress in remote sensing and GIS has enabled the execution of extensive water remediation investigations. Enhanced spectral and spatial resolution sensors, along with refined geospatial modeling techniques, now facilitate the cost-effective and highly accurate monitoring of various water quality parameters—including chlorophyll-a levels, algae blooms, turbidity, suspended sediments, and mineral content—in water bodies, including groundwater [68]. Time series datasets of water quality can be obtained from satellite images through remote sensing analysis. Furthermore, spatially distributed datasets of water quality derived from remote sensing can serve as valuable products. Apart from these key applications of remote sensing, the extracted water quality parameters can be utilized for forecasting and predictive purposes [69]. Remote sensing (RS) involves identifying Earth's surface features and estimating their geo-biophysical properties through the interaction of electromagnetic radiation. Spectral, spatial, temporal, and polarization signatures are key characteristics of the sensor/target, aiding in target discrimination. Earth surface data captured by sensors in various wavelengths (reflected, scattered, and/or emitted) undergoes radiometric and geometric corrections before spectral information is extracted [44].

GIS is an evolving technology that requires more efficient data processing and expertise solutions to create unified models for environmental simulation. Advances in methods for accessing and generating data, such as the internet, global positioning systems (GPS), and RS, are contributing to this development. Advanced computer-based and efficient methods for water management have emerged, with RS, GIS, and GPS playing crucial roles. GIS serves as a significant tool for addressing water resource issues by enabling the development of solutions to manage water resources at a regional scale. It aids in understanding the natural environment, mitigating flooding risks, assessing water availability, and evaluating water quality [70].

3.4 Detection of marine pollutants

The introduction of chemical contaminants into seawater has become a significant concern in recent years. While most pollutants originate from land-based sources like industrial activities, urban areas, and rivers, they can also enter the marine environment directly through activities such as oil and gas exploitation and shipping operations [7]. Thus, it is necessary to detect and reduce such marine pollutants. Electrochemical sensors like amperometric sensors, potentiometric sensors, electronic sensors like piezoelectric sensors, pellistors, solid-state devices [71] solid-phase extraction, spectrophotometry, gas chromatography, high-performance liquid chromatography, and mass spectrometry, electrochemical-impedance spectroscopy (EIS) are used to detect marine gases [8].

3.4.1 Methane

One of the strongest greenhouse gases, which is more potent than carbon dioxide, is methane [72]. The breakdown of organic substances in the ocean results in the evolution of methane. The methane gas evolves commonly from the natural gas hydrate layer from the seabed [6]. Fluid flow, the ebullition of gas bubbles, or pore water-seawater diffusion are the three ways that methane can move from the seafloor into the water column [73]. The increase in methane content causes several hazards to human health as well as the environment, thus making researchers focus on detection and control. The methane was analyzed in the 1970s by IR spectra showing peaks around 7425 cm^{-1} in the near-IR and around 1500 cm^{-1} in the far-IR [74] (figure 3.3). In a study by Hester *et al*, methane when dissolved in water resulted from a stretching vibrational peak at 2915 cm^{-1}, along with two weaker bands at 3017 and 3066 cm^{-1} [75]. Seismic methods are one of the widely used ways for detecting indirectly and quantification of gas hydrate in marine sediments [76]. A gas monitoring module (GMM) has been created by Marinaro *et al* to measure methane concentration in seawater continuously and over the long term at the benthic boundary layer [77].

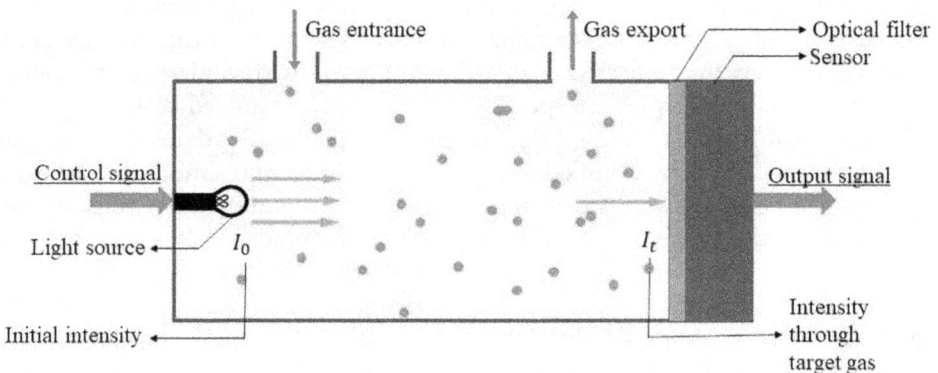

Figure 3.3. Methane gas detection system by IR (reproduced with permission from [78] CC BY 4.0).

3.4.2 Radon

Radon and the radioactive particles from it are a big cause of cancer in humans, especially lung cancer. They are the second most common reason for lung cancer after smoking [79]. It is a gas that comes from the breakdown of uranium in rocks and soil that is slightly radioactive and found underground. The amount of radon can vary depending on the type of rock below the ground. Because radon is a gas, it can seep into the air we breathe and the water we drink, giving us a small amount of radiation exposure [80]. Dulaiova *et al* improved on their own tool for finding radon gas in water. They used three RAD7 analyzers to measure radon (^{222}Rn) in air and replaced the air–water exchanger with a membrane contactor (Liquicel) for better response [81]. The RAD7 radon-in-air monitor (made by Durridge Co.) has been adapted to measure radon-222 (^{222}Rn) in water samples (figure 3.4) [82]. Key *et al* measured radon gas, radium gas (radon's parent element), and total radium in shallow ocean floor sediments. They found there was more radon in the water than one would expect based on the amount in the sediment and they took an existing idea (Broecker's model) for how radon travels in ocean mud [83].

3.4.3 Carbon dioxide

The process of capturing and storing carbon dioxide (CO_2) in deep geological formations offers an effective means of eliminating these emissions from the climate system. In various regions, storage reservoirs are positioned offshore, typically at depths exceeding a kilometer beneath shelf seas important to society [85]. Leakage from geological storage systems in marine environments can occur through infrastructure breakdown, releasing CO_2 directly into the water column, or via geological failures, allowing CO_2 to percolate through strata and sediment layers before reaching the pelagic water column [86]. Releasing a lot of CO_2 into the middle or deep parts of the ocean will create big plumes of seawater that become more acidic, with pH levels between 6 and 8 [87]. Bates *et al* showed the longest continuous

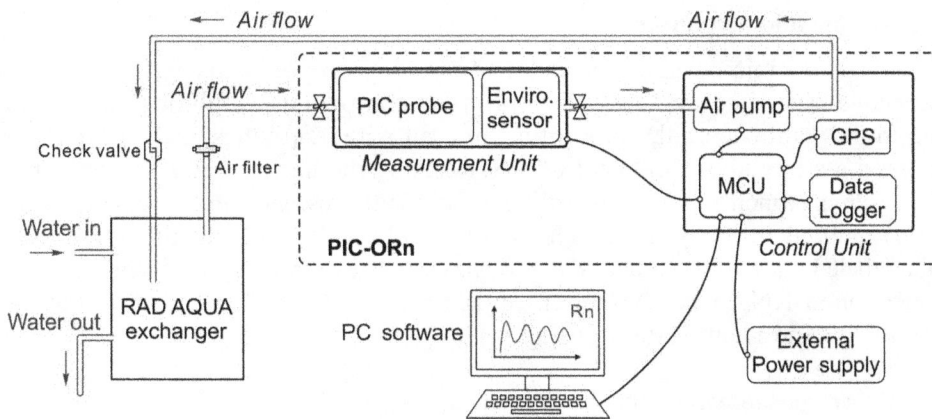

Figure 3.4. Design of a computerized system designed for continuous radon measurement in water (reproduced with permission from [84] CC BY 4.0).

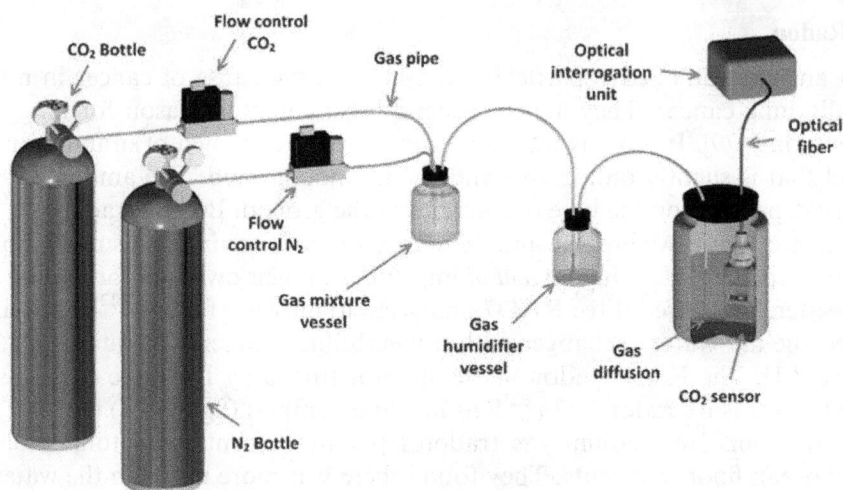

Figure 3.5. A schematic assemblage of the CO_2/dCO_2 (reproduced with permission from [89] CC BY 4.0).

record of ocean CO_2 fluctuations and ocean acidification in the North Atlantic subtropical gyre near Bermuda, spanning from 1983 to 2011, indicating a significant rise in dissolved inorganic carbon (DIC) and partial pressure of CO_2 (pCO_2) levels in surface seawater (figure 3.5). DIC increased by approximately 40 μmol kg^{-1}, and pCO_2 rose by around 50 μatm, marking an approximately 20% increase during this time [88].

Johnson *et al* used an *in situ* method involving placing a non-dispersive infrared (NDIR) sensor inside a water-impermeable but gas-permeable polytetrafluoro-ethylene (PTFE) membrane, enabling direct CO_2 concentration measurements at a designated depth in freshwater environments, eliminating the need for pumps or reagents [90]. Zilberman *et al* developed a novel composite material and an optoelectronic sensor for detecting dissolved CO_2 in water samples [91].

3.4.4 Marine microorganism

Various marine microorganisms such as fungi, myxomycetes, bacteria, and micro-algae have been isolated, and they produce compounds exhibiting antioxidant, antitumoral, antibacterial, apoptotic, and antiviral activities [92]. Linda *et al* presented a study about the field of RS detecting the harmful algal blooms in the aquatic environments [93]. Groben *et al* used fluorescence *in situ* hybridization (FISH) method to apply molecular probes to phytoplankton. In this technique, a probe labeled with fluorescence is introduced into the cell and specifically attaches to the ribosomal RNA (rRNA) present in the ribosomes and helps to find the cell's morphology and identification purposes [94].

3.4.5 Other marine pollutants

There is an urgent need for *in situ* monitoring of phenolic compounds in marine waters due to their inherent toxicity, with these organic compounds primarily

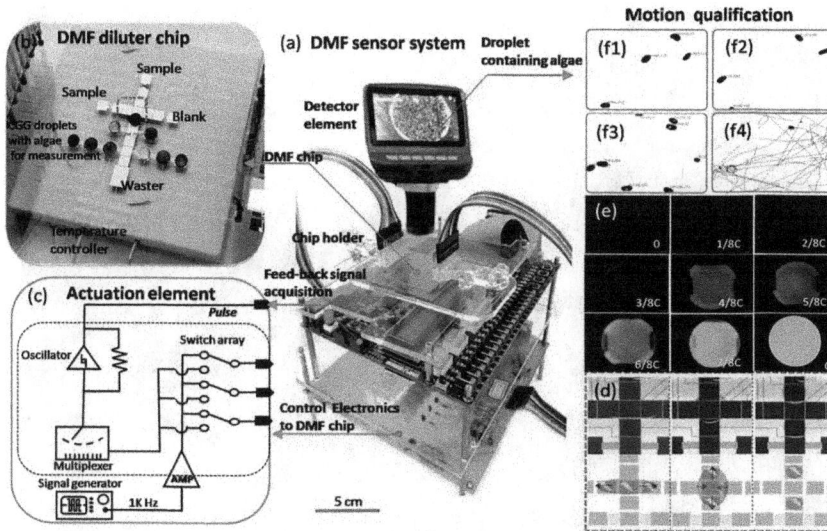

Figure 3.6. (a) Control-engaged DMF sensor, (b) DMF diluter chip, (c) droplet control system, (d) droplet cross-split way, (e) dilution performance validation and (f) alga motion qualification by Image-J (reproduced from [35] with permission from Elsevier).

originating from industrial activities and entering aquatic environments through runoff from industrial and agricultural processes [95]. Anirudhan and Alexander have designed an electrochemical sensor using potentiometric principles to detect hexachlorocyclohexane in diverse real samples, including seawater samples artificially contaminated with this pollutant [96]. Péron *et al* have devised a surface-enhanced Raman scattering (SERS) sensor to detect trace levels of naphthalene in seawater, achieving a detection limit of 10 parts per billion (ppb) [96]. Copper is recognized as a water contaminant that can be found in seawater due to both natural processes and human activities like mining and petroleum refining [97]. Twomey *et al* introduced an electrochemical sensor utilizing microelectrode arrays for detecting copper within the range of 1–5 millimolar (mM), achieving a detection limit of 157 nanomolars [98]. A whole-algae biosensor was framed for low-cost and quick detection of toxic contaminants in seawater using a digital microfluidic (DMF) diluter chip, an actuation element, a detector element, and a microalgae bioreporter, as shown in figure 3.6.

3.4.6 Detecting marine nutrients

Elevated concentrations of nutrients like nitrite, nitrate, ammonium, phosphate, and silicate in marine environments are associated with increased algae growth, leading to potential disruptions in water quality [99]. Different types of nutrient sensors, such as electronic sensors, chemical sensors, and biosensors, have been extensively utilized for continuous monitoring in estuaries and seawater [100]. The two important nutrients in the ocean are nitrite and nitrate. It is important to regularly check their levels in the ocean because they are needed for phytoplankton to grow.

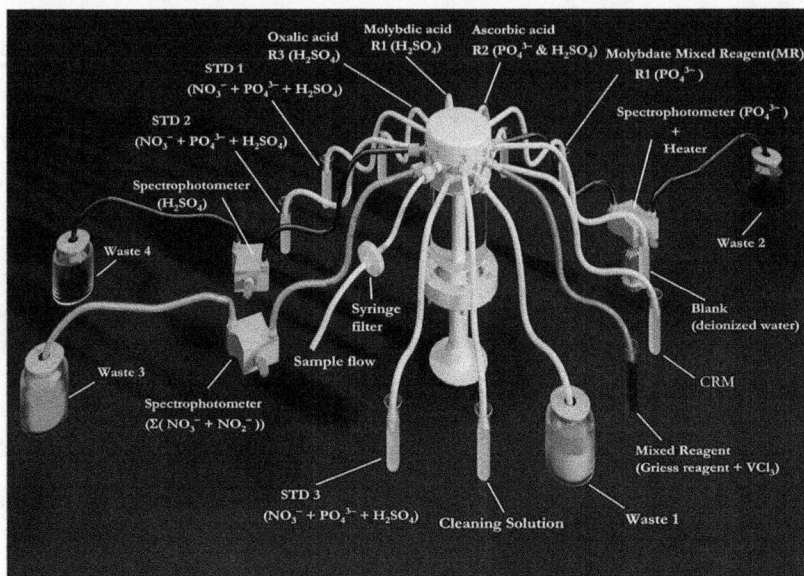

Figure 3.7. A three-dimensional schematic representation of the AutoLab auto analyzer (reproduced with permission from [104] CC BY 4.0).

However, we only consider them harmful if they come from human sources in excessive amounts, causing issues like eutrophication [5]. Cosnier *et al* demonstrated that by immobilizing nitrate reductase on an electrode's surface using a unique electropolymerizable viologen $C_{12}V_2^+$ developed an effective biosensor that is specifically designed to detect nitrate [101]. Due to their low sample and power usage, high sensitivity, quick response, and ease of operation and miniaturization, electrochemical sensors have become widely used as effective analytical tools for rapidly detecting nitrate in environmental samples [100]. Lacombe *et al* [102] and Aguilar *et al* [103] have introduced electrochemical sensors utilizing voltammetry for silicate detection, achieving limits of detection below 1 mM. The automated sensor for the spot detection of the existence of micronutrients present in seawater with good accuracy was found to be advantageous (figure 3.7).

3.5 Conclusions

The ever-increasing population and intensified human activities have elevated pollution levels to a critical threshold. We must now implement regulatory measures and rigorous monitoring processes to manage this pollution. Marine pollution is mainly due to solid wastage and oil/chemical spills deposited in the coastal areas. By proper utilization of spaceborne sensors and geospatial technology, it is possible to locate the level of contaminants. Spaceborne sensors with excellent resolution like hyperspectral sensors find more accurate sources for monitoring marine and ocean pollutants. Satellite sensors are widely used in oceanography to measure the depth and color of the ocean from which the concentration of the pollutants was measured.

The extensive progression in computer-aided technology, increases and made easy for pollutant tracking mechanisms. Government should employ the full benefits of RS systems like GIS and decision support systems to overcome marine pollution. Still, the research is progressing on the development of novel sensors to detect floating pollutants and algae growth in water resources. A few technical problems associated with RS techniques involve misleading radar signals due to poor reflection, varied accuracy in vertical direction, and small cross-sections. In spite of all technical difficulties, spaceborne sensors play a pivotal role in the detection/monitoring of organic/inorganic pollutants in marine water.

References

[1] Justino C I L, Freitas A C, Duarte A C and Santos T A P R 2015 Sensors and biosensors for monitoring marine contaminants *Trends Environ. Anal. Chem.* **6–7** 21–30

[2] Hafeez S, Sing Wong M, Abbas S *et al* 2019 Detection and monitoring of marine pollution using remote sensing technologies *Monitoring of Marine Pollution* (London: IntechOpen)

[3] Yuan S, Li Y, Bao F *et al* 2023 Marine environmental monitoring with unmanned vehicle platforms: present applications and future prospects *Sci. Total Environ.* **858** 159741

[4] Zaidi Farouk M I H, Jamil Z and Abdul Latip M F 2023 Towards online surface water quality monitoring technology: a review *Environ. Res.* **238** 117147

[5] Kröger S, Piletsky S and Turner A P F 2002 Biosensors for marine pollution research, monitoring and control *Mar. Pollut. Bull.* **45** 24–34

[6] Liu Y, Lu H and Cui Y 2023 A review of marine *in situ* sensors and biosensors *J. Mar. Sci. Eng* **11** 1469

[7] da Costa Filho B M, Duarte A C and Rocha-Santos T A P 2022 Environmental monitoring approaches for the detection of organic contaminants in marine environments: a critical review *Trends Environ. Anal. Chem.* **33** e00154

[8] Hojjati-Najafabadi A, Mansoorianfar M, Liang T, Shahin K and Karimi-Maleh H 2022 A review on magnetic sensors for monitoring of hazardous pollutants in water resources *Sci. Total Environ.* **824** 153844

[9] Gavrilaş S, Ursachi C Ş, Perţa-Crişan S and Munteanu F-D 2022 Recent trends in biosensors for environmental quality monitoring *Sensors* **22** 1513

[10] Chevalier P, Piccardo M, de Naurois G-M, Gabay I, Katzir A and Capasso F 2018 In-water fiber-optic evanescent wave sensing with quantum cascade lasers *Sens. Actuators B* **262** 195–9

[11] Han S, Zhou X, Tang Y *et al* 2016 Practical, highly sensitive, and regenerable evanescent-wave biosensor for detection of Hg^{2+} and Pb^{2+} in water *Biosens. Bioelectron.* **80** 265–72

[12] Jiao L, Zhong N, Zhao X, Ma S, Fu X and Dong D 2020 Recent advances in fiber-optic evanescent wave sensors for monitoring organic and inorganic pollutants in water *TrAC, Trends Anal. Chem.* **127** 115892

[13] Shaban A, Eddaif L and Telegdi J 2023 Sensors for water and wastewater monitoring *Advanced Sensor Technology* (Amsterdam: Elsevier) pp 517–63

[14] Rout D R, Jena H M, Baigenzhenov O and Hosseini-Bandegharaei A 2023 Graphene-based materials for effective adsorption of organic and inorganic pollutants: a critical and comprehensive review *Sci. Total Environ.* **863** 160871

[15] Aslam A A, Irshad A, Nazir M S and Atif M 2023 A review on covalent organic frameworks as adsorbents for organic pollutants *J. Clean. Prod.* **400** 136737

[16] Zaheen B, Ahmad A, Luque R, Hussain S and Noreen R 2023 Inorganic pollutants and their degradation with nanomaterials *Sodium Alginate-Based Nanomaterials for Wastewater Treatment* (Amsterdam: Elsevier) pp 57–95

[17] Ghaffar I, Hussain A, Hasan A and Deepanraj B 2023 Microalgal-induced remediation of wastewaters loaded with organic and inorganic pollutants: an overview *Chemosphere* **320** 137921

[18] Skålvik A M, Saetre C, Frøysa K-E, Bjørk R N and Tengberg A 2023 Challenges, limitations, and measurement strategies to ensure data quality in deep-sea sensors *Front. Mar. Sci.* **10** 1152236

[19] Nigam V K and Shukla P 2015 Enzyme based biosensors for detection of environmental pollutants-a review *J. Microbiol. Biotechnol.* **25** 1773–81

[20] Conroy P J, Hearty S, Leonard P and O'Kennedy R J 2009 Antibody production, design and use for biosensor-based applications *Semin. Cell Dev. Biol.* **20** 10–26

[21] Haleem A, Javaid M, Singh R P, Suman R and Rab S 2021 Biosensors applications in medical field: a brief review *Sens. Int.* **2** 100100

[22] Wang Q, Yang Q and Wu W 2022 Ensuring seafood safe to spoon: a brief review of biosensors for marine biotoxin monitoring *Crit. Rev. Food Sci. Nutr.* **62** 2495–507

[23] Vilariño N, Fonfría E, Louzao M C and Botana L 2009 Use of biosensors as alternatives to current regulatory methods for marine biotoxins *Sensors* **9** 9414–43

[24] Herrera-Domínguez M, Morales-Luna G, Mahlknecht J, Cheng Q, Aguilar-Hernández I and Ornelas-Soto N 2023 Optical biosensors and their applications for the detection of water pollutants *Biosensors* **13** 370

[25] Norizan M N, Moklis M H, Ngah Demon S Z *et al* 2020 Carbon nanotubes: functionalisation and their application in chemical sensors *RSC Adv.* **10** 43704–32

[26] Mills G and Fones G 2012 A review of *in situ* methods and sensors for monitoring the marine environment *Sens. Rev.* **32** 17–28

[27] Tang R, Shi Y, Hou Z and Wei L 2017 Carbon nanotube-based chemiresistive sensors *Sensors* **17** 882

[28] Campanella L, Cubadda F, Sammartino M and Saoncella A 2001 An algal biosensor for the monitoring of water toxicity in estuarine environments *Water Res.* **35** 69–76

[29] Frense D, Müller A and Beckmann D 1998 Detection of environmental pollutants using optical biosensor with immobilized algae cells *Sens. Actuators B* **51** 256–60

[30] Turemis M, Silletti S, Pezzotti G, Sanchís J, Farré M and Giardi M T 2018 Optical biosensor based on the microalga-paramecium symbiosis for improved marine monitoring *Sens. Actuators B* **270** 424–32

[31] Tsopela A, Laborde A, Salvagnac L *et al* 2016 Development of a lab-on-chip electrochemical biosensor for water quality analysis based on microalgal photosynthesis *Biosens. Bioelectron.* **79** 568–73

[32] Zhang Q, Zhang X, Zhang X *et al* 2019 A feedback-controlling digital microfluidic fluorimetric sensor device for simple and rapid detection of mercury (II) in costal seawater *Mar. Pollut. Bull.* **144** 20–7

[33] Tang X, Zuo J, Yang C *et al* 2023 Current trends in biosensors for biotoxins (mycotoxins, marine toxins, and bacterial food toxins): principles, application, and perspective *TrAC, Trends Anal. Chem.* **165** 117144

[34] Tian Y, Du L, Zhu P *et al* 2021 Recent progress in micro/nano biosensors for shellfish toxin detection *Biosens. Bioelectron.* **176** 112899

[35] Han S, Zhang Q, Zhang X *et al* 2019 A digital microfluidic diluter-based microalgal motion biosensor for marine pollution monitoring *Biosens. Bioelectron.* **143** 111597

[36] Albaladejo C, Sánchez P, Iborra A, Soto F, López J A and Torres R 2010 Wireless sensor networks for oceanographic monitoring: a systematic review *Sensors* **10** 6948–68

[37] Kako S, Isobe A and Magome S 2012 Low altitude remote-sensing method to monitor marine and beach litter of various colors using a balloon equipped with a digital camera *Mar. Pollut. Bull.* **64** 1156–62

[38] Cajaraville M P, Bebianno M J, Blasco J, Porte C, Sarasquete C and Viarengo A 2000 The use of biomarkers to assess the impact of pollution in coastal environments of the Iberian Peninsula: a practical approach *Sci. Total Environ.* **247** 295–311

[39] Lian Z, Zhao M, Wang J and Yu R-C 2021 Dual-emission ratiometric fluorescent sensor based molecularly imprinted nanoparticles for visual detection of okadaic acid in seawater and sediment *Sens. Actuators B* **346** 130465

[40] Sarkar A, Ray D, Shrivastava A N and Sarker S 2006 Molecular biomarkers: their significance and application in marine pollution monitoring *Ecotoxicology* **15** 333–40

[41] Fallati L, Polidori A, Salvatore C, Saponari L, Savini A and Galli P 2019 Anthropogenic marine debris assessment with unmanned aerial vehicle imagery and deep learning: a case study along the beaches of the Republic of Maldives *Sci. Total Environ.* **693** 133581

[42] Jaywant S A and Arif K M 2019 A comprehensive review of microfluidic water quality monitoring sensors *Sensors* **19** 4781

[43] Haji Gholizadeh M, Melesse A M and Reddi L 2016 Spaceborne and airborne sensors in water quality assessment *Int. J. Remote Sens.* **37** 3143–80

[44] Adjovu G E, Stephen H, James D and Ahmad S 2023 Overview of the application of remote sensing in effective monitoring of water quality parameters *Remote Sens.* **15** 1938

[45] Cao Q, Yu G, Sun S, Dou Y, Li H and Qiao Z 2021 Monitoring water quality of the haihe river based on ground-based hyperspectral remote sensing *Water* **14** 22

[46] Pettorelli N, Laurance W F, O'Brien T G, Wegmann M, Nagendra H and Turner W 2014 Satellite remote sensing for applied ecologists: opportunities and challenges *J. Appl. Ecol.* **51** 839–48

[47] Carreno-Luengo H, Lowe S, Zuffada C, Esterhuizen S and Oveisgharan S 2017 Spaceborne GNSS-R from the SMAP mission: first assessment of polarimetric scatterometry over land and cryosphere *Remote Sens.* **9** 362

[48] Amediek A and Wirth M 2017 Pointing verification method for spaceborne lidars *Remote Sens.* **9** 56

[49] Novák D, Gregor L and Veselý J 2022 Capability of a ground-based passive surveillance system to detect and track spaceborne SAR in LEO orbits *Remote Sens.* **14** 4586

[50] Velasco-Garcia M N 2009 Optical biosensors for probing at the cellular level: a review of recent progress and future prospects *Semin. Cell Dev. Biol.* **20** 27–33

[51] Huang C, Chen Y, Zhang S and Wu J 2018 Detecting, extracting, and monitoring surface water from space using optical sensors: a review *Rev. Geophys.* **56** 333–60

[52] Sansone L, Macchia E, Taddei C, Torsi L and Giordano M 2018 Label-free optical biosensing at femtomolar detection limit *Sens. Actuators B* **255** 1097–104

[53] Mizaikoff B 2003 Infrared optical sensors for water quality monitoring *Water Sci. Technol.* **47** 35–42

[54] Ullah N, Mansha M, Khan I and Qurashi A 2018 Nanomaterial-based optical chemical sensors for the detection of heavy metals in water: recent advances and challenges *TrAC, Trends Anal. Chem.* **100** 155–66

[55] King D, Lyons W B, Flanagan C and Lewis E 2004 An optical-fiber sensor for use in water systems utilizing digital signal processing techniques and artificial neural network pattern recognition *IEEE Sens. J.* **4** 21–7

[56] Hamza M E, Othman M A and Swillam M A 2022 Plasmonic biosensors: review *Biology* **11** 621

[57] Ripka P and Janosek M 2010 Advances in magnetic field sensors *IEEE Sens. J.* **10** 1108–16

[58] Ripka P and Závěta K 2009 Magnetic sensors: principles and applications *Handbook of Magnetic Materials* (Elsevier) ch 3 pp 347–420

[59] Gooneratne C, Li B and Moellendick T 2017 Downhole applications of magnetic sensors *Sensors* **17** 2384

[60] Khan M A, Sun J, Li B, Przybysz A and Kosel J 2021 Magnetic sensors-a review and recent technologies *Eng. Res. Express* **3** 022005

[61] Kanoun O, Lazarević-Pašti T, Pašti I *et al* 2021 A review of nanocomposite-modified electrochemical sensors for water quality monitoring *Sensors* **21** 4131

[62] Cho G, Azzouzi S, Zucchi G and Lebental B 2021 Electrical and electrochemical sensors based on carbon nanotubes for the monitoring of chemicals in water—a review *Sensors* **22** 218

[63] Sunaina , Kaur H, Kumari N, Sharma A, Sachdeva M and Mutreja V 2022 Optical and electrochemical microfluidic sensors for water contaminants: a short review *Mater. Today Proc.* **48** 1673–9

[64] Jang A, Zou Z, Lee K K, Ahn C H and Bishop P L 2010 Potentiometric and voltammetric polymer lab chip sensors for determination of nitrate, pH and Cd(II) in water *Talanta* **83** 1–8

[65] Kopparthy V L and Guilbeau E J 2017 Highly sensitive microfluidic chip sensor for biochemical detection *IEEE Sens. J.* **17** 6510–4

[66] Zhou B, Bian C, Tong J and Xia S 2017 Fabrication of a miniature multi-parameter sensor chip for water quality assessment *Sensors* **17** 157

[67] Sharma B, Tyagi S, Singh P, Dobhal R and Jaiswal V 2015 Application of remote sensing and gis in hydrological studies in india: an overview *Natl. Acad. Sci. Lett.* **38** 1–8

[68] Ramadas M and Samantaray A K 2018 Applications of remote sensing and gis in water quality monitoring and remediation: a state-of-the-art review *Water Remediation* (Springer) pp 225–46

[69] Sheppard D, Tsegaye T D, Tadesse W, McKay D and Coleman T L The application of remote sensing, geographic information systems, and global positioning system technology to improve water quality in northern Alabama *IGARSS 2001. Scanning the Present and Resolving the Future. Proc.. IEEE 2001 Int. Geoscience and Remote Sensing Symp. (Cat. No. 01CH37217)* 3 (Piscataway, NJ: IEEE) pp 1291–3

[70] Dandge K P and Patil S S 2022 Spatial distribution of ground water quality index using remote sensing and GIS techniques *Appl. Water Sci.* **12** 7

[71] Kamieniak J, Randviir E P and Banks C E 2015 The latest developments in the analytical sensing of methane *TrAC, Trends Anal. Chem.* **73** 146–57

[72] Roberts D A, Bradley E S, Cheung R, Leifer I, Dennison P E and Margolis J S 2010 Mapping methane emissions from a marine geological seep source using imaging spectrometry *Remote Sens. Environ.* **114** 592–606

[73] Lohrberg A, Schmale O, Ostrovsky I, Niemann H, Held P and Schneider von Deimling J 2020 Discovery and quantification of a widespread methane ebullition event in a coastal inlet (Baltic Sea) using a novel sonar strategy *Sci. Rep.* **10** 4393

[74] Werle P, Slemr F, Maurer K, Kormann R, Mücke R and Jänker B 2002 Near- and mid-infrared laser-optical sensors for gas analysis *Opt. Lasers Eng.* **37** 101–14

[75] Hester K C, Dunk R M, White S N, Brewer P G, Peltzer E T and Sloan E D 2007 Gas hydrate measurements at hydrate ridge using Raman spectroscopy *Geochim. Cosmochim. Acta* **71** 2947–59

[76] Pecher I A and Holbrook W S 2003 Seismic methods for detecting and quantifying marine methane hydrate/free gas reservoirs *Natural Gas Hydrate* (Springer) pp 275–94

[77] Marinaro G, Etiope G, Gasparoni F *et al* 2004 GMM? A gas monitoring module for long-term detection of methane leakage from the seafloor *Environ. Geol.* **46** 1053–8

[78] Aldhafeeri T, Tran M-K, Vrolyk R, Pope M and Fowler M 2020 A review of methane gas detection sensors: recent developments and future perspectives *Inventions* **5** 28

[79] Yu C, Sun Y and Wang N 2023 Outdoor radon and its progeny in relation to the particulate matter during different polluted weather in Beijing *Atmosphere (Basel)* **14** 1132

[80] Ismail N F, Hashim S, Sanusi M S M, Abdul Rahman A T and Bradley D A 2021 Radon levels of water sources in the southwest coastal region of peninsular Malaysia *Appl. Sci.* **11** 6842

[81] Dulaiova H, Camilli R, Henderson P B and Charette M A 2010 Coupled radon, methane and nitrate sensors for large-scale assessment of groundwater discharge and non-point source pollution to coastal waters *J. Environ. Radioact.* **101** 553–63

[82] Seo J and Kim G 2021 Rapid and precise measurements of radon in water using a pulsed ionization chamber *Limnol. Oceanogr. Methods* **19** 245–52

[83] Key R M, Guinasso N L and Schink D 1979 Emanation of radon-222 from marine sediments *Mar. Chem.* **7** 221–50

[84] Li C, Zhao S, Zhang C *et al* 2022 Further refinements of a continuous radon monitor for surface ocean water measurements *Front. Mar. Sci.* **9** 1047126

[85] Blackford J, Stahl H, Bull J M *et al* 2014 Detection and impacts of leakage from sub-seafloor deep geological carbon dioxide storage *Nat. Clim. Chang.* **4** 1011–6

[86] Blackford J, Widdicombe S, Lowe D and Chen B 2010 Environmental risks and performance assessment of carbon dioxide (CO_2) leakage in marine ecosystems *Developments and Innovation in Carbon Dioxide (CO_2) Capture and Storage Technology* (Amsterdam: Elsevier) pp 344–73

[87] Huesemann M H, Skillman A D and Crecelius E A 2002 The inhibition of marine nitrification by ocean disposal of carbon dioxide *Mar. Pollut. Bull.* **44** 142–8

[88] Bates N R, Best M H P, Neely K, Garley R, Dickson A G and Johnson R J 2012 Detecting anthropogenic carbon dioxide uptake and ocean acidification in the North Atlantic Ocean *Biogeosciences* **9** 2509–22

[89] Mendes J P, Coelho L, Kovacs B *et al* 2019 Dissolved carbon dioxide sensing platform for freshwater and saline water applications: characterization and validation in aquaculture environments *Sensors* **19** 5513

[90] Johnson M S, Billett M F, Dinsmore K J, Wallin M, Dyson K E and Jassal R S 2010 Direct and continuous measurement of dissolved carbon dioxide in freshwater aquatic systems—method and applications *Ecohydrology* **3** 68–78

[91] Zilberman Y, Ameri S K and Sonkusale S 2014 Microfluidic optoelectronic sensor based on a composite halochromic material for dissolved carbon dioxide detection *Sens. Actuators B* **194** 404–9

[92] Ameen F, AlNadhari S and Al-Homaidan A A 2021 Marine microorganisms as an untapped source of bioactive compounds *Saudi J. Biol. Sci.* **28** 224–31

[93] Medlin L and Orozco J 2017 Molecular techniques for the detection of organisms in aquatic environments, with emphasis on harmful algal bloom species *Sensors* **17** 1184

[94] Groben R, Eller G, Lange M and Medlin L K 2004 Using fluorescently-labelled rRNA probes for hierarchical estimation of phytoplankton diversity: a mini-review *Nov. Hedwigia* **79** 313–20

[95] Malzahn K, Windmiller J R, Valdés-Ramírez G, Schöning M J and Wang J 2011 Wearable electrochemical sensors for *in situ* analysis in marine environments *Analyst* **136** 2912

[96] Anirudhan T S and Alexander S 2015 Design and fabrication of molecularly imprinted polymer-based potentiometric sensor from the surface modified multiwalled carbon nanotube for the determination of lindane (γ-hexachlorocyclohexane), an organochlorine pesticide *Biosens. Bioelectron.* **64** 586–93

[97] Zhou N, Li J, Chen H, Liao C and Chen L 2013 A functional graphene oxide-ionic liquid composites–gold nanoparticle sensing platform for ultrasensitive electrochemical detection of Hg^{2+} *Analyst* **138** 1091

[98] Herzog G, Moujahid W, Twomey K, Lyons C and Ogurtsov V I 2013 On-chip electrochemical microsystems for measurements of copper and conductivity in artificial seawater *Talanta* **116** 26–32

[99] Kröger S and Law R J 2005 Biosensors for marine applications *Biosens. Bioelectron.* **20** 1903–13

[100] Cao L, Xiang H, Xu J *et al* 2022 Nutrient detection sensors in seawater based on ISI Web of Science Database *J. Sens.* **2022** 1–12

[101] Cosnier S, Innocent C and Jouanneau Y 1994 Amperometric detection of nitrate via a nitrate reductase immobilized and electrically wired at the electrode surface *Anal. Chem.* **66** 3198–201

[102] Lacombe M, Garcon V, Thouron D, Lebris N and Comtat M 2008 Silicate electrochemical measurements in seawater: chemical and analytical aspects towards a reagentless sensor *Talanta* **77** 744–50

[103] Aguilar D, Barus C, Giraud W *et al* 2015 Silicon-based electrochemical microdevices for silicate detection in seawater *Sens. Actuators B* **211** 116–24

[104] Altahan M F, Esposito M and Achterberg E P 2022 Improvement of on-site sensor for simultaneous determination of phosphate, silicic acid, nitrate plus nitrite in seawater *Sensors* **22** 3479

IOP Publishing

Sensors for Marine Biosciences
Next-generation sensing approaches
Shyam S Pandey, Rout George Kerry and Kshitij RB Singh

Chapter 4

Polymeric matrix-based biosensors for the detection of pollutants in the marine ecosystem

Cansu İlke Kuru-Sumer and Fulden Ulucan-Karnak

Biosensors combine the extreme sensitivity and specificity of biological systems with the processing capacity of microelectronics to provide simple, low-cost measuring instruments for use in the field or *in situ* deployment. While their potential for use in the marine environment is considerable, much of the published research to date has concentrated on applications in freshwater and wastewater. Polymers' intrinsic biocompatibility, as well as their distinctive optical, electrical, and mechanical characteristics, have piqued the interest of researchers, especially in biosensors. Measurement of biological, climatic, and anthropogenic changes and pollutants is the foundation for developing appropriate management plans for the long-term use and conservation of the maritime environment. Sensors have historically been employed to identify physical characteristics in marine investigations, but there is a rising demand for real-time information concerning chemical and biological aspects. These parameters are presently measured in seawater samples and then examined in the laboratory. Marine applications provide a significant barrier in terms of the resilience necessary for remote application, but recent advances in portable medical equipment and receptor design show that these needs may now be realistically met. In this chapter, these concepts will be investigated in detail.

4.1 Introduction
Being a primary biodiversity source, the marine environment is crucial to the management of the global climate. In order to address the issues facing our planet, such as a sustainable energy and food supply, new processes, materials, bioactive compounds and medical treatments, it may be possible to create new goods and services using marine resources. Nevertheless, the maritime ecosystem is also becoming more susceptible to climate change, and human activity has an impact on urban, industrial, and tourism growth [1].

doi:10.1088/978-0-7503-5999-3ch4

Massive volumes of saltwater are found in the ocean, and the ocean's general circulation of saltwater allows it to take part in many ecological processes, including the regulation of global heat, carbon, and climate cycles. Human activities such as trash disposal, harbour development in coastal zones, and mineral resource exploration can have a substantial impact on marine habitats. Long-term ocean management and protection depend on an understanding of all the variables and mechanisms that affect the health of the marine ecosystem. The main purpose is the awareness of the marine environment and its resources, including methane, radon, ferrous ions, carbon dioxide, bacteria, nutrients, and seafood, requiring an awareness of a wide range of chemical substances and animals. The utilization of marine resources and the sustained advancement of human civilization depend on the investigation and measurement of the substances and creatures found in the ocean [2, 3].

The Marine Strategy Framework Directive (MSFD), implemented in 2008, was the most significant effort of the European Union in this field [4]. It outlined all the steps required to create a marine environment that is safe, conserved, and healthy. The MSFD strives to attain or maintain a good environmental status in the maritime environment, which is defined as the ecologically diversified and dynamic oceans and seas that are clean, healthy, and productive. This safeguards the marine environment's potential for both current and future generations by enabling its sustainable use. To preserve the marine environment, MSFD recommends developing and putting into practice marine strategies that minimize pollution and stop it from deteriorating. It also makes sure that there are no major effects or risks to marine ecosystems and biodiversity, human health [1, 5].

By mid-July 2023, the European Commission is required by Article 23 of the directive to assess the directive and, if applicable, suggest any necessary adjustments. Adopted in May 2020, the EU's biodiversity plan for 2030 emphasizes the necessity of taking more aggressive measures to safeguard and restore marine ecosystems. Recalling the significance of attaining 'good status' for achieving all aquatic habitats' zero pollution goal under the Green Deal the European Commission stated that it would review and, if necessary, revise the marine strategy framework directive in the zero pollution action plan for air, water, and soil, which was adopted a year later [4].

4.1.1 Pollutants that should be avoided for marine waters to have a healthy environment

Chemical elements or compounds that are persistent, poisonous and likely to bioaccumulate, as well as substances groups that raise concerns of a similar nature, are classified as contaminants. In addition, the following categories are frequently used to categorize chemical contaminants: (a) stable trace elements, (b) organic substances, (c) hydrocarbon pollutions, and (d) radionuclides. However, monitoring the marine environment requires taking into account a number of criteria, including the type of contamination, its origins, its distribution, its concentration, its persistence, its uptake by the biota, and its effects on the ecosystem [6].

Although they are necessary for survival and are found in the environment naturally, heavy metals can become dangerous when they build up inside living things. Mercury, cadmium, arsenic, chromium, nickel, copper, and lead are a few of the heavy metals that poison the environment most frequently (figure 4.1). [7]

A classification of substances that may be discharged into the marine environment as a result of anthropogenic sea-based activities may be found in the literature. It contains 276 compounds together with the primary sea-based sources and their CAS identification number: There are 19 metals and metalloids, 10 organometallic compounds, 24 inorganic, 204 organic, and 19 radionuclides in all. The largest number of substances are contributed by offshore oil and gas operations, which are followed by shipping and mariculture operations [5]. In table 4.1, the most common hazardous compounds in marine environments are given.

Ocean pollution is a global concern that surpasses the boundaries of individual countries and is derived from multiple sources. In most nations, there is insufficient regulation, leading to a worsening of the situation. Approximately 80% of ocean pollution originates from terrestrial sources. Plastic waste is the most prominent form of ocean pollution and has attracted significant attention. Its detrimental effects include the harm it causes to marine life such as fish, dolphins, whales, and seabirds. Moreover, plastic waste decomposes into fibers containing various harmful and carcinogenic substances, as well as micro- and nanoparticles. These chemical-

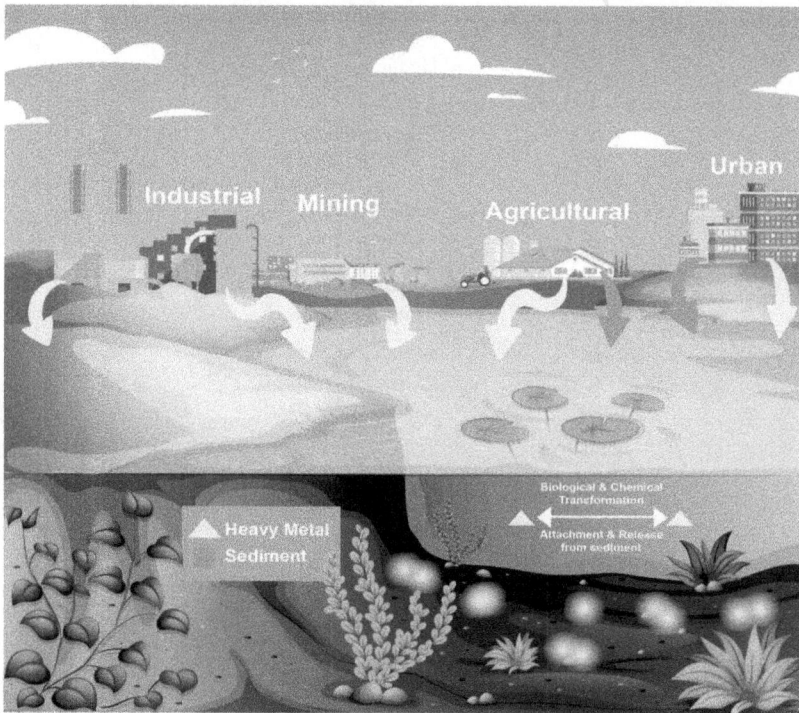

Figure 4.1. An explanation of environmental heavy metals via a diagram (reproduced with permission from [7] CC BY 4.0).

Table 4.1. Priority hazardous compounds in marine environment [1–3].

Heavy metals	Copper
	Cadmium
	Nickel
	Zinc
	Lead
	Mercury
Pesticides	Endosulfan
	Atrazine
	Trifluralin
	Chlorfenvinphos
	Simazine
	Chlorpyrifos
	İsoproturon
	Tributyltin-cation
Organotin compounds	Tributyltin compounds
Alkilphenolic compounds	Nonylphenol
	Ctylphenol
Organochlorine compounds	Chloroalkanes
	Brominated diphenyl ether
	Alachlor
	Di-(2-ethylhexyl) phthalate
	Dichloromethane–trichloromethane
	Chlorobenzenes
	Hexachlorobutadiene
	Pentachlorophenol
	Hexachlorocyclohexane
Aliphatic and polycyclic aromatic hydrocarbons	Benzene
	Fluoranthene
	Naphthalene
	Polyaromatic hydrocarbons
	Anthracene
	Benzopyrene
	Benzofluoranthene
	Benzoperylene
	Indenopyrene
Other priority hazardous substances	Bifenox
	Aclonifen
	17a-Ethinylestradiol
	17b-Estradiol
	Cybutryne
	Dichlorvos
	Cypermethrin
	Dicofol
	Terbutryn
	Diclofenac
	Quinoxyfen
	Perfluorooctane sulfonic acid
	Dioxins
	Hexabromocyclododecane Heptachlor/heptachlor epoxide

laden particles are absorbed by fish and shellfish, eventually entering the marine food chain and potentially being consumed by humans. The potential health risks posed by these contaminants are currently being assessed.

Further elements of marine pollution encompass mercury emitted from burning coal and small-scale gold mining, petroleum releases from oil spills and pipeline ruptures, enduring organic pollutants like PCBs and DDT, numerous synthetic chemicals with uncertain effects, pesticides, nitrogen, and phosphorus from agricultural runoff and animal feces, as well as sewage discharges with various microbial impurities. Alongside ocean warming and acidification, ocean pollution contributes to more frequent and severe harmful algal blooms, coral reef degradation, and the proliferation of life-threatening diseases. Therefore, new generation technologies are always needed to overcome these pollutions and their harmful effects [8–10].

4.2 Methods used to detect pollutants in marine waters

For a sustainable blue economy, scientists and researchers have been focusing on extensive ocean monitoring. Traditional methods like autonomous underwater vehicles (AUVs), gliders, profiling floats, volunteer measurements from ships, drifters, and sensing nodes with cable networks are now accessible for ocean monitoring. All of these methods of marine monitoring often assess temperature, conductivity, pH, salinity, dissolved oxygen, chlorophyll fluorescence, turbidity, and colour dissolved organic matter (CDOM) [11]. Before any analysis, sample integrity must be preserved during sampling, transportation, and storage. Correct sample handling, processing, and pre-treatment, analysis setup, instrument operation, accuracy, and calibration, and lastly data analysis, interpretation, and reporting are all necessary for an accurate analysis [12].

The most frequent method for measuring marine pollution is to collect *in situ* water samples from various depths of sea using boats/ships equipped with water samplers. In the laboratory, water samples are analysed to identify the physical and chemical qualities of the water. Such procedures are precise yet time-consuming and geographically limited, necessitating the use of qualified professionals and laboratory analysis. However, real-time or near-real-time measurements of marine pollutants and toxins at several spatial scales are required for monitoring and managing environmental impacts as well as understanding the processes governing their spatial distribution. Researchers also studying water resources and decision-makers can monitor waterbodies more successfully with the use of remotely sensed data. The qualitative characteristics of waterbodies have been extensively measured using remote sensing techniques [11, 13].

Since data collection and availability provide the biggest challenges to target setting, trend monitoring, and marine environment assessment, a variety of tactics and instruments are used in these areas. These include sampling, observation, and marine measurements and analysis. In the past, samples were taken aboard ships and examined there or when they got back to the lab. For the purpose of doing automated marine measurements, a number of platforms and equipment have been created, including drifter, submersible, satellite, remote-operated vehicle, towed body and 'SmartBuoy.' These tactics include direct sampling, utilization of electromagnetic waves for distant sensing and acoustic techniques, hydrological measures utilizing probes for temperature, depth, conductivity, and aerial and satellite photography [5].

Zampoukas *et al* list moorings and buoys, ships, gliders, underwater video and imagery, underwater acoustics, continuous plankton recorders, remote sensing, autonomous underwater vehicles, and other techniques as potential aids for an effective monitoring of the spatial scale pertinent to the MSFD. Nonetheless, the majority of autonomous or remote measurements rely on sensors to examine oceanographic characteristics like depth, turbidity, conductivity, and temperature [14].

The fact that marine pollutants can exist in small enough concentrations to have an adverse effect on marine ecosystems makes monitoring them one of the biggest challenges. As a result, the analytical techniques ought to be capable of identifying and detecting these contaminants at the pg l^{-1} or ng l^{-1} level. For instance, the levels of steroid estrogens have been measured by gas chromatography with tandem mass spectrometry (GC–MS/MS) or liquid chromatography-tandem mass spectrometry with electrospray ionization in negative mode (LC/ESI-MS/MS). Alternative analytical methods, such as the combination of triple quadrupole linear ion-trap MS (UHPLC-QTRAP1MS) and ultrahigh-pressure LC (UHPLC), enable the detection of polar organic pollutants in marine waters. However, such analytical approaches needed sample preparation and sometimes preconcentration processes in order to undertake the quantitative assessment of trace quantities of marine pollutants. These requirements could result in a lengthy analysis process that would cost more money and resources than they would have in the case of *in situ* continuous monitoring. Thus, rapid alert systems should not be used with lengthy result interpretation times [15].

As well as biosensors for real-time monitoring of biohazards and man-made chemical pollutants in the marine environment, the development of novel multi-functional sensors for *in situ* monitoring of the marine environment and related maritime operations are two areas where ocean management challenges are being addressed by joining forces in research. Sensing technologies are needed to enhance accurate measurements of important water characteristics. Therefore, it is vitally necessary to develop early-warning systems, like sensor systems, in order to enable sensitive and selective marine contaminant detection [1–4, 6].

4.3 Polymeric matrix-based biosensors for marine determinants

The growing population and disputes brought on by industrial activity along the shore have an effect on the maritime environment. Activities that affect the water quality in the coastal zone include garbage disposal, port development, dredging, and extraction procedures. Because of the negative impacts on health and socio-economic situations caused by these marine toxins, it is critical to identify them on-site before they reach the market. In order to foresee the effects of change, plan for mitigation, and direct adaptation, ocean monitoring and study are crucial. While physical parameters are often determined by sensors in marine research, real-time chemical, and biological parameter information is becoming more and more in demand. At the moment, these characteristics are assessed in samples that are taken at sea and then examined in a lab. We are now able to collect data on every facet of marine health because of the expanding scientific capacity to combine biological identification with platform engineering. Researchers use this capacity to watch the ocean in real time, making essential decisions and protecting this vital resource, by combining a hierarchy of measuring devices from satellite to *in situ* [5, 11, 16, 17].

A wide range of innovative biosensor designs, including molecularly imprinted polymers and nano-biosensors, provide insight into the problems associated with developing and deploying marine biosensors. Molecularly imprinted polymers (MIPs) have proven to be a viable option for antibodies in ELISA biosensors, which rely on optical and electrochemical detection. Electrochemical sensors combined with biomimetic recognition components, such as synthetic MIPs or aptamers, have the potential to provide very sensitive and specific analysis (figure 4.2) [12, 13, 18].

Coatings are often employed to increase [1] selectivity and [2] biocompatibility of biosensors [14]. Using polymers as a coating with a combination of biosensors is an increasing demand in the biosensor development field. Polymers are helpful tools to obtain low detection limits, sensitivity, efficient electron transfer, good stability and conductivity. The current advancements in polymer materials have a significant impact on the electron transport kinetics of enzyme electrodes and redox reactions. It is clear that research on new assembly techniques is still concentrated on achieving high specificity and sensitivity, fast reaction times, and application flexibility [20, 21].

Natural polymers such as alginate, cellulose, starch and chitosan are commonly employed in combination with sensors. The versatility of biopolymers stems from their features, such as their capacity to collect and aggregate analytes on surfaces of sensor and the ability to produce and use them in a variety of forms (film, sponge, hydrogel). Their composites are also advaced to imrpove durability of the material, mechanical strength, yield, quality, stability, hardness and stifness [22].

Polymeric-based biosensors could be used to detect many ingredients in the marine environment. Nutrients (nitrate, nitrit, phosphate, silicate etc), anti-

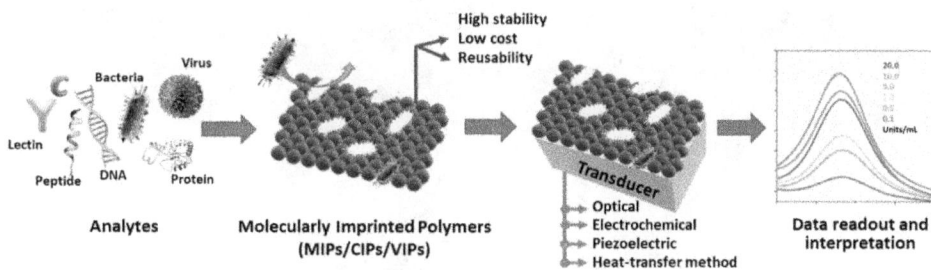

Figure 4.2. Biosensor development schematics using MIPs as bio-recognition components for a range of bioanalytes. Biomarkers such as toxins, tiny compounds, surface or secreted proteins, or even entire cells can be targeted for the purpose of disease detection. MIPs created specifically for viruses are known as virus imprinted polymers (VIPs), and those created specifically for complete cells, such as bacteria, are referred to as cell imprinted polymers (CIPs) (reproduced with permission from [19], copyright 2022 The Authors, published by Elsevier B.V.).

biofouling agents, pesticides (biocides), endocrine disruptors are the main examples of these bioanalytes (figure 4.3).

In table 4.2, polymeric-based-material-used biosensors for the marine environment are listed, especially in last 5 years. Different polymeric materials can be found for different bioanalytes detection.

Other types of biosensors can be found for the purpose of water contaminant detection. Although their recognition mechanisms, transducers and technologies are different, the aim remains stable. Because of the fast industrialization, growing urbanization, rising living standards, and reliance on irrigated agriculture, there is an acute increase in the need for clean, fresh water. There is a severe shortage of clean drinking water in the world, not just due to dwindling supply but also to declining quality. Because of this, a lot of governments are prioritizing lowering the amount of harmful effluents that contaminate water and enter the ecosystem [35].

Biosensor technologies are increasingly demanded in the detection and monitoring of marine-based pollutants, nutrients and other chemicals. The robustness needed for remote and *in situ* applications is a significant barrier for marine applications, but new advances in portable medical equipment and receptor design indicate that these demands can now be met.

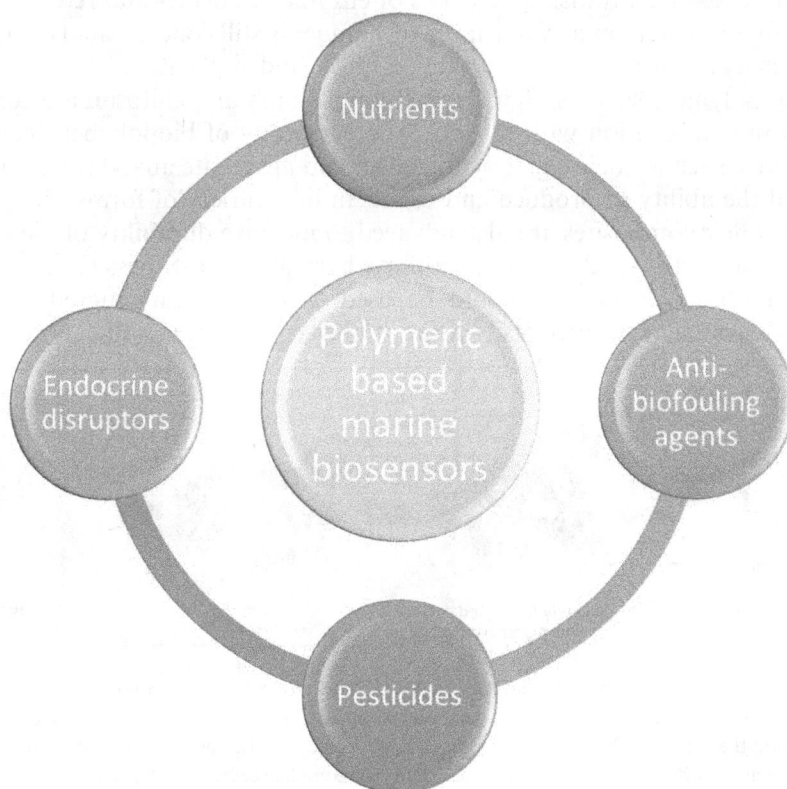

Figure 4.3. Subfields of bioanalytes that are detected with polymeric-based marine biosensors.

Table 4.2. Polymeric material-based marine biosensor studies in the literature.

Polymeric material	Analyte	References
Poly (diallyl dimethyl ammonium chloride) and Chitosan	Heavy metal ions (Zn^{2+}, Cu^{2+}, $Cr_2O_7^{2-}$ and Ni^{2+}) and pesticides	[23]
Peptide modified magnetic beads embedded polymeric membrane	*Staphylococcus aureus*	[2]
Aptamer imprinted polymer (chitosan-graphene)	Lead ion (Pb(II))	[24]
Microarray glass chips	Triazine biocide, chloramphenicol and sulfonamide antibiotics, polybrominated diphenyl ether flame-retardant, hormone, algae toxin	[25]
poly(methyl methacrylate) (PMMA) based microfluidic disc	Heavy metal content (Cu^{2+}, Zn^{2+}) and Ah3w fluorescent-peptide	[26]
PET-binding peptide modified AuNp	Polyethylene terephthalate (PET)	[27]
Anti-fouling polymers onto Au layers	Biotin-avidin	[28]
Escherichia coli pBAV1K-ACU-lucFF cells	Poly (acrylic acid)	[29]
Aptamer-HRP complex	Palytoxin	[30]
Algae based digital microfluidic	Copper, lead, phenol, and Nonylphenol (NP)	[31]
poly(3,4-ethylenedioxythiophene): poly(styrene sulfonic acid) (PEDOT:PSS)	*Navicula* sp. and *amphiprore* sp.	[32]
hydrophobin based chimera binded polystyrene multiwell plates	Mercury (II) ions	[33]
Aptamer binded divinyl sulfone beads	Brevetoxins (Btxs)	[34]

4.4 Conclusions

Water sustainability and marine conservation, also referred to as ocean conservation, are closely related. After all, the global ecosystem of our world depends on a healthy ocean. The specific goal of marine conservation is to save and maintain the ecosystems found in the world's water systems. But just as human population growth is having a detrimental effect on the climate problem, the ocean is also suffering. Marine life has been directly damaged by overexploitation and irresponsible water management practices. In a similar vein, when glaciers melt, global warming has raised ocean temperatures and raised sea levels. Ocean habitats are being harmed by all of these factors, which puts the survival of biodiversity worldwide in jeopardy [3].

Healthy oceans and seas are necessary for human survival and life on Earth. We as a society have to be alert to create sustainable marine systems and oceans for the future. It is possible that future generations will have very different expectations for their seas than do we. Even while we are unable to foresee these needs, we can evaluate whether the way the sea is now used will prevent future generations from making the most of the oceans [32, 36].

The creation of better scientific understanding is of utmost importance, as stated in the Commission Decision on GES criteria. This is especially true through the EU Marine and Maritime Research agenda, which is part of the Europe 2020 agenda. In light of the aforementioned factors, the European Commission's Seventh Framework Programme and the forthcoming Horizon 2020 offer potential for the start of pertinent marine research. A number of European initiatives are now being developed to support EU laws and regulations like MSFD. The FP7 suggested the creation of novel multifunctional sensors for *in situ* monitoring of the marine environment and associated maritime activities, as well as biosensors for real-time monitoring of natural and man-made chemical contaminants in the marine environment [1].

It should be clear that developing biosensors for *in situ* measurements is a challenging but highly valuable task, and that efforts are being made to overcome some of the concerns brought up. Even if the technology has a lot of potential for use in future observational tactics, there are a few obstacles that need to be taken into account in order to prevent widespread adoption. For *in situ* high-frequency measurements to be performed successfully, the instruments used must be extremely durable against potential physical impacts and able to function in an extremely corrosive environment. The ultimate goal of most improvements is to be able to make such observations, but even so, a major advancement that shouldn't be disregarded is the ability to perform decentralized measurements in the field rather than needing to process collected materials in specialized laboratories. This is the domain in which biosensors are expected to find their first meaningful uses: the accessible, low-cost test that can be used to pre-screen samples, potentially assisting in guiding the sample collection effort for more in-depth chemical analysis [6].

The growing use of marine resources has brought attention to the critical issue of marine biotoxin contamination, which has become a significant concern in recent years. Protecting both human health and marine ecosystems requires continuous monitoring and identification of marine biotoxins (MBs). At present, biosensor detection has become a crucial method for detecting marine biotoxins, providing high sensitivity and specificity, and effectively contributing to the protection of marine environments and human health. Although substantial progress has been achieved in marine biotoxin detection technology, practical obstacles remain, underscoring the need for further exploration of the advantages and drawbacks of new detection technologies. Furthermore, the exploration of using multiple technologies together shows potential for improving detection efficiency, accuracy, and ensuring food safety.

The preceding discussion illustrates the complexity and significance of developing biosensors for on-site measurements, as well as ongoing efforts to address associated challenges. While the primary aim of advancements is to enable such observations, a significant development worth noting is the ability to conduct decentralized field measurements rather than relying on specialized laboratory analysis. It is anticipated that biosensors will initially serve in this capacity, providing affordable and accessible preliminary screening of samples to facilitate more extensive chemical analysis through informed sample collection.

4.5 Challenges and future perspectives

The fact that several biosensing techniques have been developed to date for the identification of various marine toxins, nutrients and hazardous substances indicates how important this issue is to our daily existence. To clarify, while the development of traditional techniques might extend researchers' understanding of the detection platform concept, some obstacles may restrict their use in bioanalytes determination. For this aim to be achieved, biosensors must be more practical for rapid screening and on-site use in order to produce trustworthy low-carbon footprint environmental monitoring data with a lower amount of hazardous waste. One of the most significant drawbacks among them is the mobility of biosensing. In order to provide point-of-care (POC) analysis of these bioanalytes, microfluidics devices, lateral flow assays, and smartphone-based sensors have garnered a lot of attention lately. These biosensors have several advantages, such as mobility, sensitivity, and selectivity. In addition, even non-expert users may readily utilize them. It is evident how important bioreceptors and nanomaterials were to their development. In terms of nano-materials, scientists tried to enhance the characteristics of these platforms by taking use of the advantages of metallic, carbon-based, polymeric nanoparticles. However, there is a deficiency in the use of various kinds of nanomaterials, particularly those based on silica, including mesoporous silica nanoparticles. Additionally, emerging nanomaterials including metal–organic frameworks (MOFs) and a family of two-dimensional inorganic compounds (MXene) may offer fresh perspectives on how to build reliable, transportable electrochemical and optical biosensors. Antibodies and aptamers were used in the majority of manufactured portable biosensors for measurement in order to boost the selectivity and sensitivity of the bioreceptors; nonetheless, the application of alternative bioreceptor kinds is severely lacking. To identify potential storage options and, more crucially, to allow the commercial use of portable optical and electrochemical biosensors for POC examination, it is necessary to investigate the portable biosensors stability, which are either related to nanomaterials [37–40].

Overall, the portable biosensors that have been described are still in their infancy, and the majority of their testing has taken place in lab environments. Therefore, controlled settings employing real-life samples with little-to-no user input are one of the key challenges for these biosensors in the future.

Listed below are the current trends and challenges encountered by biosensors in the detection of marine biotoxins (MBs).

(1) Miniaturization and portability: There is a growing trend towards smaller and portable biosensors, enabling real-time and on-site MB detection. This advancement allows for rapid and automated identification of MBs using minimal sample quantities and processing steps. Additionally, it facilitates easier field monitoring and the establishment of early-warning systems for MB outbreaks.

(2) Multiplexed detection: The increasing popularity of biosensors capable of simultaneously detecting multiple toxins, known as multiplexed detection, is driven by the need for comprehensive monitoring of maritime environments and the desire to reduce the time and cost associated with such studies.

(3) Nanotechnology integration: The enhancement of sensitivity and selectivity in biosensor designs can be achieved through the incorporation of nanotechnology. The utilization of nanomaterials, including nanoparticles and nanocomposites, is currently under examination to improve the performance of biosensors in detecting marine biotoxins.

(4) The attachment of biomolecules (e.g., antibodies and aptamers) to sensor surfaces is facilitated by advancements in surface functionalization techniques, thereby enhancing the stability and specificity of biosensors for marine biotoxin detection.

(5) Biosensors are increasingly combined with data analytics and Internet of Things platforms to enable real-time data analysis and monitoring. This integration allows for continuous surveillance of maritime habitats and swift response to toxicological incidents.

(6) Despite advancements, challenges such as cross-reactivity, sample matrix interference, standardization and validation issues, concerns about field deployment, and reliability hinder the use of biosensors for MB detection. Overcoming these challenges and leveraging new advancements will drive the continual evolution of biosensors for MB detection, thereby enhancing public health and marine environment monitoring [41].

The ultimate goal in the fabrication and evolution of a sensor or biosensor is to achieve the commercialization of sensing systems, potentially stemming from the success of their features.

References

[1] Justino C I L, Freitas A C, Duarte A C and Santos T A P R 2015 Sensors and biosensors for monitoring marine contaminants *Trends Environ. Anal. Chem.* **6–7** 21–30

[2] Liu Y, Yang Y, Fan Y, Zhao Q, Gao G and Zhi J 2023 Feasibility investigation and development of microbial electrochemical biosensors for marine pollution monitoring *Talanta* **255** 124204

[3] Ward D, Melbourne-Thomas J, Pecl G T, Evans K, Green M, McCormack P C *et al* 2022 Safeguarding marine life: conservation of biodiversity and ecosystems *Rev. Fish Biol. Fish.* **32** 65–100

[4] Directive—2008/56—EN—EUR-Lex. https://eur-lex.europa.eu/eli/dir/2008/56/oj (accessed 21 August 2024)

[5] Tornero V and Hanke G 2016 Chemical contaminants entering the marine environment from sea-based sources: a review with a focus on European seas *Mar. Pollut. Bull.* **112** 17–38

[6] Kröger S and Law R J 2005 Biosensors for marine applications *Biosens. Bioelectron.* **20** 1903–13

[7] Mitra S, Chakraborty A J, Tareq A M, Emran T B, Nainu F, Khusro A *et al* 2022 Impact of heavy metals on the environment and human health: novel therapeutic insights to counter the toxicity *J. King Saud. Univ. Sci.* **34** 101865

[8] Trégarot E *et al* 2024 Effects of climate change on marine coastal ecosystems—a review to guide research and management *Biol. Conserv.* **289** 110394

[9] Yuan Z, Nag R and Cummins E 2022 Human health concerns regarding microplastics in the aquatic environment—from marine to food systems *Sci. Total Environ.* **823** 153730

[10] Landrigan P J, Stegeman J J, Fleming L E, Allemand D, Anderson D M, Backer L C *et al* 2020 Human health and ocean pollution *Ann. Glob. Health* **86** 1–64

[11] Hafeez S, Wong M S, Abbas S, Kwok C Y T, Nichol J, Lee K H *et al* 2018 Detection and monitoring of marine pollution using remote sensing technologies *Monitoring of Marine Pollution* (IntechOpen) ch 2

[12] Reichelt-Brushett A 2023 Collecting, measuring, and understanding contaminant concentrations in the marine environment *Marine Pollution – Monitoring, Management and Mitigation* (Springer) pp 23–51

[13] Gholizadeh M H, Melesse A M and Reddi L 2016 A comprehensive review on water quality parameters estimation using remote sensing techniques *Sensors* **16** 1298

[14] Zampoukas N, Piha H, Bigagli E, Hoepffner N, Hanke G and Cardoso A C 2012 Monitoring for the marine strategy framework directive: requirements and options *Publications Office of the European Union* https://publications.jrc.ec.europa.eu/repository/handle/JRC68179 (accessed 21 August 2024)

[15] Ma Z, Meliana C, Munawaroh H S H, Karaman C, Karimi-Maleh H, Low S S *et al* 2022 Recent advances in the analytical strategies of microbial biosensor for detection of pollutants *Chemosphere* **306** 135515

[16] Bano K, Khan W S, Cao C, Khan R F H and Webster T J 2020 Biosensors for detection of marine toxins *Nanobiosensors* (New York: Wiley) ch 14 pp 329–56

[17] Regan F and Hansen P D 2023 Biosensors for the marine environment: introduction *Biosensors for the Marine Environment* (Berlin: Springer) pp 1–9

[18] Kurbanoglu S, Yarman A, Zhang X and Scheller F W 2023 Electrochemical MIP sensors for environmental analysis *Biosensors for the Marine Environment* (Berlin: Springer) pp 139–64

[19] Rajpal S and Mishra P 2022 Next generation biosensors employing molecularly imprinted polymers as sensing elements for *in vitro* diagnostics *Biosens. Bioelectron.* X **11** 100201

[20] Davis F and Higson S P J 2007 Polymers in biosensors *Biomedical Polymers* (Elsevier) pp 174–96

[21] Alam M W, Islam Bhat S, Al Qahtani H S, Aamir M, Amin M N, Farhan M *et al* 2022 Recent progress, challenges, and trends in polymer-based sensors: a review *Polymers (Basel)* **14** 2164

[22] Wang X and Uchiyam S 2013 Polymers for biosensors construction *State of the Art in Biosensors—General Aspects* (Rijeka: InTech)

[23] Madej-Kiełbik L, Gzyra-Jagieła K, Jóźwik-Pruska J, Dziuba R and Bednarowicz A 2022 Biopolymer composites with sensors for environmental and medical applications *Materials* **15** 7493

[24] Lv E, Li Y, Ding J and Qin W 2021 Magnetic-field-driven extraction of bioreceptors into polymeric membranes for label-free potentiometric biosensing *Angew. Chem. Int. Ed.* **60** 2609–13

[25] Zhu N, Liu X, Peng K, Cao H, Yuan M, Ye T *et al* 2022 A novel aptamer-imprinted polymer-based electrochemical biosensor for the detection of lead in aquatic products *Molecules* **28** 196

[26] Sanchis A, Salvador J P, Campbell K, Elliott C T, Shelver W L, Li Q X *et al* 2018 Fluorescent microarray for multiplexed quantification of environmental contaminants in seawater samples *Talanta* **184** 499–506

[27] Maguire I and Regan F 2019 A design pipeline for development of a multi-analyte marine bio-sensor lab-on-a-disc platform *OCEANS 2019—Marseille* (Piscataway, NJ: IEEE) pp 1–8

[28] Behera A, Mahapatra S R, Majhi S, Misra N, Sharma R, Singh J *et al* 2023 Gold nanoparticle assisted colorimetric biosensors for rapid polyethylene terephthalate (PET) sensing for sustainable environment to monitor microplastics *Environ. Res.* **234** 116556

[29] Jesmer A H, Huynh V, Marple A S T, Ding X, Moran-Mirabal J M and Wylie R G 2021 Graft-then-shrink: simultaneous generation of antifouling polymeric interfaces and localized surface plasmon resonance biosensors *ACS Appl. Mater. Interfaces* **13** 52362–73

[30] Puhakka E and Santala V 2022 Method for acrylic acid monomer detection with recombinant biosensor cells for enhanced plastic degradation monitoring from water environments *Mar. Pollut. Bull.* **178** 113568

[31] Gao S, Zheng X, Hu B, Sun M, Wu J, Jiao B *et al* 2017 Enzyme-linked, aptamer-based, competitive biolayer interferometry biosensor for palytoxin *Biosens. Bioelectron.* **89** 952–8

[32] Han S, Zhang Q, Zhang X, Liu X, Lu L, Wei J *et al* 2019 A digital microfluidic diluter-based microalgal motion biosensor for marine pollution monitoring *Biosens. Bioelectron.* **143** 111597

[33] Liao J, Lin S, Liu K, Yang Y, Zhang R, Du W *et al* 2014 Organic electrochemical transistor based biosensor for detecting marine diatoms in seawater medium *Sens. Actuators B* **203** 677–82

[34] Pennacchio A, Giampaolo F, Piccialli F, Cuomo S, Notomista E, Spinelli M *et al* 2022 A machine learning-enhanced biosensor for mercury detection based on an hydrophobin chimera *Biosens. Bioelectron.* **196** 113696

[35] Laad M and Ghule B 2023 Removal of toxic contaminants from drinking water using biosensors: a systematic review *Groundw. Sustain. Dev.* **20** 100888

[36] Bailey D M and Hopkins C R 2023 Sustainable use of ocean resources *Mar Policy* **154** 105672

[37] Shati A A, Al-dolaimy F, Alfaifi M Y, Sayyed R Z, Mansouri S, Aminov Z *et al* 2023 Recent advances in using nanomaterials for portable biosensing platforms towards marine toxins application: up-to-date technology and future prospects *Microchem. J.* **195** 109500

[38] Kröger S and Law R J 2005 Biosensors for marine applications *Biosens. Bioelectron.* **20** 1903–13

[39] Huang C W, Lin C, Nguyen M K, Hussain A, Bui X T and Ngo H H 2023 A review of biosensor for environmental monitoring: principle, application, and corresponding achievement of sustainable development goals *Bioengineered* **14** 58–80

[40] Rodriguez-Mozaz S, Alda M, Marco M and Barcelo D 2005 Biosensors for environmental monitoring a global perspective *Talanta* **65** 291–7

[41] Zhu X, Zhao Y, Wu L, Gao X, Huang H, Han Y *et al* 2024 Advances in biosensors for the rapid detection of marine biotoxins: current status and future perspectives *Biosensors* **14** 203

IOP Publishing

Sensors for Marine Biosciences
Next-generation sensing approaches
Shyam S Pandey, Rout George Kerry and Kshitij RB Singh

Chapter 5

Remote sensing and satellite biological sensors for sustainability of the marine ecosystem

Merve Asena Özbek, Gaye Ezgi Yılmaz, Aykut Arif Topçu and Adil Denizli

Oceans, containing high amounts of seawater, are the home of the marine ecosystem and play a crucial role in maintaining the marine ecosystem biodiversity; hence, marine environment safety is a global issue around the World. Remote sensing (RS) technology provides new opportunities for monitoring biodiversity and measuring some health indicator levels without being in physical contact with the area of interest and opens a new avenue for maintaining the ecosystem. In this context, we emphasize the usability of RS technology for marine environment safety and summarize the basic concepts of RS using some of indicators used in environment health.

5.1 Introduction

Oceans contain huge amounts of seawater and cover over 70% of the Earth's surface. These macroscopic fluid systems are the home of marine ecosystems and play an essential role in protecting marine biodiversity [1]; furthermore, oceans participate in some important life circulation cycles such as carbon cycles, global heat cycles, and climate systems [2]. So, the amounts of ocean constituents are paramount to protecting marine ecosystems and biodiversity; for instance, the higher uptake of CO_2 by the oceans leads to an increase in ocean acidification and results in some problems for marine ecosystems [1]. On the other hand, the discharge of industrial waste into coastal areas, human-caused global warming, rising sea levels, and others have detrimental effects on marine ecosystems [3, 4]; hence, some of the precautions such as the monitoring of ocean ingredients, the construction of marine protected areas (MPAs), the use of mangroves in coastal zones, and RS have been taken in protecting the marine ecosystem health, biodiversity, and our planet as well (figure 5.1).

doi:10.1088/978-0-7503-5999-3ch5
5-1

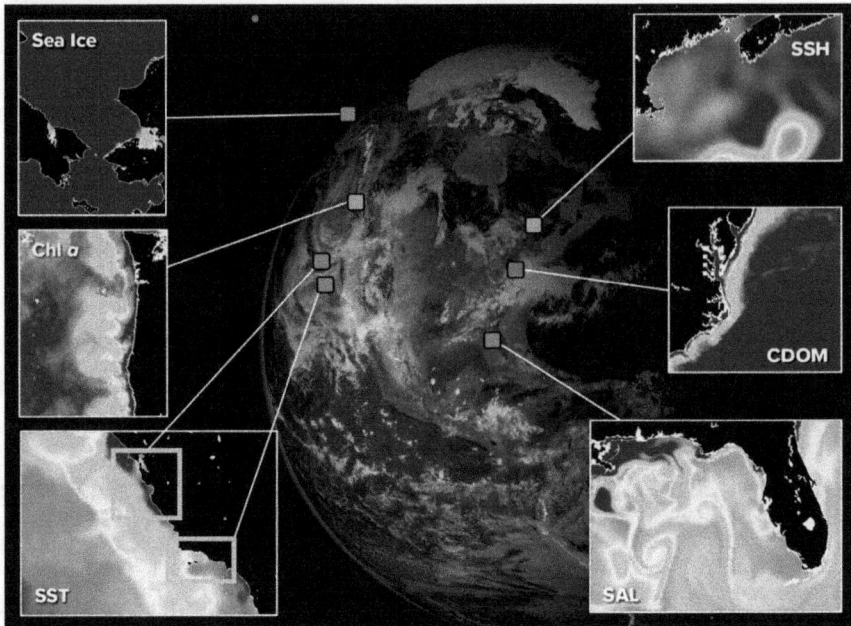

Figure 5.1. An example of RS study on different locations where were illustrated with red, blue, and green colors using some biological and physical parameters (reproduced with the permission from [4] © National Oceanic and Atmospheric Administration, United States Department of Commerce, CC BY 4.0).

Biodiversity is an important indicator reflecting the ecosystem health and the observation of biodiversity with RS. In particular, observing ocean color gives some valuable information about the marine ecosystem health such as dominant taxa or the biomass [4]. With RS systems, the color devices measure the ultraviolet, visible, and near-infrared light where the top of the Earth's atmosphere and the remote-sensing reflectance (R_{rs}), that is, the measure of the Earth's surface color, is gained using the atmospheric correction algorithms [4]; however, the constitutions of the marine ecosystem such as the amount or the concentration of the particles, chlorophyll-a, which could absorb or the scatter of the light change the measurement level of light. For instance, the coral diversity, algae diversity, benthic habitat characteristics, and plankton communities were monitored by RS platforms, including, suborbital sensors [5–8], nanosatellites [9], the Coastal Zone Color Scanner (CZCS) [4] using the color and turbidity of the ocean.

From this perspective, the complexity of the ocean ecosystem, and the process, of the change of environmental conditions such as global warming, urbanization, sea-level rise, etc, have affected the marine ecosystem balance; so, the monitoring of these changes with the recent technologies could be useful for maintaining the marine ecosystem, biodiversity, human health, and our planet as well.

In this section, we will first address the development of RS and then, the applicability and limitations of RS for marine ecosystem balance will be discussed.

5.2 Satellite remote sensing technology on the marine ecosystem

Sustaining biodiversity, supporting livelihoods, regulating climate, and providing key ecosystem services, marine ecosystems are crucial; however, marine ecosystems are threatened in many ways, including by pollution, overfishing, climate change, and habitat degradation. Thus, monitoring of ecosystem in an effective way is highly crucial to ensure long-term sustainability.

RS and satellite biological sensors have emerged as powerful tools for monitoring environment safety and these technologies offer new opportunities to improve our comprehension of the dynamics of marine ecosystems by supplying valuable data on ocean conditions and marine life from space-based platforms.

Satellite remote sensing (SRS) civilian applications are rapidly expanding due to the improvements in platform and sensor technology, data transmission and storage, and rising consumer demand for satellite data products. A variety of practical industries, including hydrology, land use, and agriculture, use RS satellite data that is provided by the integration of communications, meteorology, positioning, and aviation. Nowadays, satellites are one of the effective tools used in environmental observation [10, 11]. SRS of the marine environment is a potential technology for conservation issues and has become essential for environmental monitoring of ecology and determining the detrimental effects on habitats. Besides, some of the key data sources or the keynotes, e.g., high-resolution photos used in understanding the relationship between marine animals and their habitats have been systematically collected via satellites around the world to protect the marine environment's safety [12, 13].

Both on-site and RS data sources can be used to gathering information about the maritime environment. Although localized sensors are *in situ* data sources and can produce accurate point measurements, their geographic coverage is constrained and their results may not be always transmitted in real time. In contrast, RS provides valuable data about the ocean surface without requiring physical contact with the concerned environment [14, 15].

5.3 Important parameters used in monitoring of marine ecosystem health

One of the main purposes of RS technologies is to maintain the marine ecosystem, marine diversity and minimize the detrimental effects of environmental changes on marine ecosystem health using the observation and collection of spatial, spectral, radiometric, and temporal resolutions of an area of interest with the help of aircraft and satellites [16]. Additionally, monitoring of marine health indicators such as salinity, water turbidity, and color of the ocean gives valuable information about the marine ecosystem balance [16]. As a result, the input of the ecosystem constitutes the above-mentioned that are observed by RS systems to give some predictions about the ecosystem's future.

Herein, the important parameters used in monitoring marine ecosystem health including ocean color, habitat mapping, the measurement of physical parameters, and oil spills with the RS will be just summarized; but, in the following part, the monitoring of these important parameters with the help of RS platforms will be exemplified with the recent articles and the latest approaches.

5.3.1 Ocean color

Ocean color is one of the important parameters to determine the marine ecosystem balance and the change of its color depending on absorption, reflection, or scattering is used to predict the ecosystem health; for instance, the light in the red wavelength is absorbed by the uncontaminated water reflecting the blue color that is the indicator of marine ecosystem health and it is possible to calibrate the sensor that is used in mapping studies of the area of interest using the ocean color [17].

The determination of chlorophyll-a that is responsible for photosynthesis and coral reefs that is the indicator of the highly productive areas is used in predicting the marine ecosystem health via change of the color of the marine ecosystem [16]. For instance, the determination of green-yellow color reflected by chlorophyll-a is the indicator of marine ecosystem health, whereas ocean bleaching is a sign of deterioration of the marine ecosystem depending on coral dying due to changes in the ecosystem conditions. So, the determination of coral reefs and chlorophyll-a with satellite remote systems via color change is useful to predict the marine ecosystem's safety and maintain biodiversity.

5.3.2 Mapping studies

Mapping studies of the marine ecosystem is another approach to give some valuable information about the habitat of the area of interest using satellite and the wetlands and mangroves that are the habitats of the various species; so, these habitats have been mapped out in long or short time periods with the help of satellites to obtain information about the changes of the surface of the habitats and their dynamics; moreover, the ecosystem health and the biodiversity (bloom algae community, sediment bacteria, seagrass) of the area of interest can be imaged by using satellite color mapping as well [17].

5.3.3 Determination of physical parameters and oil spills

The climate of the marine ecosystem is the key parameter for maintaining biodiversity and sea surface temperature (SST), which is one the main indicators used in determining the climate water quality of the coastal region [18] and monitoring the productivity, climate change, and pollution of the target area [17] (figure 5.2). Salinity is another indicator used in the evaluation of water balance, evaporation rate and ocean currents, and its low level provides the quality of freshwater sources [16]. The combination of satellite color images and algorithms are utilized to measure its level [18]. Sea currents, waves, and surface winds have detrimental effects on marine ecosystems, for instance, these parameters could exchange the air at the sea surface resulting in sea-level rise, and changing the environmental conditions of the polar zones [19]. Moreover, the transfer of heat, salt, and pollutants by the combination of sea currents and winds could change

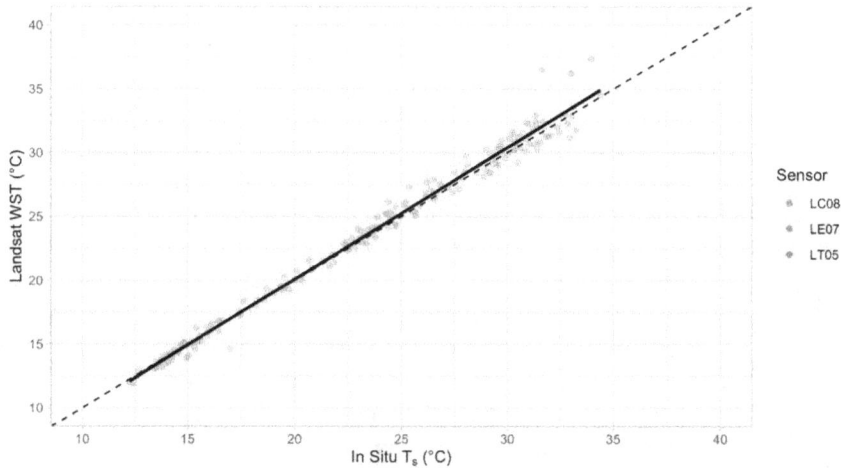

Figure 5.2. The comparisation results of water surface temperature (WST) and surface temperature (T_s) from the Salton Sea Validation Site via Scatter-plot of Landsat (reproduced with permission from [18], copyright © 2021 California Institute of Technology, Gov't sponsorship acknowledged, published by American Chemical Society).

the ecosystem dynamic and lead to erosion, and sediment transport [19]. Sea oil pollution is the other major pollutant for marine ecosystems and the discharge of oil derivatives from ships or oil tankers without treatment and the oil slicks from the oil drilling or the oil pipes are the main sources of these environmental pollutants [20].

5.4 Some of the limitations of remote sensing

Before monitoring or collecting some information using satellite data or ground-based information, the functions, locations, limits, and biodiversity of the ecosystem of interest should be investigated. For this purpose, the Red List of Ecosystems provides a thorough structure for categorizing and tracking ecosystems and could be used in tracking and reporting on the condition of ecosystems around the world [21].

Since temporal, spatial, and spectral resolutions are intrinsic limits of SRS, it is reasonable to expect that data product characteristics may affect mapping accuracy and monitoring prospects for particular ecosystem processes in particular locations. As long as data and product needs are well defined, integrated utilization of numerous RS sources and enhanced RS capability can assist in overcoming many of these well-known obstacles [22, 23]. It should be clear in discussions on ecosystem function monitoring which processes are being watched over for each function under consideration. To accurately depict the required ecosystem function, field observations must frequently be supplemented with RS proxies. Combining satellite data analysis with on-site measurements or laboratory process measurements can be necessary for improving the capability and value of satellite-based indicators for monitoring ecosystem function. This is particularly crucial in highly dynamic environments like coastal waters and the seafloor [24–26].

Because the majority of the data gathered to monitor ecosystem processes will be substituted, it is crucial to evaluate and acknowledge the predicted advantages and restrictions of the quantity being monitored in terms of sensitivity, representativeness, cost, and accuracy. As knowledge and technology develop, data access and processing costs fall, and ecosystem function monitoring becomes more and more important, the list of satellite RS products that can be used to do so is expected to evolve quickly. Information sharing between data providers, ecosystem modelers, RS specialists and ecologists is necessary to enhance the use of SRS data for monitoring ecosystem functions and to take advantage of present and future opportunities; so, a clear and shared platform for discussion and communication of data products that include well-defined terminology, conceptual translation across disciplines, support for version controls, and data sharing must be identified immediately. In developing such a platform, it is necessary to regularly update the information provided and ensure its sustainability in the long term, as well as to allocate consistent and ongoing funding for the development and maintenance of the platform [27, 28].

Another deficiency in this regard is that there are very few ecologists and conservation biologists with theoretical and/or practical training, especially in the sciences of ecology and geophysics, in addition to knowledge of RS. Conceptual models of ecological functions that explicitly describe the variables supporting ecosystem functions and the interactions among them might highlight potential variations between disciplines. As a result, it is important to consider the need for centralized platforms that were created in collaboration, as well as for a shared language and set of tools [29].

5.5 Satellite remote sensor applications for environment safety

Changes in this environment are arguably one of the most important areas in ecology, directly changing the biodiversity and temperature conditions and agricultural activities as well [30]. The possibility of planetary-scale observations of the sea surface from satellites has been a topic of considerable interest since the first spacecraft mission returned and gave color images of the ocean structure. After that period, the enhancement of satellite data offered new opportunities and provided scientific information from space [31].

Today advances in RS technology have produced non-destructive and rapid methods for conducting restoration work, especially in sensitive marine environments [32]. Satellite sensors are very effective in monitoring many ocean and coastal parameters, as well as water quality [33]; so, in this section, we highlighted the usabilities of RS technologies by monitoring the valuable indicators used in marine environment safety.

5.6 Determination of valuable ocean parameters via ocean color sensing

In the 1970s, the first passive RS technique was developed to monitor land and surface water conditions; but, the first ocean color sensor, the proof-of-concept

Coastal Color Scanner (CZCS), was launched in 1978 and measured the ocean color from space until 1986. Following this sensor development, the other RS platforms have opened a new avenue for RS technologies [34]. For instance, Fan *et al* [35] developed a new sensing platform, named Marine and Aerosol Acquisition Tool (OC-SMART) to analyze the global ocean color products received from satellite sensors under complex environmental conditions. In another case study, Nagamani and colleagues [36] developed the Oceansat-2 Ocean Color Monitor (OCM-2) to monitor the phytoplankton blooms on which East of India and the results of developed remote sensor technology were examined with the help of optical and *in situ* biological parameters such as chlorophyll measurements, phytoplankton taxonomy, and pigments. In the next study [37], five different ocean color sensor systems, namely the Terra satellite (Terra MODIS), Moderate Resolution Imaging Spectroradiometer aboard the Aqua satellite (Aqua MODIS), Medium Range Imaging Spectrometer aboard the Environmental Satellite (Envisat MERIS), Medium Resolution Spectral Imager aboard the FY-3 satellite (FY-3 MERSI), and Geostationary Ocean Colour Imager (GOCI), were developed by the authors and their ocean color monitoring capabilities dependent on measuring of suspended particulate matter (SPM) retrieval were conducted between 2004 and 2012 in China. Moreover, Doxaran *et al* [38] aimed to measure the particulate organic carbon (POC) level found in SPM at different locations of the Mackenzie River using an ocean colour algorithm. For that purpose, the MODIS satellite imaginary system was used to measure the R_{rs} signals of the different locations to examine the differences of the locations dependent on ocean color change. In another research study [39], the semivariogram approach from geostatistics was used in determining the global patterns of mesoscale ocean variability with the help of the Sea-viewing Wide Field-of-view (SeaWiFS) by measuring ocean color data.

The monitoring of chlorophyll-a (chl-a) level is a key parameter giving some valuable information about the water quality; for example, Bresciani *et al* [40] aimed to monitor chl-a level of the European perialpine region using Medium Resolution Imaging Spectrometer (MERIS) satellite images and compared the chl-a levels of 12 perialpine lakes (small and large lakes) between the years of 2003–2009.

Many research articles have been reported to measure chl-a levels in different regions and some of them, especially the different sensing platforms, are shown in table 5.1. Ocean transport is a key issue for global transportation and has an economic return around the world; however, the oil spill with its detrimental effects is one of the key proxies for evaluating environmental health, and various research studies have been reported to minimize the detrimental effects of this global problem. Hence, in this part, we aimed to highlight the usability of RS for oil spill monitoring using recent articles. The first case study of this section, reported by Zhang *et al* [48], was based on oil spill determination caused by an unpredictable explosion in the Gulf of Mexico in the Macondo seafloor well, and to monitor the oil spill, the researchers selected the optical polarization method, which is a common approach for oil spill detection using open-source POLDER/ PARASAL polarization time-series data. Additionally, the other RS method, synthetic aperture radar (SAR), is capable of working in all weather conditions

Table 5.1. The sensing platforms for chl-a monitoring.

Sensing platform	Study area	Duration of the study	References
MERIS	The costral area in the Baltic Sea	2008–2010	[41]
Landsand-8 OLI Sentinel	Timsah Lake	2014–2020	[42]
Hyperion	Lake Atillan	No exact data	[43]
TSI	Permanent lakes and reservoirs across China	2016–2021	[44]
SeaWifs MOIS CZCS	The Artic Ocean	1979–2016	[45]
CZCS	Southern and Baja California	1979–1986	[46]
BST and MODIS	Subregions of the Nortwestern Pasific	2007–2018	[47]

Abbreviations: Coastal Zone Color Scanner data (CZCS) and SeaWiFS, based Extreme Gradient Boosting (BST), Medium Resolution Imaging Spectrophotometer (MERIS), Operational Land Image (OLI), Trophic State Index (TSI), Moderate Resolution Imaging Spectroradiometer (MODIS).

and at night during an oil spill detection, for instance, the researchers [49] used polarimetric SAR to monitor an oil spill caused by an accident near the central Philippines island of Guimaras in 2006. In the next study, Trivero *et al* [50] developed a system called Oil Leak Automatic Detector (OSAD) that was originally designed for C-band SAR images and was later adapted to ENVISAT data. OSAD is integrated into complex hardware and software architecture for operational marine monitoring and oil spill drift estimation, and this developed system is also capable of processing L and X-band images from various satellites. Additionally, laser-induced fluorescence (LIF) sensing platforms have been employed not only to monitor oil spills on the sea surface but also to determine the oil types, their thickness, and their volumes [51]; from these properties of LIF systems, a portable LIF light detection and ranging (LiDAR) platform with 5 kg weight was developed for the unmanned aerial vehicle (UVA) and used in real-time monitoring of the aquatic environment in China. In addition, Moon *et al* [52] utilized small-size LiDAR sensor and developed an algorithm to investigate the feasibility of the LiDAR sensor to detect the three different types of oils, light crude, bunker A, and bunker C and their volumes at 905 nm. Chaudhary and Kumar [53] utilized SAR platforms to separate oil spills from water using the quad-pol and hybrid-pol datasets, and during the separation of the oil and water, various parameters were examined to check the separation capability of the datasets.

Salinity is an important parameter affecting the surface, heat transport and deep circulation of the oceans and the change of its level could affect the marine environment and marine biodiversity as well; thus, the monitoring of its level is of great importance to preserve environment safety and biodiversity. The first study about the information on salinity level measurement was reported by Borovskaya *et al* [54], and during the determination of salinity level in hypersaline East Sivash Bay, the RS platform was combined with machine learning using 93 *in situ* samples and six Sentinel-2 datasets that were evaluated with some models. The next work was reported by Zhao *et al* [55], which was a long-term study between 2015–2020 in the Bering Sea comparing to previous works. During the examination of salinity measurements, and the data that were periodically collected monthly were analyzed with the help of SMAP, SMOC, and Optimally Interpolated Sea Surface Salinity (OIS) satellites. Some researchers combined with machine learning approaches and MODIS [56], (Soil Moisture Active and Passive (SWAP), MODIS [57] RS data to measure the salinity levels in China, but, especially, in the second study, Zhang *et al* [57] used three different machine learning methods, Random Forest, Particle Swarm Optimization Support Vector Regression (PSO-SVR), and Automatic Machine Learning (TPOT) and compared their accuracy performances. Moreover, Tian and coworkers [58] aimed to assess the salinity level of the areas with no observation; so, in that study, Tian *et al* employed machine learning that offers some predictions about the output using historical inputs and the researchers used the synthetic dataset including original dataset and noised dataset and adopted for two machine learning approaches. Nowadays, RS platforms are paving the way for studying salinity levels in desired areas and many RS platforms have been constructed for this purpose, so, in table 5.2, we summarize some RS platforms used in salinity level measurements.

Table 5.2. Some of the sensing platforms for salinity measurements of different locations.

Sensing platform	Study area	References
SMOS and SWAP	The Gulf of St Lawrence	[59]
Combination of ML and satallite observations	The South China Sea	[60]
Combination of saildrones and SMAP	The West Coast of North America	[61]
Combination of multisensor satellite and multivariate OI algorithm	The Mediterranean Sea	[62]
SMAP and JPLSMAP	The California Coast	[63]

Abbreviations: SMOS (soil moisture ocean salinity), SMAP (soil moisture active and passive), JPLSMAP [(The Jet Propulsion Laboratory (JPL) SMAP].

5.7 Conclusion and prospects

Protecting the marine ecosystem is a key issue for marine environment safety, marine biodiversity, and our planet as well; so, one of our duties is to protect our ecosystem and biodiversity. However, in most cases, it is not possible to immediately protect or control the areas of interest like marine habitats thanks to the geographic locations. In such circumstances, monitoring of marine ecosystem health indicators, e.g., ocean color, salinity, sea surface temperature, surface winds, oil spills, etc, give valuable information about the marine ecosystem balance.

RS from sea drones, satellites, or the combinations of other platforms mentioned above is an effective approach for monitoring some marine ecosystem health indicators or measuring their levels, and it provides new opportunities without being in contact with the area of interest for marine ecosystem health.

Recently, RS and satellite biological sensors with the combination of machine learning or some algorithms have paved the way for monitoring of marine ecosystem health indicators and have been used in tackling some problems and enhancing the performance of RS platforms; so, in this context, we aimed to focus on the usability of RS for monitoring of marine ecosystem health with the combination of the other sensing platforms and some useful algorithms using recent articles.

We hope and trust that the combination of advanced deep learning, machine learning, and RS platforms will open a new avenue for protecting our ecosystem and our planet in the near future.

References

[1] Fukuba T and Fujii T 2021 Lab-on-a-chip technology for: *in situ* combined observations in oceanography *Lab Chip* **21** 55–74

[2] Liu Y, Lu H and Cui Y 2023 A review of marine *in situ* sensors and biosensors *J. Mar. Sci. Eng* **11** 1469

[3] Mills G and Fones G 2012 A review of *in situ*/IT methods and sensors for monitoring the marine environment *Sens. Rev.* **32** 17–28

[4] Kavanaugh M T, Bell T, Catlett D, Cimino M A, Doney S C, Klajbor W *et al* 2021 Satellite remote sensing and the marine biodiversity observation network: current science and future steps *Oceanography* **34** 62–79

[5] Collin A, Lambert N and Etienne S 2018 Satellite-based salt marsh elevation, vegetation height, and species composition mapping using the superspectral worldview-3 imagery *Int. J. Remote Sens.* **39** 5619–37

[6] Johnston D W 2019 Unoccupied aircraft systems in marine science and conservation *Annu. Rev. Mar. Sci.* **11** 439–63

[7] Bell T W, Okin G S, Cavanaugh K C and Hochberg E J 2020 Impact of water characteristics on the discrimination of benthic cover in and around coral reefs from imaging spectrometer data *Remote Sens. Environ.* **239** 111631

[8] Monteiro J G, Jiménez J L, Gizzi F, Přikryl P, Lefcheck J S, Santos R S *et al* 2021 Novel approach to enhance coastal habitat and biotope mapping with drone aerial imagery analysis *Sci. Rep.* **11** 1–13

[9] Aragon B, Ziliani M G, Houborg R, Franz T E and McCabe M F 2021 CubeSats deliver new insights into agricultural water use at daily and 3 m resolutions *Sci. Rep.* **11** 1–12

[10] Turner W, Spector S, Gardiner N, Fladeland M, Sterling E and Steininger M 2003 Remote sensing for biodiversity science and conservation *Trends Ecol. Evol.* **18** 306–14

[11] Mumby P J, Skirving W, Strong A E, Hardy J T, LeDrew E F, Hochberg E J *et al* 2004 Remote sensing of coral reefs and their physical environment *Mar. Pollut. Bull.* **48** 219–28

[12] Chassot E, Bonhommeau S, Reygondeau G, Nieto K, Polovina J J, Huret M *et al* 2011 Satellite remote sensing for an ecosystem approach to fisheries management *ICES J. Mar. Sci.* **68** 651–66

[13] Polovina J J and Howell E A 2005 Ecosystem indicators derived from satellite remotely sensed oceanographic data for the North Pacific *ICES J. Mar. Sci.* **62** 319–27

[14] Roemmich D, Johnson G C, Riser S, Davis R, Gilson J, Owens W B *et al* 2009 The Argo Program: observing the global ocean with profiling floats *Oceanography* **22** 34–43

[15] Williamson M J, Tebbs E J, Dawson T P and Jacoby D M P 2019 Satellite remote sensing in shark and ray ecology, conservation and management *Front. Mar. Sci.* **6** 135

[16] Klemas V 2011 Remote sensing techniques for studying coastal ecosystems: an overview *J. Coast. Res.* **27** 2–17

[17] Sheppard C 2019 *World Seas: An Environmental Evaluation: Volume III: Ecological Issues and Environmental Impacts* (Elsevier)

[18] Halverson G H, Lee C M, Hestir E L, Hulley G C, Cawse-Nicholson K, Hook S J *et al* 2022 Decline in thermal habitat conditions for the endangered delta smelt as seen from landsat satellites (1985–2019) *Environ. Sci. Technol.* **56** 185–93

[19] Hauser D, Abdalla S, Ardhuin F, Bidlot J R, Bourassa M, Cotton D *et al* 2023 Satellite remote sensing of surface winds, waves, and currents: where are we now? *Surv. Geophys.* **44** 1357–446

[20] Krestenitis M, Orfanidis G, Ioannidis K, Avgerinakis K, Vrochidis S and Kompatsiaris I 2019 Oil spill identification from satellite images using deep neural networks *Remote Sens.* **11** 1–22

[21] Berryman A A 2002 Population: a central concept for ecology? *Oikos* **97** 439–42

[22] Paganini M, Leidner A K, Geller G, Turner W and Wegmann M 2016 The role of space agencies in remotely sensed essential biodiversity variables *Remote Sens. Ecol. Conserv.* **2** 132–40

[23] Pettorelli N, Laurance W F, O'Brien T G, Wegmann M, Nagendra H and Turner W 2014 Satellite remote sensing for applied ecologists: opportunities and challenges *J. Appl. Ecol.* **51** 839–48

[24] Stephens P A, Pettorelli N, Barlow J, Whittingham M J and Cadotte M W 2015 Management by proxy? The use of indices in applied ecology *J. Appl. Ecol.* **52** 1–6

[25] Tilstone , Mallor-Hoya S G, Gohin F, Couto A B, Sá C, Goela P *et al* 2017 Which ocean colour algorithm for MERIS in North West European waters? *Remote Sens. Environ.* **189** 132–51

[26] Racault M F, Platt T, Sathyendranath S, Ağirbaş E, Martinez Vicente V and Brewin R 2014 Plankton indicators and ocean observing systems: support to the marine ecosystem state assessment *J. Plankton Res.* **36** 621–9

[27] Pettorelli N, Schulte to Bühne H, Tulloch A, Dubois G, Macinnis-Ng C, Queirós A M *et al* 2018 Satellite remote sensing of ecosystem functions: opportunities, challenges and way forward *Remote Sens. Ecol. Conserv.* **4** 71–93

[28] Queirós A M, Bruggeman J, Stephens N, Artioli Y, Butenschön M, Blackford J C *et al* 2015 Placing biodiversity in ecosystem models without getting lost in translation *J. Sea Res.* **98** 83–90

[29] Cabello J, Fernández N, Alcaraz-Segura D, Oyonarte C, Piñeiro G, Altesor A *et al* 2012 The ecosystem functioning dimension in conservation: insights from remote sensing *Biodivers. Conserv.* **21** 3287–305

[30] Trasviña-Moreno C A, Blasco R, Marco Á, Casas R and Trasviña-Castro A 2017 Unmanned aerial vehicle based wireless sensor network for marine-coastal environment monitoring *Sensors (Switzerland)* **17** 1–22

[31] Apel J 1980 Satellite sensing of ocean surface dynamics *Annu. Rev. Earth Planet Sci.* **8** 303–42

[32] Ridge J T and Johnston D W 2020 Unoccupied Aircraft Systems (UAS) for marine ecosystem restoration *Front. Mar. Sci.* **7** 1–13

[33] O'Connor E, Smeaton A F, O'Connor N E and Regan F 2012 Investigation into the use of satellite remote sensing data products as part of a multi-modal marine environmental monitoring network *Proc. Remote Sens Ocean Sea Ice, Coast Waters Large Water Regions 2012 SPIE Remote Sensing* 8532 85320F

[34] Chang N B, Imen S and Vannah B 2015 Remote sensing for monitoring surface water quality status and ecosystem state in relation to the nutrient cycle: a 40-year perspective *Crit. Rev. Environ. Sci. Technol.* **45** 101–66

[35] Fan Y, Li W, Chen N, Ahn J H, Park Y J, Kratzer S *et al* 2021 OC-SMART: a machine learning based data analysis platform for satellite ocean color sensors *Remote Sens. Environ.* **253**

[36] Nagamani P V, Latha T P, Bhavani I V G, Rao Y U, Raman M, Amminedu E *et al* 2017 Optical detection of diatom bloom in the coastal waters of Bay of Bengal using Oceansat-2 OCM *Proc. Natl Acad. Sci. India A—Phys. Sci.* **87** 867–78

[37] Shen F, Zhou Y, Peng X and Chen Y 2014 Satellite multi-sensor mapping of suspended particulate matter in turbid estuarine and coastal ocean, China *Int. J. Remote Sens.* **35** 4173–92

[38] Doxaran D, Ehn J, Bélanger S, Matsuoka A, Hooker S and Babin M 2012 Optical characterisation of suspended particles in the Mackenzie River plume (Canadian Arctic Ocean) and implications for ocean colour remote sensing *Biogeosciences* **9** 3213–29

[39] Doney S C, Glover D M, McCue S J and Fuentes M 1998 Mesoscale variability of Sea-viewing Wide Field-of-view Sensor (SeaWiFS) satellite ocean color: global patterns and spatial scales *J. Geophys. Res. Ocean* **108** 1–15

[40] Bresciani M, Stroppiana D, Odermatt D, Morabito G and Giardino C 2011 Assessing remotely sensed chlorophyll-a for the implementation of the Water Framework Directive in European perialpine lakes *Sci. Total Environ.* **409** 3083–91

[41] Harvey E T, Kratzer S and Philipson P 2015 Satellite-based water quality monitoring for improved spatial and temporal retrieval of chlorophyll-a in coastal waters *Remote Sens. Environ.* **158** 417–30

[42] Seleem T, Bafi D, Karantzia M and Parcharidis I 2022 Water quality monitoring using Landsat 8 and Sentinel-2 satellite data (2014–2020) in Timsah Lake, Ismailia, Suez Canal Region (Egypt) *J. Indian Soc. Remote Sens.* **50** 2411–28

[43] Flores-Anderson A I, Griffin R, Dix M, Romero-Oliva C S, Ochaeta G, Skinner-Alvarado J *et al* 2020 Hyperspectral satellite remote sensing of water quality in Lake Atitlán, Guatemala *Front. Environ. Sci.* **8**

[44] Liu Y, Ke Y, Wu H, Zhang C and Chen X 2023 A satellite-based hybrid model for trophic state evaluation in inland waters across China *Environ. Res.* **225** 115509

[45] Oziel L, Massicotte P, Babin M and Devred E 2022 Decadal changes in Arctic Ocean Chlorophyll a: bridging ocean color observations from the 1980s to present time *Remote Sens. Environ.* **275** 113020

[46] Barocio-León O A, Millán-Núñez R, Santamaría-del-Ángel E and González-Silvera A 2007 Productividad primaria del fitoplancton en la zona eufócotica del Sistema de la Corriente de California estimada mediante imágenes del CZCS *Cienc. Mar.* **33** 59–72

[47] Xing M, Yao F, Zhang J, Meng X, Jiang L and Bao Y 2022 Data reconstruction of daily MODIS chlorophyll-a concentration and spatio-temporal variations in the Northwestern Pacific *Sci. Total Environ.* **843** 156981

[48] Zhang Z, Yan L, Jiang X, Ding J, Zhang F, Jiang K *et al* 2022 Exploring the potential of optical polarization remote sensing for oil spill detection: a case study of deepwater horizon *Remote Sens.* **14** 1–18

[49] Ma L and Chen G 2013 Method for oil spill monitoring by polarimetric SAR *Environ. Forensics* **14** 294–300

[50] Trivero P, Adamo M, Biamino W, Borasi M, Cavagnero M, De Carolis G *et al* 2016 Automatic oil slick detection from SAR images: results and improvements in the framework of the PRIMI pilot project *Deep. Res. Part II Top. Stud. Oceanogr.* **133** 146–58

[51] Sun L, Zhang Y, Ouyang C, Yin S, Ren X and Fu S 2023 A portable UAV-based laser-induced fluorescence lidar system for oil pollution and aquatic environment monitoring *Opt. Commun.* **527** 128914

[52] Moon J H and Jung M 2020 Geometrical properties of spilled oil on seawater detected using a LiDAR sensor *J. Sens.* **2020**

[53] Chaudhary V and Kumar S 2020 Marine oil slicks detection using spaceborne and airborne SAR data *Adv. Sp. Res.* **66** 854–72

[54] Borovskaya R, Krivoguz D, Chernyi S, Kozhurin E, Khorosheltseva V and Zinchenko E 2022 Surface water salinity evaluation and identification for using remote sensing data and machine learning approach *J. Mar. Sci. Eng.* **10** 257

[55] Zhao J, Wang Y, Liu W, Bi H, Cokelet E D, Mordy C W *et al* 2022 Sea surface salinity variability in the Bering Sea in 2015–2020 *Remote Sens.* **14** 1–19

[56] Wang Z, Wang G, Guo X, Hu J and Dai M 2022 Reconstruction of high-resolution sea surface salinity over 2003–2020 in the South China Sea Using the Machine Learning Algorithm LightGBM Model *Remote Sens.* **14** 6147

[57] Zhang X, Wu M, Han W, Bi L, Shang Y and Yang Y 2022 Sea Surface Salinity Inversion Model for Changjiang Estuary and Adjoining Sea Area with SMAP and MODIS data based on machine learning and preliminary application *Remote Sens.* **14** 5358

[58] Tian T, Leng H, Wang G, Li G, Song J, Zhu J *et al* 2022 Comparison of machine learning approaches for reconstructing sea subsurface salinity using synthetic data *Remote Sens.* **14** 5650

[59] Dumas J and Gilbert D 2023 Comparison of SMOS, SMAP and *in situ* sea surface salinity in the Gulf of St. Lawrence *Atmos—Ocean* **61** 148–61

[60] Dong L, Qi J, Yin B, Zhi H, Li D, Yang S *et al* 2022 Reconstruction of subsurface salinity structure in the South China Sea Using Satellite observations: a LightGBM-based deep forest method *Remote Sens.* **14** 1–19

[61] Tang W, Yueh S H, Fore A G, Vazquez-Cuervo J, Gentemann C, Hayashi A K *et al* 2022 Using saildrones to assess the SMAP sea surface salinity retrieval in the coastal regions *IEEE J. Sel. Top. Appl. Earth Obs. Remote Sens.* **15** 7042–51

[62] Sammartino M, Aronica S, Santoleri R and Nardelli B B 2022 Retrieving Mediterranean sea surface salinity distribution and interannual trends from multi-sensor satellite and *in situ* data *Remote Sens.* **14** 2502

[63] Vazquez-Cuervo J, García-Reyes M and Gómez-Valdés J 2023 Identification of sea surface temperature and sea surface salinity fronts along the California Coast: application using saildrone and satellite derived products *Remote Sens.* **15** 484

IOP Publishing

Sensors for Marine Biosciences
Next-generation sensing approaches
Shyam S Pandey, Rout George Kerry and Kshitij RB Singh

Chapter 6

Artificial intelligence-, machine learning- and Internet of Things-based sensors for detection of marine pollutants

Chansi, Karan Hadwani and Tinku Basu

Marine pollution is a serious issue that requires global attention. Early detection and monitoring of marine pollutants are critical for protecting our oceans and ensuring a healthy future. Traditional methods, such as laboratory analysis and visual observations, are often time-consuming, labor-intensive, and limited in their spatial and temporal coverage. Fortunately, the emergence of artificial intelligence (AI), machine learning (ML), and the Internet of Things (IoT) presents a revolutionary approach to addressing these limitations and revolutionizing the detection of marine pollutants. AI is used to improve the accuracy and sensitivity of biosensors, ML-based new algorithms help in data analysis, and IoT is utilized to collect and transmit data from biosensors in real time. With this background, the present chapter highlights the basics of AI, ML algorithms, IoT-based dataset storage techniques, and their applicability in the design and development of biosensors with specific reference to marine pollutant detection. The development of AI, ML, and IoT-based sensors for the detection of marine pollutants is still in its early stages. However, the potential benefits of this technology are great, hence a section is devoted to future prospects and challenges faced for their development.

6.1 Introduction

Marine pollution is defined as the presence of a mixture of chemicals and junk in water bodies at high levels leading to contamination of the environment and ecosystems. Marine pollution is a serious problem affecting the environment and ecosystems. It comes from different sources generating varied groups of pollutants. The most common types of ocean pollution include plastic pollution, chemical pollution, high noise levels, and food waste [1]. The source of pollution in the sea is mainly due to human activities, and the most important source of pollution is

doi:10.1088/978-0-7503-5999-3ch6

6-1

non-point sources. Non-point polluting sources include many smaller sources such as sewage, cars, trucks, and boats, as well as larger sources such as construction, shipping spills, and offshore oil and gas production [2]. The commonest marine pollution is caused by plastic. A whopping 80% of the trash littering these vast waters, from the sun-dappled surface to the inky depths, is plastic. This plastic plague is wreaking havoc on marine life and the delicate balance of ocean ecosystems. Every year, a staggering 14 million tons of plastic find its way into the oceans, primarily from land sources like overflowing cities, storm drains carrying waste, and irresponsible dumping. As plastic breaks down under the relentless assault of sunlight, wind, and waves, it crumbles into tiny, invisible fragments called microplastics and even smaller nanoplastics. These minuscule particles are a silent threat, easily mistaken for food by marine animals [3]. Marine pollution is a planetary scale problem and has noticeable effects on the environment. Early detection and monitoring of marine pollution are essential steps to prevent and diminish the uncontrolled harmful effects. In fact, early detection of marine pollution allows an extra window for quick interception. Monitoring also holds crucial importance as it deals with the data which reflects the extent and intensity of any type of pollution, this can help identify environmental hazards and instantaneous actions [4]. Monitoring pollutants can also allow tracking progress in reducing the concentration of the pollutants, which can increase the effectiveness of the reinforcement steps taken. Conventional marine pollution monitoring methods often prove to be time-consuming, costly, and labour-intensive, hindering the timely detection of pollutants and their detrimental effects. AI, ML, and IoT-based biosensors [5] offer a promising solution to this challenge. These innovative technologies pave the way for the development of real-time, *in situ*, monitoring systems capable of continuous data collection on a wide spectrum of marine pollutants. AI and ML algorithms [6] can then analyze this data to identify patterns indicative of pollution presence. This approach offers several advantages over traditional methods. Firstly, it is significantly faster, with biosensors capable of real-time pollutant detection. Secondly, they are quite cost-effective due to the relatively low deployment and maintenance costs of biosensors. Thirdly, it is highly scalable, allowing biosensor networks to be deployed over vast areas for comprehensive marine environment coverage. While AI, ML, and IoT-based biosensors [7] are still in their infancy, they hold immense potential to revolutionize marine pollution reduction. These modern tools have a unique capacity for acquiring real-time data, statistical analysis of acquired data, and interpretation of data based on real-time scenarios to provide early warnings of pollution events, safeguarding marine ecosystems and human health. The subsequent sections provide a basic overview (figure 6.1) of the fundamentals of AI- and ML-based techniques and provide an outline as they have evolved over the course of time in detection and monitoring of marine pollutants, contributing significantly towards the protection of ocean [8]. As these technologies mature, we can anticipate even more innovative applications for marine pollution monitoring, hence the future prospects of these techniques are described.

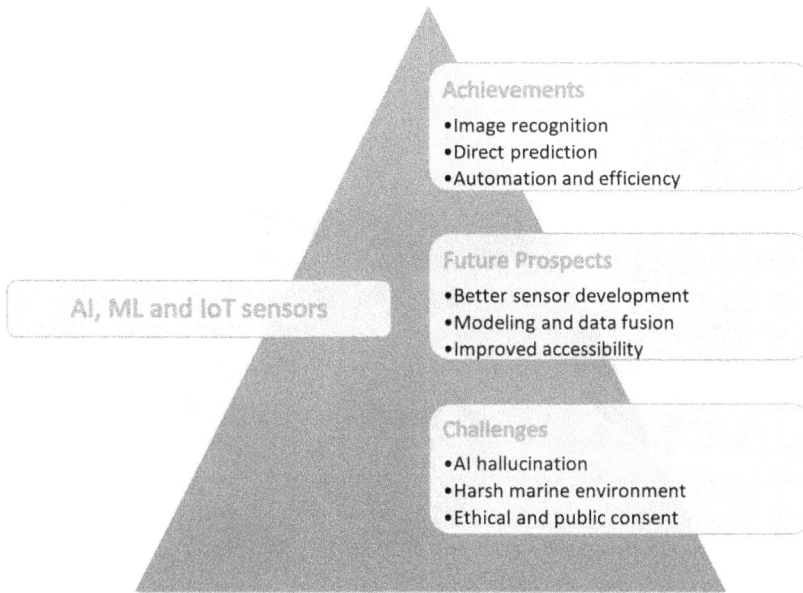

Figure 6.1. Overview of IoT, ML and AI-based sensors.

6.2 Fundamentals of AI

The current section is devoted towards basic understanding of AI.

6.2.1 Definition and concept

AI is a field of computer science that studies intelligent machines that can think, learn, and act independently. The concept of intelligence was first put forward by John McCarthy at the Dartmouth Conference in 1955 [9]. AI has no invariably acquired explanation because of its wide span and volatile nature. The lack of criteria for the creation of AI and the ambiguity of the word 'intelligence' leads to contradiction. AI aims to use complex computers to simulate human behavior and requires a lot of analysis before it is considered intelligent. AI has a history of less than a century and has faced three challenges. The first wave, called 'symbolic AI,' involves the completion of tasks. The second wave is 'data-driven' technology that bypasses human expertise and replaces learning algorithms. The third wave, 'strong' or 'advanced' intelligence, refers to machines that can be intelligent at many tasks. The future of AI includes artificial general intelligence (AGI), quantum computing, and brain simulation, which are not achievable with current technology and require an evolutionary transformation.

Here are five crucial elements of AI [10] arranged in the order of least-to-highest priority (this ordering is based on what AI needs to do to comprehend the environment around it to learn from it, solve complex problems, analyze reasoning and logic, and then provide a result, as shown in figure 6.2. These elements are:

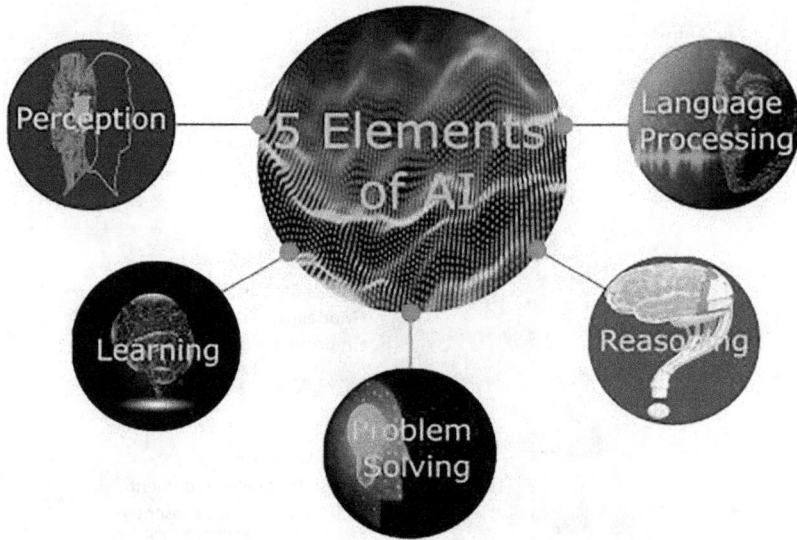

Figure 6.2. Elements of AI.

1. Perception—Perception is the ability of an intelligent being to perceive and understand its environment. This can be done using a variety of sensors, including cameras, microphones, and accelerometers. Any AI that interacts with a real-world environment, such as a self-driving car or robot, requires perception [11].
2. Learning—The ability of an AI system to improve its performance over time by being exposed to new information or experience is called learning. This can be achieved using a variety of ML methods, including supervised learning, unsupervised learning, and additive learning. Learning is essential for any intellectual skill that needs to be changed or transferred to a new task [12].
3. Problem solving—Problem solving refers to the ability of cognitive skills to solve complex problems. This requires the use of many techniques, including search algorithms, planning algorithms, and game theory. Problem-solving is essential for all cognitive skills that need to achieve goals in a complex environment [13].
4. Reasoning—The ability of an AI system to conclude its data and beliefs is called reasoning. This requires a lot of good ideas and math to extract new information from existing information. Any AI system that needs to make decisions or act by understanding its environment must be able to reason [14].
5. Language processing—Language processing refers to the intellectual mind's ability to understand and produce human language. Many methods are used, such as natural language processing (NLP) and machine translation. Language is essential for all cognitive abilities that have to process human-generated information or communicate with people [15].

6.2.2 Types of AI

There are a variety of means to allocate AI, one highly relied-on way is by dividing it into three categories:

1. Narrow AI (ANI)—This term was coined by John McCarthy in 1956. ANI is the conventionally used AI as well as for the most part a limited AI system. ANI requires restructuring as soon as it experiences even slight changes to maintain its degree regarding making smart decisions. Unlike natural intelligence, like humans who can adjust to new goals or environments through learning, ANI focuses on specific tasks. Image recognition, filtering spam emails, and providing virtual assistance are all examples of ANI's capabilities [16].

2. General AI (AGI)—British mathematician and philosopher Irving John Good coined the term 'general artificial intelligence' in 1965 [17]. AGI is a type of intelligence that can do all the tasks that humans can do. AGI systems will be able to teach themselves new tasks and expand their knowledge into new areas. Although AGI systems do not yet exist, they are the goal of many AI researchers.

3. Artificial superintelligence (ASI)—A Swedish philosopher called Nick Bostrom created this term in 2001 [18]. ASI is predicted to be better and smarter than humans. ASI systems are not known to exist, and it is unknown whether they are even possible to build. ASI is a contentious topic, with some scientists feeling it poses a severe threat to humanity and others believing it could benefit the planet. It is vital to highlight that ASI is currently speculative, and it is too soon to predict its influence. This type of intelligence will be successful in areas such as art, decision making, and emotional relationships. These features are now part of what distinguishes machines from humans. In other words, these things are considered unique to humans [19].

6.2.3 Application of AI in pollutant monitoring

Monitoring pollution is an important part of environmental management, especially in marine ecosystems. It involves the collection, analysis, and interpretation of data on the presence, concentration, and distribution of pollutants in the environment. Monitoring pollution is important for many reasons, such as measuring the amount and severity of pollution, identifying pollution sources, evaluating the effectiveness of pollution control, protecting human and animal health, and caring for the environment and biodiversity [20]. Assessing the extent and severity of pollutants is important for developing pollution control and management strategies. For example, supervising levels of heavy metal ions in marine ecosystems can help to identify sources of pollution and develop strategies to reduce their release into the environment. Pollution control plans can be articulated according to the origin. Investigation of urban and industrial water pollution can help to locate sources and develop strategies to reduce environmental pollution. Assessment of the extent of pollution is crucial for checking the effectiveness of pollution control. Observing air pollution can identify new and emerging pollution problems which guide to develop pollution and quality control strategies. Pollution monitoring can be done using a variety of technologies and solutions, such as sensors, satellite tracking systems, and other types of monitoring. The best pollution monitoring services are those that use various technologies to complete the picture of pollution.

AI has been explored as a useful tool in pollution monitoring. AI is attempted in air, water, soil, and industrial waste management, as well as environmental policy and decision-making. AI can analyze large datasets, identify patterns, and predict potential risks. By training AI models on vast amounts of environmental data, it is possible to create predictive models that aid in understanding the dynamics of pollution and its impact. AI is still a relatively new technology, but it has the potential to revolutionize the way we monitor and manage pollution. AI can assist us with faster, smarter, and more intelligent detection and solve pollution problems faster and better by providing real-time data and insights.

Almalawai *et al* [21] have employed AI to observe air pollution and forecast the extent of pollutants such as CO_2, SO_2, NO_2, and other particle-sized matter present in the atmosphere by using several AI models such as linear regression, support vector regression, and gradient boosted decision tree to fulfill the purpose and make a forecasting AI-based air quality index model. This proposed model can work with large data and it can easily detect the usual trends. The main advantage of using this model is that it can also foresee the potential health hazards associated with the rise in the amount of pollutants [22]. Neo *et al* have presented an AI-assisted air quality monitoring system that can be used in smart cities to reduce environmental pollution and promote sustainability during urbanization. Feature-optimized Long Short-Term Memory (LSTM) is a promising AI model for predicting PM10, PM2.5, CO, and O_3 concentration [23]. A real-life application of AI has been made by Dr Daniel J Jacob and his team where they have created a global 3D chemical transport model known as GEOS (Goddard Earth Observing System)-Chem. The GEOS-Chem from NASA can simulate the transport and chemical reactions of air pollutants in the atmosphere. This model can deal with everything from studying the sources to their culmination points and everything happening in between [24]. De Vries *et al* have created a novel AI-based detection algorithm to assess floating plastic parts and pieces present in the sea with the help of vessel-based optical data. The developed algorithm is capable of accurately differentiating plastic debris of different sizes and types in the North Atlantic Ocean [24]. Dhanapal *et al* have invented clean-water AI based on an IoT device that can differentiate and identify harmful bacteria and other harmful particles. This work uses a deep learning neural network and helps in boosting devices that fulfill the purpose [25]. Bruzzese *et al* developed a real-time monitoring of marine aquaculture by making a smart buoy aided with a spectro-photometer to monitor the water quality. The proposed buoy is connected to the internet by a LoRaWAN link that transfers data to the surveillance stations. There is an EU-funded AQUACROSS project [26] that developed a drone-based water quality monitoring system in Europe that helps in identifying pollution sources and tracks pollutants. An African startup named HydroIQ developed an AI-powered smart water quality assessing system to improve and keep a check on water quality. The system gathers data from a series of sensors deployed in the local water bodies.

AI combines technology and innovation to determine the quality of our air, water, and soil with incredible accuracy, providing vital information to protect our ecosystems and human health. AI systems can monitor pollutants in real time, quickly identifying sources, analyzing trends, and predicting potential risks. Traditional methods of pollutant monitoring rely on complex numerical models and physical sensors, which can be expensive and often provide limited spatial coverage. By leveraging AI's capabilities in data analysis and pattern recognition, we can create more advanced and cost-effective ways to monitor environmental quality. AI, specifically ML methods like deep learning, has shown promise in analyzing large datasets and identifying patterns that may not be easily detected by human intelligence. By training AI models on vast amounts of environmental data, we can develop predictive models that aid in understanding the dynamics of pollution and its impact, hence these techniques are discussed in detail in subsequent sections.

6.3 Machine learning

Arthur Samuel, an American computer scientist in 1959 introduced the term ML [27]. He is said to be the founder of computer gaming and ML. ML is focused on making predictions and decisions based on algorithms and statistical models developed by the experts. This is a data-driven concept in which machines are trained to recognize patterns, see patterns, and use large amounts of data to create accurate predictions. The main idea behind ML is to enable computers to learn and adapt to human-like experiences. ML assists in increasing efficiency, detecting fraud, predicting customer behavior, and improving decision-making in many industries including healthcare, commerce, transportation, and business. Bekkar *et al* used a deep learning hybrid model by combining a convolutional neural network (CNN) and LSTM which was employed to forecast PM2.5 pollutants in Beijing city. This study found that ML methods are highly favorable for air pollution monitoring. This model can analyze data from the pre-existing surveillance stations and correlate the data with this model by leveraging the strengths of ML. In the study done by Jia *et al* [28] a visible and infrared reflectance spectroscopy-based detection system was developed which helps in mapping of soil pollution. Infrared reflectance spectra proved to be a suitable candidate for mapping soil pollution. Priyadarshini *et al* [29] developed an ML model to forecast the water quality and the same can also detect the origin of the pollutants in the metropolitan areas. It is claimed that this model can anticipate the water quality based on water quality parameters decided by the government and ML algorithms, which is highly beneficial as it can help responsible authorities to develop strategies to cope and reduce the impact of water pollution. A very recent study by Pan *et al* [30] represents a comparison of different ML methods for forecasting ozone pollution levels. The final results showed that the Random Forest model had surpassed other models such as artificial neural networks and support vector regression models. The Random Forest model can predict pollutant

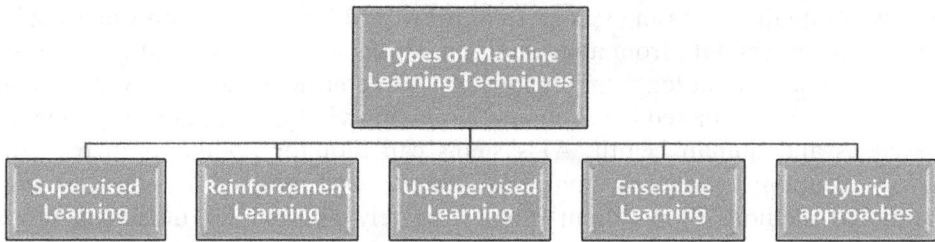

Figure 6.3. Types of ML techniques.

concentrations from surveillance stations and can easily transform outdated surveillance stations with virtually controllable surveillance stations. Booth *et al* [31] developed a highly accurate density mapping method for garbage and plastic present in the ocean by using satellite imagery.

This mapping method can also be used for detecting floating garbage and the distribution in the ocean which can be later used to develop strategies to combat the issue of rising marine pollution. Cartwright *et al* [32] reported noise pollution monitoring in New York City based on ML to analyze the noise from different hotspots. They have adapted this idea from Sounds of New York City project (SONYC), where they use deployable noise sensors and self-supervised audio learning to combine these with the ML algorithms to track down the sources of noise pollution in different parts of the city. This method could help the city planners and authorities to make new policies to reduce noise pollution occurrences. ML learning techniques are classified into five categories, as shown in figure 6.3.

6.3.1 Supervised learning

In ML, supervised learning acts like a trainee learning from an instructor's guidance. Imagine showing a computer program tons of examples with both questions and answers. Through this 'training' on paired data, the program learns to recognize patterns and make predictions for entirely new situations it hasn't encountered before [33]. The algorithm then uses these functions to make predictions on new, previously unseen data (figure 6.4). Training auditing is one of the most widely used technology methods and has a strong track record of creating unique models that are effective at the tasks taught.

In general, supervised learning algorithms can be used for two sorts of tasks, named classification and regression. Classification is when a task requires the prediction of a set category where a new data point is to be coordinated. Regression, on the other hand, functions by prediction of the numerical value of a result. For example, a supervised learning type algorithm can be applied to predict the price of an automobile based on its features and specifications [34].

There are five different types of supervised learning algorithms—linear regression, logistic regression, decision trees, support vector machines, ensemble methods.

Linear regression is a category of supervised ML that is able to do predictions based on a ceaseless output variable, the main application of linear regression is in

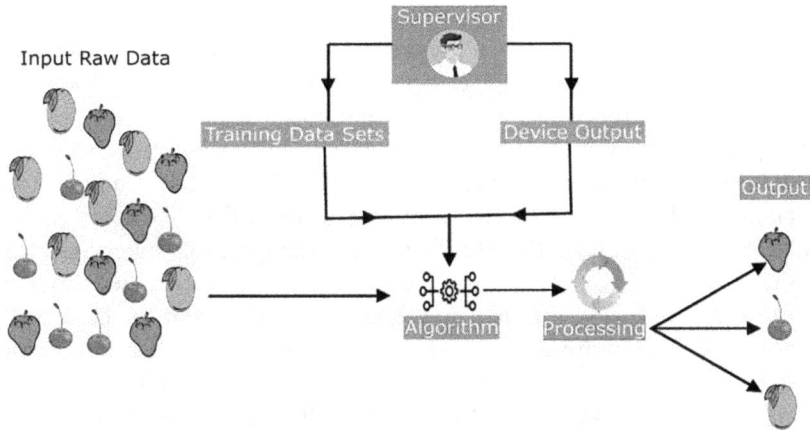

Figure 6.4. Supervised learning-based ML model.

the stock market where predictions play a major role in forecasting weather of a particular area based on the trends [35]. Logistic regression holds the prime importance for majority of the crucial analytical tools. Logistic regression makes the foundation of natural language processing and is closely related to neural networks. It can be used to categorize tasks [36]. Logistic and linear regression work in very identical ways, the latter produces a ceaseless output variable, whereas the former generates binary output variable. This algorithm can be of use if the user requires to analyze the chances of occurrence of certain incident. Decision trees is a type of algorithm that is capable of doing both classification and regression tasks in one. Decision trees [37] are a simple but effective way to model the interaction between input and output. Regression network development and categorization of network creation are comparable processes. As we did in the regression case, we use recursive binary splitting to construct the distribution tree. However, in the case of classification, mean squared deviation (MSE) cannot be used as a model for binary segmentation. Classification error rate is a natural alternative to MSE [38].

A kind of technique that may prove beneficial for both the classification and regression applications is the support vector machine (SVM). SVMs can help in the augmentation of the margin around the separation of hyperplane. The decision is declared by support vectors which are a division of training samples. These algorithms learn a hyperplane that separates one data into two divisions [39]. The ensemble method is an ML algorithm that combines different models to make predictions. Algorithms in cluster learning combine the predictions of multiple models to arrive at a final prediction. ML models can benefit from cluster learning based on accuracy and robustness. Bagging, lifting, and stacking are examples of collaborative learning [40]. It is used in image recognition, voice recognition and natural language use. Nowadays, ensemble method is used for several diseases such as to predict cardiovascular diseases, to presage a variety of cancer too.

6.3.1.1 *Applications of supervised learning*

As we know supervised learning is a powerful technique that is highly versatile and can relieve stupendous amounts of effort and hours. There are a lot of instances where supervised learning can be of extreme help, such as in prediction of diseases. It has been applied to predict the onset of diseases such as diabetes, cancer and heart disease. These algorithms can observe fixed or unfixed sequences and analyze factors that can endanger human life in vast amounts of data for prediction and can find probability of occurrence of a disease. For image recognition, supervised learning is used to identify faces, animals and vehicles in images. Supervised learning algorithms can learn patterns and features specific to different objects by training patterns on multiple domain images [41]. Another major application is commonly referred to as recognition of speech and its purpose is to convert it to text. Supervised learning algorithms can be trained to learn specific features and motifs and adapt themselves to different speeches from huge text collected [42]. Banks also use supervised learning to separate duplicitous proceedings which helps them protect their account holders and other customers from fraud [43]. Finally, supervised learning is a powerful learning machine that is widely used in fields such as disease prediction, image recognition, voice recognition, speech recognition, NIA credit scoring, fraud detection and weather forecasting. Recently, Basu *et al* [44] created two models for the detection of marine microplastics. A spectral signature profile was introduced for discriminating the floating plastic from other marine debris.

Understanding learning management practices allows us to choose the best methods for the job and build more accurate and powerful ML models.

6.3.2 Unsupervised learning

Unsupervised learning is an ML technique in which the algorithm learns from unlabeled input. Unlabeled data indicates that there are no corresponding output values, as shown in figure 6.5. Without any prior understanding of the data, the algorithm learns to find pattern and relationships in it.

Unsupervised learning is used for two main types of functions [45]—clustering and dimensionality. Clustering is the process of putting data points together based on their commonalities. For example, an unsupervised learning algorithm may be used to categorize customers based on their purchasing history. Dimensionality

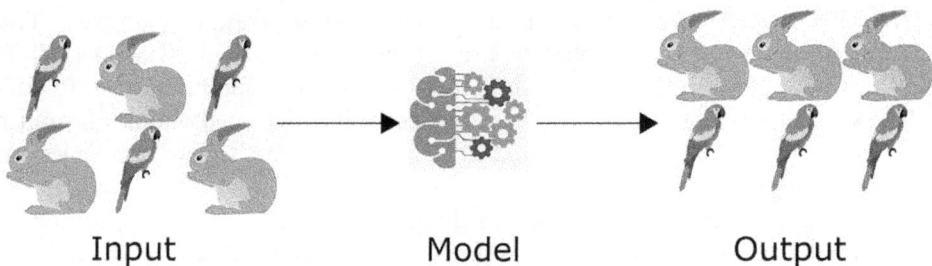

Figure 6.5. Unsupervised learning-based ML model. Image credit: Tinku Basu and Mr. Karan.

reduction problems entail lowering the amount of characteristics in a dataset without losing too much information. For example, an unsupervised learning approach could be used to lower the dimensionality of an image dataset so that it can be handled more simply by a computer.

There are five types of unsupervised learning algorithms—K-means clustering, hierarchical clustering, Gaussian mixture models (GMMs), principal component analysis (PCA) and t-distributed stochastic neighbor embedding. K-means clustering is a straightforward yet effective approach for grouping data points based on similarities. It operates by dividing the data into a predetermined number of clusters, such as three or five. Each data point is then assigned to the cluster that is closest to it by the algorithm [46]. Hierarchical clustering is another prominent approach for grouping data points based on similarities. It operates by establishing a cluster hierarchy in which each cluster is a subset of its parent cluster. Based on their closeness, the algorithm then merges or separates clusters [47]. GMMs are a sort of probabilistic clustering algorithm. They operate on the assumption that the data points come from a mixture of Gaussian distributions. The programme then learns the GMMs and assigns each data point to the Gaussian distribution from which it is most likely to have originated [48]. PCA operates by identifying the data's principle components, which are the directions of highest variance. The data is then projected onto the principle components using this method, which reduces the dimensionality of the data while preserving the maximum amount of information [49]. Stochastic neighbor embedding with t-distribution (t-SNE) is a dimensionality reduction approach. It operates by embedding the data points in a lower-dimensional space while preserving the data's local structure. As a result, t-SNE is an excellent technique for visualizing high-dimensional data [50].

6.3.2.1 Application of unsupervised learning

Unsupervised learning is an effective ML technique that is applied in a wide range of applications. Unsupervised learning algorithms use unlabeled data to uncover intrinsic patterns and structures without the use of explicit direction. Unsupervised learning is applied in customer segmentation and picture segmentation to group comparable data points together. In unsupervised learning, anomaly detection techniques such as isolation forest and local outlier factor are often used to discover uncommon or aberrant data items, such as fraud detection and network intrusion detection. Dimensionality reduction algorithms, such as PCA and t-SNE, are frequently used in unsupervised learning to minimize the amount of features in a dataset, as in picture compression and text summarization. Unsupervised learning is also used to create recommendation systems in fields such as e-commerce and social media. By evaluating user behavior and preferences, unsupervised learning algorithms can offer products or information that are likely to be of interest to the user. Unsupervised learning methods can be used in a variety of applications, such as data exploration, customer segmentation, anomaly detection, picture, and text analysis, and many more. They can reveal hidden insights in huge and complex datasets, discover patterns that may not be obvious at first glance, and assist in decision-making processes [51].

6.3.3 Reinforcement learning

Reinforcement learning (RL) [52] is a sort of ML in which the algorithm learns by interacting with its surroundings. The algorithm rewards behaviors that result in desired outcomes and penalizes those that result in undesirable consequences as shown in figure 6.6. Over time, the algorithm learns which activities are most likely to result in the intended consequences.

In most cases, RL algorithms are utilized in challenges where the algorithm must learn to control a system in order to attain a goal. An RL algorithm, for example, might be used to train a robot to walk, a self-driving car to navigate, or a trading agent to make winning trades. RL algorithms operate by interacting with the environment iteratively and learning from their experiences. The algorithm performs an action at each step and monitors the subsequent reward or penalty. This information is then used by the algorithm to update its policy, which is a function that translates states to actions. There are five different types of algorithms based on reinforcement learning—q-learning, SARSA, deep Q-Networks, policy gradient, actor-critic. Q-learning [53] is a model-free reinforcement learning algorithm that teaches itself to make decisions by maximizing the expected reward. It operates by learning an action-value function that predicts the expected reward for performing a specific action in a specific condition. Q-learning is utilized in a variety of applications, including gaming and robotics.

SARSA: SARSA [54] is yet another model-free reinforcement learning algorithm that teaches itself to make judgments by maximizing the expected reward. It operates by learning the state-action-reward-state-action (SARSA) tuple, which provides the predicted reward for doing a definite action in a specific state and transitioning to a different one. SARSA is utilized in a variety of applications, including gaming and robotics.

Deep Q Network (DQN): DQN [55] is a deep learning support system that uses neural networks for action-value prediction. It learns the optimal policy by minimizing the difference between the expected Q value and the actual Q value. DQN can be used for many applications, including gaming and robotics.

Policy Gradient (PG) [56]: A model-free reinforcement learning algorithm that learns to make decisions through direct policy development. It works by adjusting

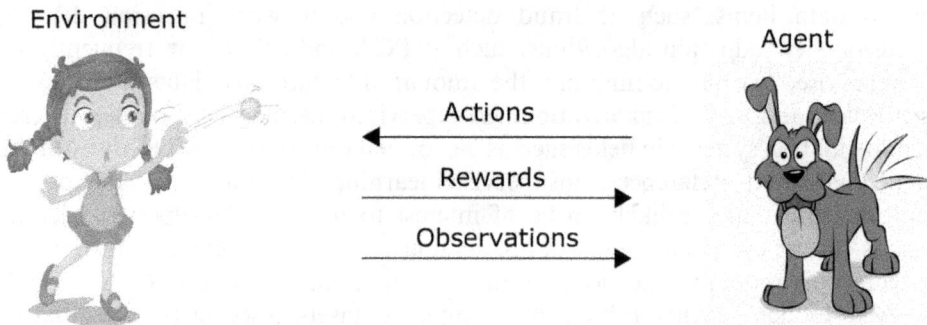

Figure 6.6. Reinforcement learning-based ML technique. Image credit: Tinku Basu and Mr. Karan.

parameters in the direction of the expected reward gradient. PG can be used in many applications, including robotics and natural language processing.

Actor-Critic: Actor-Critic [57] is a support learning model that combines the advantages of policy iteration (PG) and value iteration (Q-learning) techniques. The network will estimate the V(s) function value (how good or bad it is for a particular situation) and the policy. Actor-Critic can be used in many applications, including bots and games.

6.3.3.1 Applications of reinforcement learning

RL has wide applications in many fields. Patients in medical care can benefit from techniques learned from RL systems. RL can generate good ideas based on experience, without prior knowledge of mathematical models of biological systems. This makes the device more suitable for treatment than other control systems. RL is also used in engineering, particularly in games where AlphaGo Zero can learn the game of Go from scratch. It learns by competing with itself. After 40 days of personal training, Alpha Go Zero's performance surpassed Alpha Go Master Edition and defeated world number one Ke Jie. NLP, which uses reinforcement learning, also uses additive learning. Motivational speakers can try to test and predict what people talk about every day by examining normal language patterns. Robotics, driverless cars, finance, and business automation all use additive learning. Understanding boost learning practices allows us to choose the best algorithm for the job and build more accurate and powerful ML models [58].

6.3.4 Ensemble learning

ML can get a boost by combining the strengths of multiple models, a technique called ensemble learning. This approach is useful for both categorization tasks (classification) and continuous value prediction (regression) [59]. For instance, researchers used ensemble learning to estimate water quality in coastal areas using satellite images (Xiaotong Zhu et al [59]). Their model successfully predicted factors like chlorophyll levels, water clarity, and oxygen content. To achieve this, they identified the most informative data points from the satellite images and employed a method to explain how each piece of information influenced the model's predictions.

Bagging and boosting are two types of training used for ensemble learning [60]. Bagging leverages a technique called bootstrap aggregating to create multiple training datasets from the original data. Each of these subsets is used to train a separate model. Predictions from individual models are then combined to create the final forecast. Bagging can reduce the number of variables predicted by a single model. Boosting works by training multiple models; each model builds on the rest of the previous model. The residual is the difference between the actual value of the target variable and the prediction of the previous model.

The types of algorithms used by ensemble learning are—random forests, gradient boosting machines, XGBoost, adaBoost, and stacking. Random forests [61] are a form of ensemble learning technique that trains numerous decision trees using bagging. Bagging is a technique that divides training data into multiple subsets and trains a model on each subset. Multiple model forecasts are then integrated to produce a final projection.

Imagine a group of experts voting on a problem—that's the core idea behind random forests. This ML technique tackles both categorization (classification) and predicting values (regression) by training a multitude of decision trees. Each tree makes its own prediction, and the final output is either the most popular class (classification) or the average prediction (regression) from all the trees. This approach helps to avoid a common pitfall—models that become overly reliant on the specific training data. Random forests can also deal with missing data and maintain accuracy even when a major chunk of the data is missing. Random forests have numerous applications in sectors such as healthcare, engineering, natural language processing, robotics, self-driving cars, finance, and industry automation.

Gradient boosting machines (GBMs) [62] are another sort of ensemble learning method that employs boosting to train multiple decision trees. Boosting is a model-training strategy in which each model is trained using the residuals of the prior model. Residuals are the disparities between the target variable's actual values and the prior model's predictions.

XGBoost [63] is a GBM model optimized for speed and performance. To accelerate training, XGBoost employs a variety of approaches, including parallel computation and gradient compression. When training time is an issue, XGBoost is frequently the best solution for ensemble learning issues.

Stacking [64] is a strategy that uses a metamodel to integrate the predictions of numerous models. The metamodel is trained using the individual model predictions and the target variable. Stacking can be used to combine the predictions of any form of ML model, and it can frequently enhance the ensemble's accuracy.

AdaBoost [65] is a boosting technique that works by training a set of weak learners iteratively. A poor learner is a classifier that performs only marginally better than random guessing. AdaBoost weighs the training instances, requiring the weak learner to learn from the most difficult examples. AdaBoost is frequently used for classification tasks where the training data is unbalanced, meaning that one class has more examples than the other.

6.3.5 Hybrid approaches

Hybrid ML methods [66] use multiple methods to complement and enhance each other, resulting in more accurate and predictable results. This process is receiving increasing attention in fields such as biology, hydrology, and agriculture. The performance of an algorithm can be improved with hybrid methods. It can be employed to eliminate the shortcomings of special algorithms and combines the advantages of many algorithms. This approach is used by experts to learn to use anonymous objects. Finally, it can reduce the number of registered documents required to demonstrate the model.

There are numerous approaches for combining ML algorithms. Some common hybrid techniques are—ensemble learning, stacked learning, multi-instance learning, semi-supervised, active learning.

Ensemble learning—This entails merging many algorithms' predictions to get a more accurate prediction. Ensemble learning approaches can be used to combine any type of algorithm and are frequently used to increase supervised learning task performance.

Stacked learning—This is a sort of ensemble learning in which one algorithm's predictions are utilized as inputs to another algorithm. Stacked learning can mix any sort of algorithm, however, it is mostly employed to blend supervised and unsupervised learning methods.

Multi-instance learning: A sort of supervised learning in which labels are assigned to groups of data points rather than individual data points. Multi-instance learning is frequently employed for challenges where labeling individual data points is difficult or expensive.

6.3.5.1 Applications of hybrid approaches

Hybrid approaches have made their way into several types of applications, and problems that are based on the modern age. Traore *et al* introduced an unsupervised learning process as an improved learning algorithm for hybrid dynamical systems that can identify the modes of the system on-line. The algorithm is exploited to a real switching system multicellular converter and is found to be able to identify the modes accurately [67]. Hybrid approaches in ML can be effective in prediction of forthcoming inclinations of any relevant field [68]. To achieve accurate and interpretable outcomes, this approach combines human intelligence with ML. It is especially valuable in biomedicine, where human experience is required to evaluate complex data. Pham *et al* confirmed that employing hybrid ML approaches in the field of Marcellus Shale to have more accuracy is the highly qualified type for deeper exploration [69]. Lastly, human–machine interaction, ML automation, image and speech recognition, and AI are all hybrid ML applications.

6.4 IoT sensors

6.4.1 Introduction of IoT-based sensors

IoT sensors [70] play a pivotal role in bridging the physical and digital worlds. These intelligent devices gather real-time data from their surroundings, transforming it into actionable insights that empower various applications. IoT-based sensors are revolutionizing industries, enabling automation, optimization, and enhanced decision-making across a wide spectrum. From monitoring environmental conditions to tracking asset performance, IoT sensors are meticulously woven into the fabric of modern society. The benefits of IoT-based sensors [71] are manifold. They enable real-time data collection, providing a holistic understanding of complex systems and facilitating proactive decision-making. They promote efficiency and optimization, reducing waste, improving productivity, and minimizing costs. They enhance safety and security, alerting to potential hazards and enabling preventive measures.

However, the widespread adoption of IoT sensors also presents challenges. Data security and privacy are paramount [72], requiring robust cybersecurity measures to protect sensitive information. Data management and analytics are crucial, as the sheer volume of data generated by sensors demands sophisticated tools and techniques for processing and interpretation. Standardization and interpretability are essential to ensure seamless integration of sensors from diverse manufacturers into unified IoT ecosystems.

6.4.2 Types of IOT sensors for marine pollution detection

Marine pollution poses a significant threat to the health of our oceans and the delicate ecosystems they support [73]. To effectively combat this issue, it is crucial to develop robust monitoring systems that can accurately detect and quantify various forms of pollution in marine environments. IoT-based sensors have emerged as a powerful tool for marine pollution detection, offering real-time data collection, continuous monitoring, and a wide range of sensing capabilities.

The diverse range of marine pollutants necessitates a variety of IoT sensors tailored to detect specific contaminants and environmental parameters. Some of the key types of IoT sensors [74] employed for marine pollution detection are as follows.

6.4.2.1 Chemical sensors
These sensors detect the presence of specific chemicals in marine waters, including heavy metals, organic pollutants, and nutrients. Electrochemical sensors, optical sensors, and biosensors are commonly used for this purpose.

6.4.2.2 Turbidity sensors
Turbidity sensors measure the clarity of water, which can be an indicator of suspended sediment, algae blooms, or other contaminants. Optical sensors based on light scattering or absorption principles are widely used for turbidity measurement.

6.4.2.3 Aquatic oxygen detectors
Essential for aquatic ecosystems, aquatic oxygen detectors monitor the level of oxygen gas dissolved in water, a critical factor for marine organism survival. Optical sensors, electrochemical sensors, and membrane-based sensors are employed for dissolved oxygen measurement.

6.4.2.4 pH sensors
pH sensors measure the acidity or alkalinity of water, which can have a significant impact on marine life and ecosystems. Glass electrode sensors, field-effect transistor (FET) sensors, and optical sensors are commonly used for pH measurement.

6.4.2.5 Temperature sensors
Temperature sensors measure the temperature of seawater, which can influence the behavior and distribution of marine organisms. Thermocouples, resistance temperature detectors (RTDs), and thermistors are commonly used temperature sensors.

6.4.2.6 *Conductivity sensors*

Conductivity sensors measure the electrical conductivity of seawater, which can be an indicator of salinity, pollution, or the presence of dissolved ions. Two-electrode sensors, four-electrode sensors, and inductive conductivity sensors are widely used.

6.4.2.7 *Biofouling sensors*

Biofouling sensors detect the buildup of microorganisms on sensor surfaces, which can affect sensor performance and accuracy. Optical sensors, electrochemical sensors, and acoustic sensors are used for biofouling detection. IoT sensors for marine pollution detection are typically deployed in strategic locations, such as coastal areas, ports, and industrial sites. These sensors can be deployed on fixed platforms, such as buoys, piers, or underwater structures, or they can be integrated into mobile platforms, such as autonomous underwater vehicles (AUVs) or drones. Networking of IoT sensors is crucial for real-time data transmission and analysis. Wireless communication technologies, such as Wi-Fi, Zigbee, and cellular networks, are commonly used for data transmission from sensors to central data collection hubs or cloud platforms.

6.4.3 Data analysis and visualization for marine pollution assessment

The data collected by IoT sensors undergoes rigorous analysis and visualization to extract meaningful insights and identify potential pollution events. Advanced data analytics algorithms, ML techniques, and statistical methods are employed to analyze sensor data patterns, identify anomalies, and correlate pollution levels with environmental factors [75]. Visualization tools, such as maps, charts, and dashboards, are used to present the analyzed data in a comprehensible and informative manner. These visualizations enable researchers, environmental managers, and policymakers to make informed decisions regarding pollution prevention, remediation, and regulatory enforcement.

MARIDA is a game-changer for researchers working on automated marine debris detection using ML. This unique dataset, built from multispectral data captured by Sentinel-2 satellites, allows algorithms to differentiate marine debris from other ocean features like seaweed, ships, and different water types.

MARIDA goes beyond just data. It provides an in-depth analysis of the spectral properties of various floating materials and marine features, along with established ML methods for identifying debris. This open-access resource empowers researchers to develop new AI and deep learning solutions for tackling marine debris, including methods to pre-process satellite data for better detection.

6.5 Integration of AI, ML and IoT sensors

The convergence of IoT, ML and AI unlocks the potential of collaborative intelligence. By harnessing the power of these transformative technologies, we can unlock unprecedented levels of data-driven insights, automation, and decision-making capabilities. AI and ML provide the tools necessary to transform raw IoT data into actionable intelligence [76]. Powerful AI tools are adept at sifting through

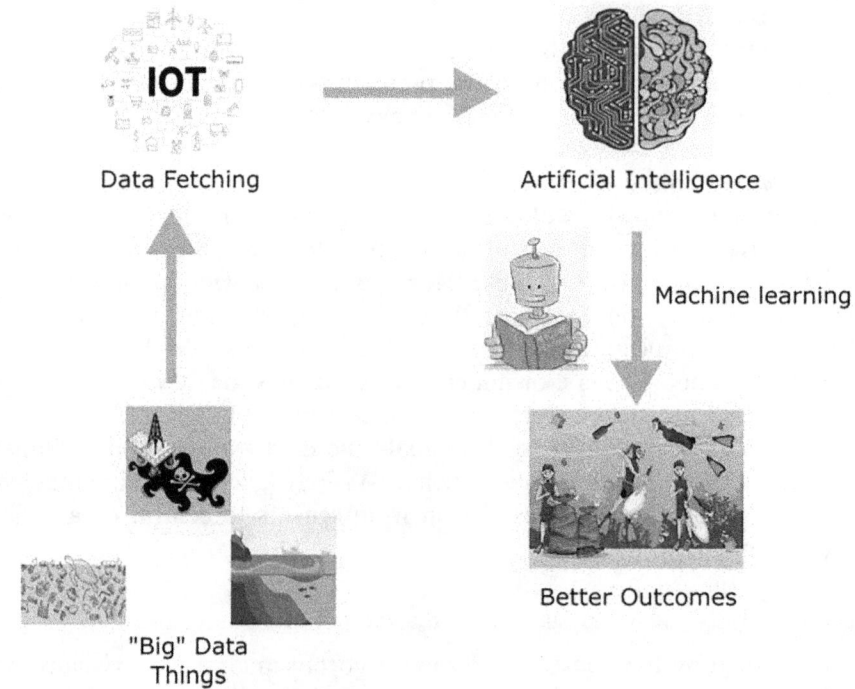

Figure 6.7. Integration of IoT, ML and AI-based technique.

enormous amounts of data to uncover hidden patterns, evolving trends, and unexpected deviations from the norm. ML can then use these insights to build predictive models, automate tasks, and optimize decision-making. The integration of IoT, ML and AI significantly alters data collection, analysis, and decision-making, as shown in figure 6.7.

However, the raw data generated by IoT sensors is often noisy, incomplete, and irrelevant for downstream AI and ML applications. Data preprocessing and feature extraction are crucial steps in transforming raw IoT data into meaningful and usable information for AI and ML models.

6.5.1 Data preprocessing and feature extraction

Unlocking the insights hidden within IoT data requires a crucial first step: data preprocessing. This process ensures the data is in the right format and quality for AI and ML algorithms to analyze it effectively [77]. It involves identifying and correcting errors, handling missing values, and removing outliers. Common data preprocessing techniques include:

Data cleaning: Identifying and correcting errors such as typos, inconsistencies, and invalid values.

Imputation: Handling missing values by filling them with estimated values using techniques like mean, median, or interpolation.

Normalization: Scaling data to a consistent range to ensure equal representation of features.

Feature extraction: Identifying relevant information.

Feature extraction involves transforming the preprocessed data into a set of relevant and informative features that capture the essence of the data for AI and ML algorithms. Common feature extraction techniques include:

Dimensionality reduction: Reducing the number of features to avoid overfitting and improve computational efficiency.

PCA: Identifying and retaining the most significant features that capture the majority of the data's variance.

Feature engineering: Creating new features by combining or transforming existing features to enhance their predictive power.

By transforming raw IoT data into a suitable format and extracting relevant features, these processes enable AI and ML algorithms to unlock the full potential of IoT-driven data, leading to transformative innovations across various industries.

6.5.2 Training and testing dataset

Training datasets are the cornerstone of AI and ML model development. They consist of labeled data that represents the specific task or domain for which the model is being trained. In the context of IoT-based systems, training datasets typically include sensor data paired with corresponding labels or target values [78]. The labels provide the context for understanding the sensor data, enabling the model to learn the relationships between sensor readings and the desired outcome. Testing datasets are crucial for evaluating the performance of trained AI and ML models. These methods offer a neutral evaluation of how well the model performs on completely new information, not just the data it was trained on. Testing datasets should be distinct from the training data to ensure a fair and accurate evaluation of the model's generalization capabilities. The collection and management of training and testing datasets for AI and ML models in IoT-based systems require careful consideration. For data handling to be truly responsible, we need to tackle issues like user privacy, data security, and the ethical implications of how information is used.

6.5.3 Model selection and optimization

The selection and optimization of appropriate AI and ML models [79] are crucial for maximizing the effectiveness of IoT-based systems. This process involves identifying the most suitable model for the specific task or domain, evaluating its performance, and refining its parameters to achieve optimal results. The selection of an appropriate AI or ML model for an IoT-based system depends on various factors, including problem type, data characteristics understanding, domain knowledge, incorporating domain expertise, computational resources, model explainability, performance benchmarks.

Once an appropriate AI or ML model has been selected, it is essential to optimize its parameters to achieve optimal performance. Optimization techniques include:

Hyperparameter tuning: Fine-tuning the model's settings, like its learning speed, control over complexity, and overall structure, can significantly boost its ability to perform well.

Regularization: To keep the model from memorizing specific details and instead focus on learning broader patterns, techniques like L1 and L2 regularization are used. This helps the model perform well on new, unseen data.

Ensemble methods: Combining multiple AI or ML models into an ensemble to enhance overall performance and reduce the impact of individual model biases.

Feature engineering: Refining and transforming features to make them more informative and relevant to the task at hand.

Data augmentation: Increasing the size and diversity of the training dataset using techniques like data augmentation to improve model robustness.

Continuous monitoring: Continuously monitoring model performance on new data to detect potential degradation and re-optimize as needed.

By carefully selecting and optimizing appropriate ML models, the power of IoT-driven data can be harnessed to solve complex problems, make informed decisions, and drive innovation across a wide range of applications.

6.6 Case studies

6.6.1 Application of AI, ML and IOT in remote and *in situ* sensor systems for oil spill response

Oil spills are a major environmental hazard that can cause significant damage to marine ecosystems, coastal communities, and the economy. Early detection of oil spills is essential for minimizing their impact, but traditional detection methods can be slow and ineffective. IoT, ML and AI sensors offer a promising new approach to oil spill detection [80]. AI-based systems are trained to map oil contamination events using statellite-based observations, aerial photographs, and other types of data. ML algorithms [81] can analyze data from IoT sensors to identify variation in the natural surroundings that could indicate presence concerning an oil spill. IoT sensors can be utilized in diverse array of locations, encompassing on coastlines, in the ocean, and on ships and aircraft. This allows for real-time monitoring of large areas for oil spills. Additionally, IoT sensors are utilized for data acquisition purposes on the characteristics of oil spills, such as their size, location, and direction of travel. de Souza *et al* [82] reported the technology of automatic identification of oil spills on the sea surface captured by remote sensing images has had a major impact on ocean monitoring. C-band SAR (ERS-1, ERS-2, Radarsat and Envisat projects) is a system based on image processing technology (filters, gradients, mathematical morphology) and artificial neural networks (ANNs). Pollution from floating oil reduces the backscatter of light. Different speckle filtering algorithms have been tested and compared with deterministic algorithms.

Yang *et al* [83] developed a deep learning model utilizing Sentinel-1 SAR imagery for detecting oil spills. The study utilized 9768 oil objects extracted from

5930 Sentinel-1 scenes spanning 2015–2018 for training and validation of the object detection system, as well as for performance evaluation. This technology holds potential for establishing an early-stage oil contamination surveillance system.

Frate *et al* [84] reported a neural network methodology employed to semi-automatically detect oil spills in ERS-SAR imagery. The network's input comprises a feature vector derived from specialized routines that assess geometric and physical attributes of identified oil spill candidates. Evaluation of the algorithm's classification efficacy was conducted using a dataset comprising confirmed instances of oil spills and visually similar phenomena.

Topouzelis *et al* [85] described the synthetic Aperture Radar images for the detection of dark formations in the marine environment utilizing two distinct neural networks: one dedicated to identifying dark formations, and another for classifying them as either oil spills or similar phenomena. This method shows considerable promise in its ability to detect dark formations with an overall accuracy of 94% and successfully differentiate oil spills from look-alikes in 89% of cases examined.

Singha *et al* [86] reported a novel methodology for SAR oil spill detection, employing a dual-stage ANN approach. The process starts with a first ANN that segments the SAR image, effectively pinpointing pixels with characteristics suggesting an oil spill. Subsequently, a second ANN utilizes extracted statistical feature parameters to classify segmented objects into either confirmed oil spills or similar-appearing phenomena. Training of the proposed algorithm involved images containing verified instances of aquatic oil disasters. Validation on a comprehensive dataset, including full-swath images, demonstrated accurate identification rates of 91.6% for reported oil spills and 98.3% for look-alike occurrences.

Wang *et al* [87] developed an innovative deep learning methodology maritime surveillance system for real-time oil spill detection. The system utilizes a CNN that can derive attributes via satellite imagery and categorize oil contamination events. The proposed system demonstrates high accuracy and robustness and fulfills its purpose.

Jiao *et al* [88] represented a comprehensive framework for oil spill detection and regulation through a remote aerial vehicle and deep learning techniques. The proposed framework employs one dual step CNN framework accurately identify and categorize petroleum discharges from UAV-captured images. The results demonstrated high efficacy of the proposed framework in detection of oil spills across varied marine ecosystems.

Wang *et al* [89] developed an IoT-based oil spill monitoring system that utilizes acoustic sensors to detect oil spills in real time. The system employs a novel signal processing algorithm to extract oil spill signatures from acoustic data.

Jin *et al* [90] introduced a Quad-polarimetric Synthetic Aperture Radar (SAR) system was utilized for the detection of oil spills, employing advanced image processing techniques such as CNNs and Simple Linear Iterative Clustering (SLIC) for analysis. The analysis employed various decomposition techniques (H/A/Alpha, Freeman 3-component, Yamaguchi 4-component) to extract features from the SAR data. Additionally, parameters like SERD, correlation, and conformity coefficients were derived for a comprehensive feature set. Among these, the

Yamaguchi parameters demonstrated superior performance, achieving a total Mean Intersection over Union (MIoU) score of 90.5%.

Yekeen *et al* [91] utilized the Mask-Region-based CNN (Mask R-CNN), a cutting-edge computer vision model, was employed to pioneer a novel deep learning approach for instance segmentation of oil spills. A dataset comprising 2882 images was meticulously curated, encompassing diverse scenarios such as land, ships, and oil spill incidents. These images were partitioned into training (2530 images, 88%) and testing (352 images, 12%) subsets following rigorous preprocessing steps. The model leveraged a Feature Pyramid Network (FPN) architecture for robust feature extraction, utilizing a learning rate of 0.001 over 30 epochs. Transfer learning involved fine-tuning a pre-trained ResNet 101 backbone trained on COCO data. Achieving impressive metrics with precision, recall, and F1-measure values of 0.964, 0.969, and 0.968, respectively, the developed instance segmentation model surpassed existing benchmarks in its ability to accurately identify and classify marine oil spills.

Song [92] integrated multiple fully polarimetric SAR feature datasets using an optimized wavelet neural network classifier (WNN) to devise an innovative method for oil spill identification. To validate this approach, they employed two sets of fully polarimetric SAR data from RADARSAT-2. Comparative analysis against unoptimized WNN and single-feature SAR methods demonstrated significant improvements in classification accuracy: a 4.96% increase over unoptimized WNN and a 7.75% increase over single-feature SAR. Overall, their proposed methodology achieved a high classification accuracy of up to 97.67%.

6.6.2 Detection of harmful algal bloom using AI, ML and IoT sensors

Toxic algal blooms [93] pose a substantial risk to marine ecosystems and human vitality. Harmful algal blooms (HABs) can cause significant environmental damage, economic losses, and even human health risks, making their timely and accurate detection crucial for effective mitigation strategies. Traditional methods for HAB detection, such as visual observations and laboratory analysis, have inherent limitations that hinder their effectiveness [94]. Visual observations are often subjective, rely on trained personnel, and are limited to surface blooms. Laboratory analysis is time-consuming, labor-intensive, and requires sample collection, which can be challenging in remote areas or during adverse weather conditions.

IoT, ML and AI emerged as promising tools for HAB detection and monitoring. To detect and track dangerous algal blooms, AI and ML algorithms can evaluate massive volumes of data from a number of sources, including satellite photography, sensor networks, and drones. [93]. IoT sensors can offer immediate information on the condition of water, based on defined parameters, such as chlorophyll-a concentration, which can be used to predict the likelihood of HAB formation. AI and ML algorithms can be trained to identify HABs in satellite imagery, aerial photography, and data from sensors deployed in the marine environment.

Ananias *et al* [95] unveiled ABD, an AI-powered algae growth detector that monitors growing algae in aquatic settings via satellite-based imagery. The developed procedure offers a reliable and precise computational process like a substitute for conventional techniques for detecting algal blooms. One completely automated approach based on data which includes preparation and collecting features is developed. by simply assuming an image time series as input which simulates an intelligent machine-based classifier that can identify algal blooms.

Mozo et al [96] proposed ML technology employed by an automated high-frequency monitoring system to develop driven by data chlorophyll-a detectors. Utilizing several affordable techniques the sensor harvests Chl-a fluorescence, which in addition can be used for devices with restricted equipment and batteries. Models using ML can become better through the integration of collection and feedback.

Tan et al [97] reported a three-step ML model for the detection of algal bloom utilizing stationary RGB camera images. Captured images were first classified based on the initial framework, and pictures containing particular dangerous growing algae were selected for additional examination. A mask that is capable of removal of interference from non-water objects is made using a second approach. Lastly, harmful algal growths were found and identified using the third model, which is particularly useful in bodies of water. The three actions used together considerably lessened the adverse impacts of the outside world. As a result, high precision final recognition can be achieved.

Srinivasan et al [98] built a CNN with 25 epochs of 900 photos to detect nine different types of algal blooms. The CNN can predict the forms of algal blooms with an accuracy of 80%. They utilized the remote sensing data to provide significant information about the coverage of chlorophyll which can be used to locate HABs.

A remote sensing tool [99] for monitoring HABs in near real time was reported by Pamula et al. The analysis leveraged a combination of regression techniques: multiple regression, support vector regression (SVR), and random forest regression (RFR), were trained and tested using Sentinel-2 and Sat-8 imagery from 2013 to 2017 and from 2015 to 2020. By contrasting the three models to estimate chl-a, phycocyanin, and turbidity, the performance was evaluated.

6.6.3 Monitoring heavy metal ion concentration in marine ecosystems

Marine ecosystems face a growing threat from anthropogenic pollution, particularly heavy metal ion contamination. These metallic pollutants, like lead, mercury, and cadmium, bioaccumulate in marine organisms, posing significant risks to ecosystem health and human well-being. Traditional monitoring approaches, relying on lab analysis and sampling expeditions, are often laborious, expensive, and limited in spatial and temporal coverage. Through several breakthroughs and advancements, AI presents itself as a high impact tool for revolutionizing heavy metal ion monitoring in marine environments.

AI algorithms, with their ability to analyze vast datasets and identify complex patterns, can significantly enhance the effectiveness of existing monitoring methods. AI-powered drones and satellites equipped with hyperspectral cameras can analyze water coloration and detect anomalies indicative of heavy metal contamination.

This data can then be combined with other environmental parameters, such as temperature and salinity, to create real-time maps of heavy metal distribution, enabling targeted monitoring efforts and rapid response to pollution events.

Moreover, AI can analyze chemical sensor data from AUVs and bio-indicators, such as fish and shellfish, to measure heavy metal concentration directly. These models can directly measure heavy metal concentration levels and predict future trends with high accuracy, enabling proactive intervention and mitigation strategies. ML leverages diverse geographical parameters collected via sensors to train itself and predict heavy metal concentration levels with high accuracy. This predictive capability enables proactive intervention and mitigation strategies before detrimental ecological effects occur.

AI also holds immense potential in data integration and analysis. Traditional monitoring generates vast amounts of data from various sources, often stored in disparate formats and locations. AI-powered data management platforms can integrate these datasets, allowing scientists to identify relationships between heavy metal pollution and other environmental factors. This holistic analysis reveals the underlying causes of heavy metal contamination and facilitates the development of targeted interventions.

The benefits of AI-based monitoring go beyond improved data collection and analysis. AI can automate routine tasks, such as data analysis and report generation, freeing up valuable time and resources for scientists to focus on research and developing effective management strategies. Additionally, AI-powered tools can be deployed in remote and inaccessible areas, providing valuable data from previously unmonitored regions.

Lu *et al* [100] reported a study that analyzed water quality and concentration of heavy metal in a water body via SVM and ANN models. This model was predominantly based on five crucial parameters which were highly efficient against particulate transition and post-transition metals. Results signify that AI models have the ability to simulate metal concentrations even with very limited data and vitality of sensitivity analysis that will enhance monitoring and management.

Yang *et al* [101] developed a tool named Artificial Mussels (AMs) for monitoring the levels of radio nuclides in the marine environment. This study focuses on absorption and buildup of ^{238}U, ^{88}Sr, and ^{133}Cs by AMs corresponding with their amounts present in water, taking up to two months to reach homeostasis with elevated concentration levels.

Park *et al* [102] acquired a SERS spectroscopic standard database of $Pb(NO_3)_2$ and applied it to investigate the effectiveness of numerous ML algorithmic computations. The suggested model achieved 84.6% of the target accuracy across the cross-batch assessment challenge.

A CNN-based model, introduced by Zhang *et al* [103] can seamlessly, accurately and effectively recognize ions of heavy metals. At first, square-wave voltage measurements provided data from metal solutions. Afterwards, cleansing of data and expansion was done and a dataset of 1200 samples was obtained. Lastly, a CNN-based recognition system named HMID-NET was established. HMID-NET leverages a two-branched architecture built upon a core foundation.

This configuration allows the network to simultaneously identify the type and concentration of metal ions within a solution. Experiment findings on multiple sets with various ionic species and concentrations demonstrated the developed HMID-NET algorithm reached 99.99% accuracy and an average comparative error of 8.85% in regards of concentrations.

Liu et al [104] have developed a fluorescent detector composed of quantum dots made from carbon driven by stepwise prediction and ML techniques for identifying metal ions (Cr^{6+}, Fe^{3+}, Fe^{2+}, and Hg^{2+}) in external samples. Xylenol orange served as a receptor for establishing the pH-regulated sensing array. Demonstrating a high accuracy of 95%, the sensor array successfully detects four types of heavy metal ions across a concentration range of 1–50 µM, as well as binary mixed materials. Overall, the SX-model-assisted fluorescence detector is an effective way to detect heavy metal ions in environmental samples.

Simionov et al [105] proposed a new approach in determining the presence of heavy metals in water using biological markers to aid the Danube River Basin's blue economy. Water and fish samples were taken from the Lower Danube, Danube Delta, and Black Sea. Cadmium, lead, iron, zinc and copper (in water and fish meat), catalase, ultrasonic acid oxide dismutase, glutathione peroxidase, and malondialdehyde are among the parameters assessed in the sample. The Pollution Index was used to determine the most contaminated ecosystems, and the results revealed that the Danube River was considerably damaged by metals, particularly zinc.

6.7 Future direction and challenges

The confluence of advancements in IoT, ML and AI offers promising avenues for developing robust and efficient detection systems for marine pollutants [106]. The growing synergy between them has ignited a revolution in marine pollution detection. By leveraging AI-powered IoT sensors, we are on the cusp of a paradigm shift towards high-frequency intelligence gathering and interpretation, enabling proactive interventions and safeguarding the health of our oceans.

Looking ahead, several promising future directions beckon. Enhanced sensor technology promises miniaturized, low-cost, and energy-efficient sensors with advanced sensing capabilities. These advancements will allow for wider deployment and longer operation in diverse marine environments, extending our reach into previously inaccessible regions. Real-time data processing and analytics will be made possible by advancements in edge AI and distributed computing, enabling on-sensor analysis and rapid response to pollution events. This will facilitate autonomous decision-making and minimize the need for centralized processing centres.

Furthermore, the fusion of multimodal data will revolutionize our understanding of marine pollution dynamics. By integrating data from diverse sources, including sensors, satellites, and citizen science initiatives, we can create a comprehensive picture of pollution distribution, sources, and impacts. These sensors will integrate diverse sensing modalities like acoustics, optics, and electrochemistry will provide richer data. This holistic approach will pave the way for more effective and targeted pollution mitigation strategies.

On the AI front, advanced ML and AI models will be developed with greater explanatory power and improved accuracy. Explainable AI [107] will allow us to understand how these models arrive at their predictions, fostering trust and confidence in their results. ML algorithms will be able to predict pollution trends, allowing for proactive mitigation strategies and resource allocation. Deep learning models are expected to play a crucial role in pollutant identification and prediction, leading to better decision-making.

AI-based sensors will improve the accessibility and deployment of sensors. AI-based techniques allow autonomous sensor deployment reducing reliance on manual interventions and extending monitoring reach to remote areas. Open-source data formats and platforms will facilitate collaboration and accelerate the development of new ML models. Citizen science initiatives will be empowered by accessible and user-friendly technology, contributing valuable data and expanding monitoring capabilities.

Several emerging applications include predictive modeling, data integration, and citizen science integration, unlocking new applications like bio-inspired sensors and environmental DNA (e-DNA) analysis. Sensors will be able to analyze e-DNA present in water samples, providing insights into the presence and distribution of specific species and potential pollution source.

However, despite the optimism, significant challenges remain. The harsh marine environment presents a formidable obstacle, demanding sensors to be robust and functional amidst varying salinity, pressure, and temperature. Data security and privacy concerns need to be addressed, ensuring unauthorized access is prevented and individual privacy protected in citizen science initiatives.

Furthermore, AI model bias and fairness are critical issues to confront [108, 109]. Training datasets can harbor biases that are amplified by AI models, leading to discriminatory outcomes. Mitigating these biases is essential to ensure fair and equitable application of AI-based solutions [110].

Beyond technological hurdles, regulatory frameworks and policy development are crucial to address the ethical and legal implications of using AI-powered sensors [111]. Clear guidelines are needed to govern data collection, use, and ownership, ensuring responsible development and deployment of this technology.

Finally, public acceptance and ethical considerations [112] cannot be overlooked. Public concerns regarding data privacy, environmental impact, and potential misuse of AI technology need to be addressed. Transparency, collaboration, and open communication will be key to ensuring public trust and ethical implementation of AI-powered marine pollution detection solutions.

6.8 Conclusion

The escalating threat of marine pollution necessitates innovative solutions for its detection and mitigation. Traditional methods, while valuable, often face limitations in spatial and temporal coverage. The confluence of AI, ML and IoT offers a revolutionary approach to marine pollution monitoring.

AI-powered IoT sensors represent a paradigm shift in our ability to detect and track pollutants in real time. These miniaturized, low-cost, and energy-efficient devices can be deployed in vast networks, providing a comprehensive picture of pollution distribution across diverse marine environments. By harnessing the power of ML algorithms, these sensors can analyze collected data to identify specific pollutants, monitor concentration levels, and even predict future trends. This real-time information empowers scientists and policymakers to take rapid and targeted action against pollution sources, minimizing their harmful impacts on marine ecosystems and human health.

The potential of this technology extends beyond mere detection. By integrating data from various sources, including sensor networks, satellites, and citizen science initiatives, a holistic understanding of the complex interactions between pollution, environmental factors, and marine life is achieved. This data-driven approach paves the way for the development of predictive models, allowing us to anticipate pollution events and proactively implement preventative measures.

However, the path to a pollution-free ocean is not without challenges. Ensuring the robustness and long-term functionality of sensors in the harsh marine environment requires technological advancements. Concerns regarding data security, privacy, and potential bias in AI models necessitate the development of ethical frameworks and responsible data governance practices. Additionally, public acceptance and trust in AI-based solutions must be fostered through transparent communication and collaborative efforts.

Despite these challenges, the potential benefits of AI, ML, and IoT-based sensors for marine pollution detection are undeniable. By harnessing the power of these technologies and addressing the associated challenges, a future will emerge where our oceans are cleaner, healthier, and teeming with life. This future, however, demands a collaborative effort from researchers, policymakers, industry stakeholders, and the public alike. Only through a unified approach can we unlock the full potential of this transformative technology and safeguard the future of our marine environment for generations to come.

References

[1] Derraik J G B 2002 The pollution of the marine environment by plastic debris: a review *Mar. Pollut. Bull.* **44** 842–52

[2] National Oceanic and Atmospheric Administration https://commerce.gov/bureaus-and-offices/noaa

[3] Karthikeyan P and Subagunasekar M 2023 Microplastics pollution studies in India: a recent review of sources, abundances and research perspectives *Reg. Stud. Mar. Sci.* **61** 102863

[4] Theodoridis T, Kraemer J and Fava M F 2022 Plastic pollution in the ocean: data, facts, consequences https://oceanliteracy.unesco.org/plastic-pollution-ocean/

[5] Pateraki M, Fysarakis K, Sakkalis V, Spanoudakis G, Varlamis I, Maniadakis M *et al* 2019 Biosensors and Internet of Things in smart healthcare applications: challenges and opportunities *Wearable and Implantable Medical Devices* (Academic) pp 25–53

[6] Sarker I H 2021 Machine learning: algorithms, real-world applications and research directions *SN Comput. Sci.* **2** 160

[7] Temitope Yekeen S and Balogun A-L 2020 Advances in remote sensing technology, machine learning and deep learning for marine oil spill detection, prediction and vulnerability assessment *Remote Sens.* **12** 3416

[8] Shakhov V, Materukhin A, Sokolova O and Koo I 2022 Optimizing urban air pollution detection systems *Sensors* **12** 4767

[9] Mijwil M 2015 *History of Artificial Intelligence* **3** pp 1–8 (unpublished)

[10] Pereira V, Hadjielias E, Christofi M and Vrontis D 2023 A systematic literature review on the impact of artificial intelligence on workplace outcomes: a multi-process perspective *Hum. Resour. Manag. Rev.* **33** 100857

[11] Alpaydin E 2010 *Introduction to Machine Learning* (Cambridge, MA: MIT Press)

[12] Poole D L and Mackworth A K 2010 *Artificial Intelligence: Foundations of Computational Agents* (New York: Cambridge University Press)

[13] Brachman R J and Levesque H J 2004 *Knowledge Representation and Reasoning* (Amsterdam: Morgan Kaufmann)

[14] Jurafsky D and Martin J 2008 *Speech and Language Processing: An Introduction to Natural Language Processing, Computational Linguistics, and Speech Recognition* **vol 2** (Prentice-Hall)

[15] Achler T 2012 *Artificial General Intelligence Begins with Recognition: Evaluating the Flexibility of Recognition BT—Theoretical Foundations of Artificial General Intelligence* ed P Wang and B Goertzel (Paris: Atlantis Press) pp 197–217

[16] Kurzweil R 2014 The singularity is near In: R L Sandler *Ethics and Emerging Technologies* (London: Palgrave Macmillan) pp 393–406

[17] Bostrom N 2014 *Superintelligence: Paths, Dangers, Strategies* 1st edn (Oxford: Oxford University Press)

[18] Gopinath N 2023 Artificial intelligence and neuroscience: an update on fascinating relationships *Process Biochem.* **125** 113–20

[19] Almalawi A, Alsolami F, Khan A I, Alkhathlan A, Fahad A, Irshad K *et al* 2022 An IoT based system for magnify air pollution monitoring and prognosis using hybrid artificial intelligence technique *Environ. Res.* **206** 112576

[20] Ullo S L and Sinha G R 2020 Advances in smart environment monitoring systems using IoT and sensors *Sensors (Basel)* **20** 3113

[21] Li Y, Guo J e, Sun S, Li J, Wang S and Zhang C 2022 Air quality forecasting with artificial intelligence techniques: a scientometric and content analysis *Environ. Model. Softw.* **149** 105329

[22] Neo E X, Hasikin K, Lai K W, Mokhtar M I, Azizan M M, Hizaddin H F *et al* 2023 Artificial intelligence-assisted air quality monitoring for smart city management *PeerJ Comput. Sci.* **9** e1306

[23] Jacob D J and Logan J A 2003 Global 3-D modeling studies of tropospheric ozone and related gases NTRS-NASA Technical Reports Server

[24] De Vries R, Egger M, Mani T and Lebreton L 2021 Quantifying floating plastic debris at sea using vessel-based optical data and artificial intelligence *Remote Sens* **13** 3401

[25] Bruzzese R, Chatzigiannakis I and Vitaletti A 2022 Monitoring water quality with a spectrophotometer: a proof of concept of a Smart Buoy In *2022 IEEE Int. Conf. on Pervasive Computing and Communications Workshops and other Affiliated Events (PerCom Workshops) (Pisa, Italy)* (Piscataway, NJ: IEEE) pp 457–62

[26] AQUACROSS https://aquacross.eu/content/about-aquacross.html

[27] Fradkov A L 2020 Early history of machine learning *IFAC-PapersOnLine* **53** 1385–90

[28] Jia X, O'Connor D, Shi Z and Hou D 2021 VIRS based detection in combination with machine learning for mapping soil pollution *Environ. Pollut.* **268** 115845

[29] Priyadarshini I, Alkhayyat A, Obaid A J and Sharma R 2022 Water pollution reduction for sustainable urban development using machine learning techniques *Cities* **130** 103970

[30] Pan Q, Harrou F and Sun Y 2023 A comparison of machine learning methods for ozone pollution prediction *J. Big Data* **10** 63

[31] Booth H, Ma W and Karakuş O 2023 High-precision density mapping of marine debris and floating plastics via satellite imagery *Sci. Rep.* **13** 6822

[32] Cartwright M, Mydlarz C and Bello J P 2023 A retrospective on monitoring noise pollution with machine learning in the Sounds of New York City project *J. Acoust. Soc. Am.* **153** A262–2

[33] Liu Q and Wu Y 2012 Supervised learning *Encyclopedia of the Sciences of Learning* ed N M Seel (Springer)

[34] Saleh H and Layous J 2022 Machine Learning—Regression (unpublished)

[35] Kumari K and Yadav S 2018 Linear regression analysis study *J. Pract. Cardiovasc. Sci.* **4** 33

[36] Salamate F E and Zahi J 2022 Supervised learning: classification using decision trees for better practice in epidemiology case study: the prevalence of tuberculosis *Procedia Comput. Sci.* **201** 783–8

[37] Jijo B and Mohsin Abdulazeez A 2021 Classification based on decision tree algorithm for machine learning *J. Appl. Sci. Technol. Trends* **2** 20–8

[38] Hearst M A, Dumais S T, Osuna E, Platt J and Scholkopf B 1998 Support vector machines *IEEE Intell. Syst. Appl.* **13** 18–28

[39] Dieterich T G 2000 *Ensemble Methods in Machine Learning BT—Multiple Classifier Systems* (Berlin: Springer) pp 1–15

[40] Nichols J A, Herbert Chan H W and Baker M A B 2019 Machine learning: applications of artificial intelligence to imaging and diagnosis *Biophys. Rev.* **11** 111–8

[41] Li L 2020 Application of deep learning in image recognition *J. Phys.: Conf. Ser.* **1693** 012128

[42] Mohamed A, yi Lee H, Borgholt L, Havtorn J D, Edin J, Igel C *et al* 2022 Self-supervised speech representation learning: a review *IEEE J. Sel. Top. Signal Process* **16** 1179–210

[43] Barlow H B 1989 Unsupervised learning *Neural Comput.* **1** 295–311

[44] Basu B, Sannigrahi S, Basu A and Pilla F 2021 Development of novel classification algorithms for detection of floating plastic debris in coastal waterbodies using multispectral sentinel-2 remote sensing imagery *Remote Sens.* **13** 1508

[45] Langone R, Alzate C and Suykens J A K 2013 Kernel spectral clustering with memory effect *Phys. A—Stat. Mech. Appl.* **392** 2588–606

[46] Na S, Xumin L and Yong G 2010 Research on k-means clustering algorithm: an improved k-means clustering algorithm *3rd Int. Symp. on Intelligent Information Technology and Security Informatics* **2010** pp 63–7

[47] Covões T F and Hruschka E R 2013 Unsupervised learning of Gaussian mixture models: evolutionary create and eliminate for expectation maximization algorithm *2013 IEEE Congress on Evolutionary Computation* pp 3206–13

[48] Wetzel S J 2017 Unsupervised learning of phase transitions: from principal component analysis to variational autoencoders *Phys. Rev.* E **96** 022140

[49] Rogovschi N, Kitazono J, Grozavu N, Omori T and Ozawa S 2017 t-Distributed stochastic neighbor embedding spectral clustering In: *2017 Int. Joint Conf. on Neural Networks (IJCNN)* (Anchorage, AK: IEEE) pp 1628–32

[50] Damilola S 2019 A review of unsupervised artificial neural networks with applications *Int. J. Comput. Appl.* **181** 22–6

[51] Silver D, Schrittwieser J, Simonyan K, Antonoglou I, Huang A, Guez A *et al* 2017 Mastering the game of Go without human knowledge *Nature* **550** 354–9

[52] Hammoudeh A 2018 A Concise Introduction to Reinforcement Learning (unpublished)

[53] Kaur A and Gourav K 2020 A study of reinforcement learning applications and its algorithms *Int. J. Sci. Technol. Res.* **9** 4223–8

[54] Ahsan W, Yi W, Liu Y and Nallanathan A 2022 A reliable reinforcement learning for resource allocation in uplink NOMA-URLLC networks *IEEE Trans. Wirel. Commun.* **21** 5989–6002

[55] Quek Y T, Koh L L, Koh N T, Tso W A and Woo W L 2021 Deep Q-network implementation for simulated autonomous vehicle control *IET Intell. Transp. Syst.* **15** 875–85

[56] Xiao L 2022 On the convergence rates of policy gradient methods *J. Mach. Learn. Res.* **23** 1–36

[57] Konda V and Tsitsiklis J 2001 Actor-critic algorithms *Advances in Neural Information Processing Systems 12 (NIPS 1999)* 42S Solla, T Leen and K Müller

[58] Chen H, Cohn A G and Yao X 2012 *Ensemble Learning by Negative Correlation Learning BT—Ensemble Machine Learning: Methods and Applications* ed C Zhang and Y Ma (New York: Springer) pp 177–201

[59] Zhu X, Guo H, Huang J J, Tian S, Xu W and Mai Y 2022 An ensemble machine learning model for water quality estimation in coastal area based on remote sensing imagery *J. Environ. Manage.* **323** 116187

[60] Bühlmann P 2012 Bagging, boosting and ensemble methods *Handbook of Computational Statistics*

[61] Cutler A, Cutler D and Stevens J 2011 Random forests In *Machine Learning—ML* pp 157–76

[62] Natekin A and Knoll A 2013 Gradient boosting machines, a tutorial *Front. Neurorobot.* **7** 21

[63] Bentéjac C, Csörgő A and Martínez-Muñoz G 2019 A Comparative Analysis of XGBoost arXiv:1911.01914

[64] Barton M and Lennox B 2022 Model stacking to improve prediction and variable importance robustness for soft sensor development *Digit Chem. Eng.* **3** 100034

[65] Chengsheng T, Huacheng L and Bing X 2017 AdaBoost typical algorithm and its application research *MATEC Web Conf.* **139** 222

[66] Megantara A A and Ahmad T 2021 A hybrid machine learning method for increasing the performance of network intrusion detection systems *J. Big Data* **8** 142

[67] McCarty D A, Kim H W and Lee H K 2020 Evaluation of light gradient boosted machine learning technique in large scale land use and land cover classification *Environments* **7** 84

[68] Li J, An X, Li Q, Wang C, Yu H, Zhou X *et al* 2022 Application of XGBoost algorithm in the optimization of pollutant concentration *Atmos. Res.* **276** 106238

[69] Xu S, Li W, Zhu Y and Xu A 2022 A novel hybrid model for six main pollutant concentrations forecasting based on improved LSTM neural networks *Sci. Rep.* **12** 14434

[70] Sehrawat D and Gill N S 2019 Smart sensors: analysis of different types of IoT sensors *2019 3rd Int. Conf. on Trends in Electronics and Informatics (ICOEI)* pp 523–8

[71] Morchid A, El Alami R, Raezah A A and Sabbar Y 2024 Applications of internet of things (IoT) and sensors technology to increase food security and agricultural sustainability: benefits and challenges *Ain. Shams. Eng. J.* **15** 102509

[72] Bertino E 2016 Data security and privacy: concepts, approaches, and research directions *2016 IEEE 40th Annual Computer Software and Applications Conference (COMPSAC)* pp 400–7

[73] Xu G, Shi Y, Sun X and Shen W 2019 Internet of things in marine environment monitoring: a review *Sensors (Basel)* **19**

[74] Sureshkumar P H and Rajesh R 2018 The analysis of different types of IoT sensors and security trend as quantum chip for smart city management *IOSR J. Bus. Manage.* **20** 55–60

[75] Nedyalkova M and Simeonov V 2022 Developing an intelligent data analysis approach for marine sediments *Molecules* **27** 6539

[76] Merenda M, Porcaro C and Iero D 2020 Edge machine learning for AI-enabled iot devices: a review *Sensors (Basel)* **20** 2533

[77] Maharana K, Mondal S and Nemade B 2022 A review: data pre-processing and data augmentation techniques *Glob. Transit. Proc.* **3** 91–9

[78] Shafique M and Hato E 2015 Formation of training and testing datasets, for transportation mode identification *J. Traffic Logist. Eng.* **3** 77–80

[79] den Boer A V and Sierag D D 2021 Decision-based model selection *Eur. J. Oper. Res.* **290** 671–86

[80] Al-Ruzouq R, Gibril M B A, Shanableh A, Kais A, Hamed O, Al-Mansoori S *et al* 2020 Sensors, features, and machine learning for oil spill detection and monitoring: a review *Remote Sens.* **12** 3338

[81] Adi E, Anwar A, Baig Z and Zeadally S 2020 Machine learning and data analytics for the IoT *Neural Comput. Appl.* **32** 16205–33

[82] de Souza D L, Neto A D D and da Mata W 2006 Intelligent system for feature extraction of oil slick in SAR images: speckle filter analysis BT *Neural Information Processing (ICONIP 2006)* Lecture Notes in Computer Science (LNTCS, volume 4233) ed I King, J Wang, L W Chan and D Wang (Berlin: Springer) pp 729–36

[83] Yang Y J, Singha S and Mayerle R 2022 A deep learning based oil spill detector using Sentinel-1 SAR imagery *Int. J. Remote Sens.* **43** 4287–314

[84] Del Frate F, Petrocchi A, Lichtenegger J and Calabresi G 2000 Neural networks for oil spill detection using ERS-SAR data *IEEE Trans. Geosci. Remote Sens.* **38** 2282–7

[85] Topouzelis K 2008 Oil spill detection by SAR images: dark formation detection, feature extraction and classification algorithms *Sensors* **8** 6642–59

[86] Singha S, Bellerby T J and Trieschmann O 2013 Satellite oil spill detection using artificial neural networks *IEEE J. Sel. Top. Appl. Earth Obs. Remote Sens.* **6** 2355–63

[87] Wang R, Zhu Z, Zhu W, Fu X and Xing S 2021 A dynamic marine oil spill prediction model based on deep learning *J. Coastal Res.* **37** 716–25

[88] Jiao Z, Jia G and Cai Y 2019 A new approach to oil spill detection that combines deep learning with unmanned aerial vehicles *Comput. Ind. Eng.* **135** 1300–11

[89] Wang Y, Chen X and Wang L 2023 Cyber-physical oil spill monitoring and detection for offshore petroleum risk management service *Sci. Rep.* **13** 4586

[90] Jin Z, Luo Q, Yu L, Hao F and Jujie W 2019 Oil spill detection using refined convolutional neural network based on quad-polarimetric SAR images *14th IEEE International Conference on Electronic Measurement & Instruments (ICEMI) (Changsha, China)* pp 528–36

[91] Yekeen S T and Balogun A-L 2020 Automated marine oil spill detection using deep learning instance segmentation model *Int. Arch. Photogramm. Remote Sens. Spatial Inf. Sci.* **XLIII-B3-2020** 1271–6

[92] Song D, Ding Y, Li X, Zhang B and Xu M 2017 Ocean oil spill classification with RADARSAT-2 SAR based on an optimized wavelet neural network *Remote Sens* **9** 1–20

[93] Khan R M, Salehi B, Mahdianpari M, Mohammadimanesh F, Mountrakis G and Quackenbush L J 2021 A meta-analysis on harmful algal bloom (HAB) detection and monitoring: a remote sensing perspective *Remote Sens.* **13** 4347

[94] Shen L, Xu H and Guo X 2012 Satellite remote sensing of harmful algal blooms (HABs) and a potential synthesized framework *Sensors (Basel)* **12** 7778–803

[95] Ananias P H M, Negri R G, Bressane A, Colnago M and Casaca W 2023 ABD: a machine intelligent-based algal bloom detector for remote sensing images *Softw. Impacts* **15** 100482

[96] Mozo A, Morón-López J, Vakaruk S, Pompa-Pernía Á G, González-Prieto Á, Aguilar J A P et al 2022 Chlorophyll soft-sensor based on machine learning models for algal bloom predictions *Sci. Rep.* **12** 13529

[97] Tan Z, Yang C, Qiu Y, Jia W, Gao C and Duan H 2023 A three-step machine learning approach for algal bloom detection using stationary RGB camera images *Int. J. Appl. Earth Obs. Geoinf.* **122** 103421 https://sciencedirect.com/science/article/pii/S1569843223002455

[98] Srinivasan K, Duvvur V and Hess D 2018 Prediction of algal blooms in the Great Lakes through a convolution neural network of remote sensing data bioRxiv (Cold Spring Harbor Laboratory). https://doi.org/10.1101/450551

[99] Pamula A S P, Gholizadeh H, Krzmarzick M, Mausbach W and Lampert D 2023 A remote sensing tool for near real-time monitoring of harmful algal blooms and turbidity in reservoirs *JAWRA J. Am. Water Resour. Assoc.* **59** 929–49

[100] Lu H, Li H, Liu T, Fan Y, Yuan Y, Xie M et al 2019 Simulating heavy metal concentrations in an aquatic environment using artificial intelligence models and physicochemical indexes *Sci. Total Environ.* **694** 133591

[101] Yang Y, Chow T W, Zhang Y Q, Yu P K N, Ko C C and Wu R S S 2023 Artificial mussels: a new tool for monitoring radionuclides in aquatic environments *J. Mar. Sci. Eng.* **11** 1309

[102] Park S, Lee J, Khan S, Wahab A and Kim M 2022 Machine learning-based heavy metal ion detection using surface-enhanced raman spectroscopy *Sensors* **22** 596

[103] Zhang J, Chen F, Zou R, Liao J, Zhang Y, Zhu Z et al 2023 A CNN-based method for heavy-metal ion detection *Appl Sci* **13** 4520

[104] Liu Y, Chen J, Xu Z, Liu H, Yuan T, Wang X et al 2022 Detection of multiple metal ions in water with a fluorescence sensor based on carbon quantum dots assisted by stepwise prediction and machine learning *Environ Chem Lett.* **20** 3415–20

[105] Simionov I A, Cristea D, Petrea S M, Mogodan Antache A, Jijie R, Todirascu-Ciornea E *et al* 2021 Predictive innovative methods for aquatic heavy metals pollution based on bioindicators in support of blue economy in the Danube River Basin *Sustainability* **13** 8936

[106] Soori M, Arezoo B and Dastres R 2023 Artificial intelligence, machine learning and deep learning in advanced robotics, a review *Cogn. Robot.* **3** 54–70

[107] Linardatos P, Papastefanopoulos V and Kotsiantis S 2020 Explainable AI: a review of machine learning interpretability methods *Entropy (Basel)* **23** 18

[108] Min A 2023 Artifical intelligence and bias: challenges, implications, and remedies *J. Soc. Res.* **2** 3808–17

[109] Ntoutsi E, Fafalios P, Gadiraju U, Iosifidis V, Nejdl W, Vidal M *et al* 2020 Bias in data-driven artificial intelligence systems—an introductory survey *WIREs Data Min. Knowl. Discov.* **10** e1356

[110] Vicente L and Matute H 2023 Humans inherit artificial intelligence biases *Sci. Rep.* **13** 1–13

[111] Henz P 2021 Ethical and legal responsibility for artificial intelligence *Discov. Artif. Intell.* **1** 2

[112] Tulchinsky T H 2018 Ethical issues in public health *Case Studies in Public Health* (Academic) ch 13 pp 277–316

IOP Publishing

Sensors for Marine Biosciences
Next-generation sensing approaches
Shyam S Pandey, Rout George Kerry and Kshitij RB Singh

Chapter 7

Biosensor-based detection of major aquatic pathogens in the marine ecosystem

Pooja Singh and Ravindra Pratap Singh

The marine ecosystem is a water-based system that is adversely affected by inevitable human activities. These activities contribute to the spread of pathogenic marine bacteria, which ultimately infect humans and devastate aquatic life. Therefore, it is necessary to control their growth and reduce their population from the marine ecosystem to maintain their well-being and prevent several future challenges, including the protection of water and food, the environment, and prevention of disease in aquatic and human life. The most reliable method for ensuring the identification and control of microorganisms is biosensor-based detection and monitoring. This system has numerous advantages over the previous one, including on-site identification, quick and simple handling, shorter scan times, environmental friendliness, and reusable materials.

7.1 Introduction

The marine environment is a key source of biodiversity and has a significant role in the management of the global climate. This system also offers potential solutions to the many issues afflicting our planet, such as a sustainable food and energy supply. It has the potential to be a significant source of novel industrial materials and procedures that produce a range of novel bioactive phytochemical substances and beneficial remedies for the health of human beings [1]. However, the ecosystem of the marine system is changing dramatically because of human activity, changes in the climate, increased industrialization, the tourism industry, and the expansion of metropolitan areas [2]. The Integrated Maritime Policy, or 'Europe 2020,' was proposed in 2007 to enhance and develop the marine economy sustainably. It included three main objectives: (a) smart economic growth through innovation and knowledge-based approaches; (b) a more competitive and environmentally friendly economy; and (c) inclusive growth that promotes high employment, and social and

doi:10.1088/978-0-7503-5999-3ch7

territorial cohesion. Moreover, the European Union decided to continue the Maritime Strategy Framework Directive (MSFD) until 2020 as an additional measure for establishing a safe, maintained, and healthy maritime environment [3]. The presence of low concentrations that are adequate to have a detrimental effect on the ecosystem is one of the main issues with monitoring and controlling marine ecosystems. As a result, numerous techniques have been developed to analyse the pollutants, even at extremely low concentrations, ranging from nanograms to picograms per Liter. For instance, gas chromatography coupled to tandem mass spectrometry (GC–MS/MS) and liquid chromatography coupled to tandem mass spectrometry with electrospray ionizing radiation in negative mode (LC/ESI-MS/MS) has been used for detection in marine waters [4, 5]. To perform the quantitative assessment of trace quantities of marine pollutants, these analytical methods necessitate sample requirements and occasionally pre-concentration processes. They also demand time-consuming, costly resources, and continual monitoring. Additionally, data that are inappropriate for a rapid alarm system must be interpreted at the same time. As a result, in these situations, an alternative method is mentioned because it converts chemical energy into a signal that may be used for analysis [6].

A sensor in maritime systems is a device that can detect changes in the surrounding physical environment, such as variations in temperature, thermal conductivity, and chemical concentration. As a result, while building a sensor or biosensor, several factors, including consistency, sensibility, specificity, precision, reliability, reusability, and reproducibility, are necessary to determine the sensor's potential application in the future [7, 8]. Implementing biosensors has several advantages, including being quick and easy, requiring minimal samples, doing away with sample preparation steps, employing remote devices, and using minimal resources. Wireless sensors are growing increasingly prevalent in small-scale biosensor networks because of their small size, affordability, and ease of use. These kinds of sensors function with the assistance of a CPU, power source, and sensor that are all connected [2]. To keep track of and regulate the marine ecosystem at suitable time and space scales, several *in situ* and remote sensing platforms must be developed. The design and development of such tools that monitor the entire marine ecosystem is a challenging endeavour, as there are numerous experts present from various disciplines, including physics, chemistry, genetics, and nanotechnology. This is a subject that is expected to receive substantial interest in the coming years. Sensors and biosensors can monitor an extensive spectrum of factors, including chemical, physical, and biological ones [2, 9]. This chapter (figure 7.1) will cover the marine ecosystem and how human interference negatively impacts it. Along with traditional methods and their limitations to recognise marine contaminants, we will also cover multiple types of marine pathogenic pollutants. Additionally, the function of biosensors in the identification of pathogenic microorganisms in marine environments and their detrimental impacts on humans and marine life will be covered in this work as well. The conclusion and prospects for the future will be covered at the end of the chapter [9, 10].

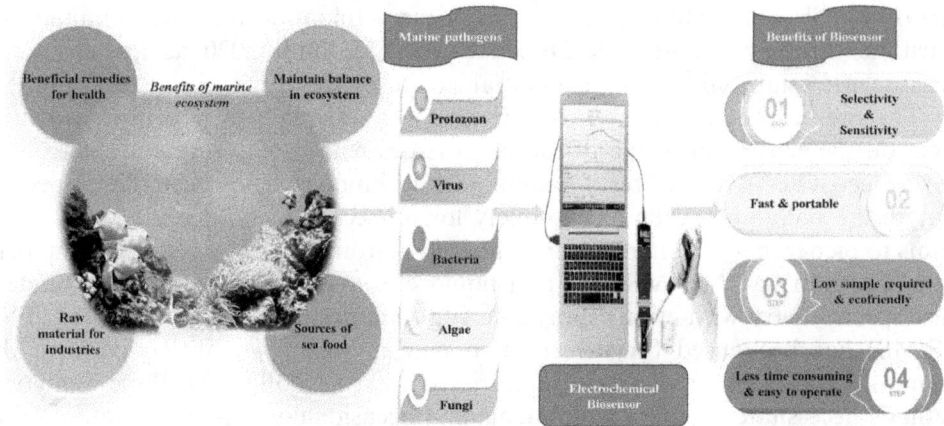

Figure 7.1. A thorough synopsis of the entire endeavour presented in an organized way.

7.2 Marine ecosystem and causes of pollution

An aquatic ecosystem is one in which the biotic and abiotic elements of the system interact in a water-based environment. This system is separated into two groups: freshwater ecosystems and marine ecosystems. More than 70% of the planet's surface is made up of the marine ecosystem, which is further separated into coral reefs, oceans, estuaries, and coastal habitats. Moreover, the 1% of the earth's surface that constitutes the freshwater ecosystem is further separated into the lentic, lotic, and wetlands ecosystems [11]. Anthropogenic activities on Earth lead to the gradual contamination of the aquatic ecosystem. These activities include the building and filling of dams, canals, roads, and bridges; deforestation; domestic and agricultural activities; and industrial settlements as thoroughly shown in the figure 7.2. Water pollution is the primary cause of pollution on the Earth and its repercussions. The primary cause of the decline in the aquatic ecosystem in many developed nations is agriculture. Nearly 38% of the water in European countries is heavily stressed by agriculture. Additionally, it is the main cause of pollution in American rivers and streams. In the same way, it is the third primary source of lakes and the second source of wetlands. China is the country that is mostly to blame for groundwater pollution and a significant amount of surface water pollution. The unrestricted volume of untreated industrial and municipal garbage poses a serious threat to aquatic ecosystems and the creatures that inhabit them in developing countries [12].

7.2.1 Agrochemicals

Population growth has put a strain on agriculture, requiring additional land that is obtained by clearing forests. This has resulted in an excess of pollutants in the aquatic ecosystem. A greater amount of agrochemicals is used to increase food production in response to the rising demand for food brought on by population growth. The use of various agrochemicals—such as fertilisers, herbicides, insecticides, and plant hormones—in an unsustainable manner leads to an increase in

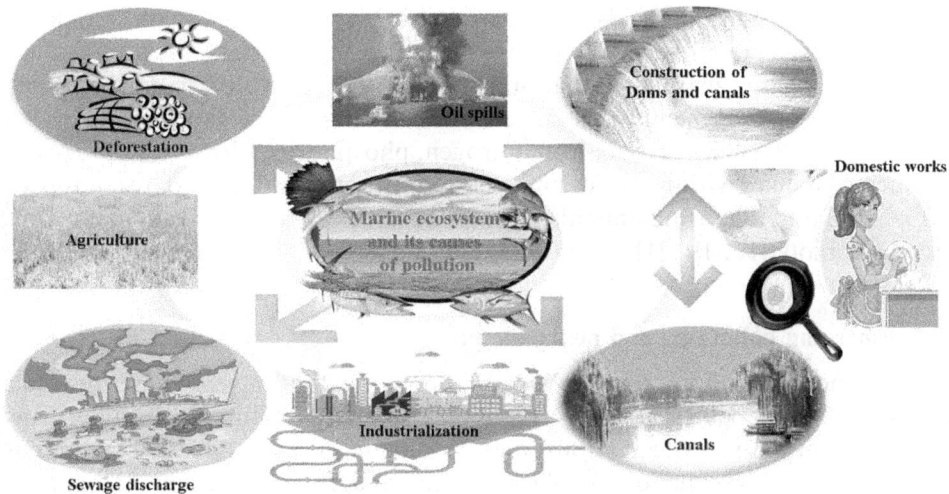

Figure 7.2. A comprehensive description of human activities that contaminate the marine ecology either directly or indirectly.

environmental contamination. Agricultural lands serve as both the primary absorbers and drains of different kinds of chemicals from surrounding fields that are gathered by water runoff, direct drift, and leaching [13–15]. High fertiliser applications are fixed by the soil and absorbed by crops through the soil; unbounded fertilisers are carried off by surface runoff from the surrounding soil into the water, causing contamination of the water. The nutrient enrichment, or 'eutrophication,' of 'lakes,' 'reservoirs,' 'ponds,' and 'coastal waters' is mostly caused by excess fertiliser that is washed by heavy rainfall and accumulated in water sources. This causes 'algae blooms,' or excessive growth of aquatic plants that kill other aquatic plants and animals. Based on the available data, 415 coastal regions have been classified as very eutrophic, out of which 169 are hypoxic. High amounts of nitrate in drinking water can result in serious health problems including 'blue-baby syndrome' from an excessive nutrient buildup. The most prevalent chemical contaminant in the world's groundwater aquifers is nitrate, which seeps into the groundwater from agriculture [16, 17]. Similarly, a lot of countries use a lot of pesticides, herbicides, and fungicides in agriculture; these chemicals are also washed and carried off by rainfall, collecting in aquatic water bodies and contaminating water supplies. They contain extremely toxic compounds that can cause cancer and other serious illnesses in humans and might perhaps wipe out marine life. They enter the food chain through absorption by aquatic life and eventually become deadly to humans. Millions of tonnes of toxic chemicals are used in agriculture, which contributes significantly to global mortality and severe human morbidity, particularly in low-income nations where impoverished farmers frequently employ extremely dangerous pesticides [15, 16, 18]. Furthermore, salinization results from the chemical's discharge through irrigation and rainfall that builds up in soils into receiving water bodies. The infiltration of salty seawater into groundwater, perhaps because of over-extraction of groundwater for coastal agricultural, can also lead to salinization. Significant saline water issues have been observed in several nations, including the United States of America, Argentina, China, Australia, India, Sudan, and other Central Asian nations. With overall effects on ecosystems, this saline water may change the geochemical cycles of important elements like carbon, iron, nitrogen, phosphorus, silicon, and sulphur. By 'producing changes within species and community composition' and 'resulting in the decline of plants, algae and animals,' this saline issue can also have an impact on the freshwater biota [16, 19–21].

7.2.2 Emerging pollutants and heavy metals

Emerging pollutants are a new class of contaminants that have entered the aquatic system during the past 20 years. Examples of these pollutants include growth hormones, antibiotics, vaccinations, and promoters in plants. These contaminants enter the water by the application of slurries and manure to agricultural land, as well as through leaching and runoff from livestock and aquafarms. Currently, the list of aquatic environments in Europe includes about 700 new pollutants, together with their metabolites and converted products, as per published literature. In addition to being the source of these pollutants, agriculture also contributes to their reintroduction into the aquatic environment by using municipality waste as fertiliser, compost, and wastewater for irrigation. An estimated 35.9 mega hectares of agricultural land are exposed to the indirect use of wastewater, which has the potential to expose people to new contaminants and cause illness through the indirect use of agricultural products [16, 22–24]. Similarly, we cover heavy metals in the paragraph that follows.

Heavy metals are aquatic pollutants that can enter the aquatic system both naturally and by human activity. There are two possible methods: the first is direct, involving the direct discharge into the fresh and marine ecosystem, and the second is indirect, using surface runoff and air deposition. Mountains and rocks are naturally occurring sources of heavy metals that are broken down by weathering and volcanic eruptions and then transported towards bodies of water by surface runoff and soil erosion. It is safe for humans to at low concentrations, but as industry and agriculture develop quickly, heavy metal concentrations are rising at an accelerated rate. Some of the most hazardous heavy metals that are typically found in aquatic environments include lead, mercury, cadmium, nickel, copper, chromium, and arsenic. Stormwater and wastewater discharged from businesses and agriculture can raise the concentrations of these heavy metals. They are among the most dangerous contaminants, because of their toxicity and ability to stay in the aquatic ecosystem. Zinc and copper are contaminants found in all fertilisers, while cadmium, mercury, and arsenic are components of some fungicides [16, 25–28]. These contaminants have a significant ecological impact because of their tenacity and difficulty being removed from the water once they build up and reach the food chain. They can amass and endanger human safety as well as produce a variety of ecological harms in specific environmental circumstances. Once they enter an aquatic ecosystem, they are drawn to the particulate matter, which will eventually cause them to settle and become absorbed into the sediments. When favourable conditions are met, such as pH and Eh, they can be readily released into the water, which can result in secondary pollution. Additionally, this pollution has the potential to bioaccumulate in living things, become persistent, very poisonous, and eventually threaten human existence on Earth by becoming magnified in food chains. These contaminants can lead to several aberrant diseases, including immunological deficiencies, poor reproductive outcomes, and improper foetal development [29–31].

7.2.3 Sewage and oil spills

The release of liquid petroleum hydrocarbons into the aquatic environment, particularly in the marine ecosystem is known as an oil spill. Pollution of the marine environment has drawn the attention of all researchers because it is harmful to both aquatic and terrestrial organisms. These pollutants are classified as hazardous waste when they unintentionally occur during transportation, damage, and other industrial mining operations. Damage from the 1% increase in oil leaks is estimated to have been US$0.178 million and they are also regarded as common organic contaminants that affect aquatic environments. These mishaps happened in pipelines, non-tanker vessels, facilities, and oil tankers, among other places. Their spills into the oceans could be the consequence of operations, ship mishaps, or deliberate releases of oily waste into the sea [32–37]. Sewage generates the most common waste product that is released into the aquatic ecosystem, making it one of the main causes of aquatic pollution. Numerous kinds of municipal, industrial, and household wastes—including bathroom, laundry, kitchen, and toilet waste—are included in the pollutants [38]. The ideal washbasin for the release of these wastes is freshwater. Research indicates that about 73% of water bodies are contaminated because of the direct discharge of 58% of wastewater from urban areas and 81% from industrial areas, with or without appropriate treatment [39]. Freshwater supplies are eventually diminished or exhausted because of these wastewater releases, which further worsen water contamination. Massive volumes of wastewater are discharged from metropolitan areas directly into neighbouring bodies of water, causing widespread ecological deterioration through decreased water availability and quality, severe flooding, extinction of species, and modifications to the biota's distribution and structure [40]. The amount and composition of wastewater, as well as how frequently and how much of it enters aquatic bodies, all affect how harmful sewage is to the environment [40]. In addition, it has high concentrations of heavy metals, bacteria, nutrients, and pharmaceutical and personal care items. Sewage is initially organic, but over time, its oxygen content decreases, earning it the appellation 'high biological oxygen demand.' when the sewage has a lot of nutrients or decomposing material, it causes eutrophication, an environment in which plants and algae proliferate in the aquatic ecosystem. When this condition occurs, the turbidity of the water grows, the biomass of plants and animals increases, the species diversity diminishes, the sedimentation rate increases, and anoxic conditions may arise, potentially leading to a shift in the dominant species within the aquatic biota

[41–43]. Many pathogenic organisms found in sewage effluent that enters surface waters have the potential to spread waterborne illnesses when polluted water is used for residential and other purposes. Around 25% of deaths globally are attributed to infectious diseases, which are brought on by pathogenic bacteria, according to UNEP data from 2006. 1400 species of pathogenic microorganisms, which include viruses, worms, bacteria, parasites, and fungi, have been identified by scientists; which can harm public health and society [44–48].

7.3 Types of marine pollutants

Numerous causative agents have been identified, including bacteria, viruses, fungi, and parasites. These agents can cause harm through various means, including animal bites, infections, penetration through rough skin and wounds, insect vectors, and animal contractions caused by inhaling particles or encountering mucous membranes. These diseases can infect humans and spread directly and indirectly through the vector. However, several illnesses are widely believed to be significant in aquatic environments [49–52]. These include illnesses like salmonellosis and campy-lobacteriosis, which are found in humans and animals far less frequently than other zoonotic diseases. This could be because of inadequate knowledge and monitoring. Furthermore, people with these diagnoses may not survive. According to other published data, 260 000 Americans fall ill each year from eating tainted fish, with fish meat being the most often linked food item to these outbreaks. Data on food-borne disease outbreaks was gathered by the Centre for Disease Control and prevention (CDC). Food-borne Disease Outbreak Surveillance System (FDOSS) show that about 857 outbreaks were linked to fish, resulting in 4815 illnesses, 359 hospitalisations, and 4815 fatalities [53]. Many studies show that there have been dangerous fish zoonotic outbreaks over the years, which further supports the need for us to keep an eye out for zoonotic infections originating from fish. The World Health Organisation (WHO) has therefore taken numerous measures to categorise the various diseases found in aquatic animals as emerging diseases, which have either recently emerged in a population or may have existed in the past but are expanding rapidly in incidence or geographic range. The fact that there are a lot of new diseases yet little information about them is available is one of the biggest drawbacks. As a result, it is crucial to make sure that information is efficiently and promptly shared with the public and other professionals; which can only occur when knowledge is swiftly and efficiently shared with the public and other experts with the question-answer format. Aetiology, geographical distribution, prevalence, incidence, eco-epidemiology, clinical symptoms, availability of diagnostic tests, assessment of zoonotic potential, possible human exposure sources, and detection of zoonotic disease potential are merely some of the many crucial assessment questions [54].

Seafood is becoming more and more popular due to its high protein content and the sustainable expansion of the fishing and aquaculture industries, which has coincided with an increase in global population. We are all aware that there is a possibility of aquatic infections spreading to humans, therefore it is not risk-free.

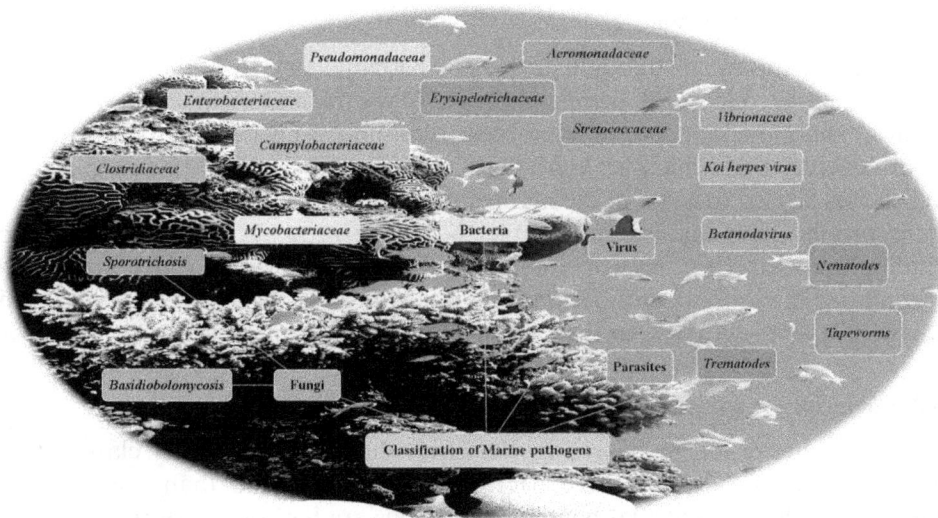

Figure 7.3. Marine pathogen classification.

On the other hand, several additional elements have demonstrated the possibility of disease transfer from fish to humans through the surrounding water [49, 55]. The immune system determines the severity of infections in aquatic environments. There are just two methods for the illness to spread. One way to contract the infection is through eating raw or undercooked fish, ingesting water or other materials contaminated with infected fish faeces or mucus, or coming into touch with the agent through open wounds, scratches, or abrasions on the skin. Even though human infections from fish are extremely rare, they should nevertheless be regarded as potential health hazards. However, zoonotic disease has been determined to be an emerging illness in humans that poses a major risk to world health and has the potential to do the most harm to the entire planet. The devastation caused by COVID-19 underscores the likelihood of human–animal interaction, especially with wildlife and livestock species that might act as viral repositories and potential hosts. As a result, it is critical to recognise the causes and mechanisms behind the genesis of different diseases and to address the spread of infectious diseases of this form. The impact of globalisation is that numerous factors constantly contribute to the spread of illness, including habitat loss, alteration in the climate, and links between the systems that support cattle and other animals [56–58]. The pathogen that infects fish can spread to humans due to several factors, including different types of human parasites, viruses, and fungi that are dependent on a person's health, including open wounds, spine penetration, immunocompromised conditions, and contaminated water in the environment. Bacteria, viruses, and parasites account for many infections caused by fish-associated pathogens and their classification is shown in figure 7.3 [49, 56].

7.3.1 Protozoan organisms

Protozoan organisms are thought to pose a zoonotic risk to humans as they can infect and cause disease in people through *Cryptosporidium species*. Worldwide, several *Cryptosporidium* species have been found in freshwater, marine, cultivated, and ornamental fish, which are pathogenic. The worldwide growth in aquaculture and fisheries has led to a rise in the number of agents that can transmit diseases from fish to people. Humans are susceptible to a wide range of pathogenic diseases, including bacterial, viral, parasitic, and fungal zoonotic illnesses, which are brought on by marine microorganisms. As a result, it is critical to draw attention to the risks to humans and, eventually, to their mitigation and prevention [59].

7.3.2 Bacterial organisms

Fish diseases are primarily caused by bacteria, which can be broadly classified into two types; wherein gram-positive bacteria are less common than gram-negative bacteria. It is also evident that bacterial infections are harboured by the bacteria in fresh fish's intestines and kidneys. Two main types of bacteria, such as *vibrio* and *mycobacterium*, are the main cause of fish production infections, which can lead to large financial losses and are frequently referred to as the limiting factor. Every year, the incidence of various zoonotic agents in seafood types varies, and it is important to regularly assess their prevalence in fish populations that are both wild and cultivated. Furthermore, decorative coloured fish can serve as a significant source of causative agents that have demonstrated elevated levels of resistance to antibiotics [60–63].

7.3.2.1 Mycobacteriaceae

Gram-positive, acid-fast, non-motile, aerobic, pleomorphic bacilli, *mycobacterium* species are members of the *Mycobacteriaceae* family, which also includes many pathogenic bacteria which are associated with fish, animals, reptiles, and humans. It is a primary factor in the demise of both farmed and wild fish. This prevalent ailment affects the fish in freshwater, brackish water, and the ocean. Non-tuberculosis *mycobacterium* infections have been discovered in over 150 fish species around the world; this presents a zoonotic danger to public health. These bacteria can spread both vertically and horizontally and affect most fish species. They are often seen in ornamental fish seen in shops and homes [64–68]. Following bacterial infection, fish may exhibit symptoms that differ depending on the pathogen species and the host species. After contracting an infection from *mycobacterium* species of bacteria, fish may exhibit symptoms like lethargy, colouration, stomach distention, exophthalmia, skin sores, and even death. The disease appears in the fish's eyes, gills, liver, kidneys, and spleen as it travels throughout the circulatory system. Additionally, enlarged liver, kidney, spleen, and lumps in internal organs can be signs of infection in fish. Long-term carriers of different types of bacteria are infected and asymptomatic fish and human illnesses frequently arise from interaction with infected aquatic environments and water. There are currently 120 identified species of *mycobacterium*, and they are significant because they are known to cause fish zoonoses, which can occasionally result in both acute and chronic illness from exposure. These include *Mycobacterium avescencs, Mycobacterium chelonae, Mycobacterium fortuitum, Mycobacterium gordonae, Mycobacterium marinum, Mycobacterium ulcerans, Mycobacterium septicum, Mycobacterium peregrinum*, and *Mycobacterium avium*, which are some of the primary causes of fish zoonoses [64, 65, 69–71]. *Mycobacterium*-induced illnesses result in deep-tissue infections, such as those in tendons and bones, severe necrotic lesions, and granulomatous skin lesions. Nevertheless, extra-respiratory and systemic respiratory illnesses are uncommon, though they can occasionally strike immunocompromised individuals. Arthritis, osteomyelitis, and bronchitis are rare conditions. In people with impaired immune systems, mycobacteriosis can develop into a systemic illness that can be fatal. The type VII secretory system, ESX genes, cytosolic accessibility, and the stimulation of host actin polymerization for motility and cell-to-cell movement are examples of *mycobacterium* virulence factors [64, 67, 69]. Four out of the 120 species of *mycobacterium* are more common than other species, including *M. marinum, M. fortuitum, M. gordonae*, and *M. chelonae*, and they are also crucial in the occurrence of outbreaks. The presence of *Mycobacterium* spp. in freshwater ornamental fish sold in pet stores poses a serious risk to the people handling them. *Piscine mycobacteriosis* has been found in ornamental fish, which are primarily used for trading in Trinidad and Tobago. To identify the existence of such zoonotic infections and restrict their propagation to human beings, a single health concept needs to be adopted [72].

7.3.2.2 Streptococcaceae

Once more, one of the main bacterial families covered in this section is Streptococcaceae, which is the source of systemic streptococcosis, a disease that has spread around the world and endangers fisheries as well as posing a risk to public health and the economy. There have been cases of meningoencephalitis and fatalities in farmed fish species due to these bacteria, which are thought to be emerging zoonotic agents, and human infections following contact with the fish [73, 74]. Both horizontal and vertical transmission of these bacteria have been documented, and they have caused sickness and mortality in fish in both fresh and saltwater. There are two different methods of transmission: indirectly through polluted water and direct contact with diseased or dead fish. A few species among these bacterial groupings are known to be highly widespread in the transmission of disease, including *Streptococcusagalactiae, Streptococcus difficile, Streptococcus difficilis, Streptococcus dysgalactiae, Streptococcus iniae*, and *Streptococcus shiloi*. Furthermore, *B. Streptococcus* (GBS) ST283 strains are one category that has been found in freshwater and marine fish, humans, and frogs. These strains exhibit clinical symptoms of the disease in fish species. Exophthalmia, stomach discomfort, loss of orientation, irregular swimming, anorexia, ocular opacity, darkening and hemorrhagic skin, and ultimately mortality are still the most common signs [69, 75–78]. Based on the clinal evidence and unstable swimming behaviour, it is also thought to be a neurotropic agent for fish. It is present in the liver, gills, kidneys, spleen, and other organs needed for fish defence against infections. Studies on its pathogenicity have revealed that *S. agalactiae* primarily entered tilapia through the gastrointestinal tract, where it was able to penetrate intestinal layers and mucosa [74]. The virulence of *Streptococcus* includes a few factor surface proteins, capsular polysaccharides, and secreted compounds. The binding of human fibrinogen to bacterial surface proteins inhibits the phagocytic activity of humans. The degradation of complement component C5A and the interleukin-8 chemokine by the bacterial peptidase C5 and protease, respectively, results in the disruption of chemotactic signals and phagocyte recruitment. Neutrophils, erythrocytes, and lymphocytes are all destroyed by the bacterium streptolysin. Increased adherence is facilitated by the production of extracellular exopolysaccharides and polysaccharides surrounding the cell and tolerance to harmful chemicals. Bacterial a-enolase breaks down fibrin clots and promotes bacterial dissemination [79]. Handling infected live and dead fish can cause cellulite, endocarditis, meningitis, severe systemic infections, suppurating ulcers, septicemia, arthritis, lympha-denitis, and in extreme cases, even death in humans. *S. iniae*, a marine pathogenic bacterium that was first isolated from freshwater dolphins in the Amazon in the 1970s, is widely recognised as a major threat to marine aquaculture due to its high prevalence and approximately 10% overall prevalence in wild marine fish and crustaceans that have been sampled from the Mediterranean Sea [80, 81].

7.3.2.3 Erysipelotrichaceae
Fish illness is caused by gram-positive bacteria called zoonoses, of which *Erysipelothrix rhusiopathiae* is the most significant component. It is associated with mammals and is responsible for several disorders, including acute sepsis, that affect the skin, particularly the connective tissues and vascular walls. It also affects animals. Clinical signs of other disorders, including cellulitis, dermatitis, and myositis, are also included. Before 2014, it was thought that *E. rhusiopathiae* was a common fish bacterium. However, some studies from publications around the nation have indicated that it has also been linked to fish mortality. One new species of ornamental fish, *Erysipelothrix piscisicarius*, has been reported in the literature [69, 82]. In warmer seasons, these bacteria can easily develop within fish and cause the disease of erysipeloidin in sellers and handlers. Fish mucus contains bacteria, which is spread by encountering either fresh or dead fish, and it harms them because of its prolonged existence in fish mucus or outside muscles. They are also linked to soil saprophytes and they can transmit their infection to humans and cause severe disorders. Thus, exposure to diseased fish, animals, and their waste products is the source of many diseases. A few of the illnesses that are listed here, including endocarditis, septicemia, and skin infections mostly affecting the hands, pose a serious risk to fishermen and veterinarians. The first case of endocarditis in humans was documented in 2017, and it was linked to research conducted off the coast of Norway and caused by a specific bacterium [69, 73, 83, 84].

7.3.2.4 Vibrionaceae
A different kind of bacteria that is covered here is known as *vibrio*, and it is extremely perilous for those who engage in aquaculture and purchase aquatic products since it can cause vibriosis in humans. This bacterium can infect fish and produce genes that confer antibiotic resistance when it is treated with an antibiotic. They are found in areas with both fresh and brackish water, and they are effective in

spreading disease to people by contaminating or infecting fish and causing skin blisters on humans. This article discusses a few of the vibrio's most harmful species including *Vibrio cholerae*, *Vibrio metschnikovi*, *Vibrio hollisae*, *Vibrio damselae*, *Vibrio vulnificus*, *Vibrio alginolyticus*, and *Vibrio parahaemolyticus* [69, 85, 86]. Additional dangerous pathogen species, including *Vibrio alginolyticus, Vibrio anguillarum, Vibrio campbellii, Vibrio harvey, Vibrio vulnificus*, and *Vibrio parahaemolyticus*, have been identified and detected in the infected fish. This species is responsible for several fatal illnesses, some of which are usually unknown and manifest as lethargy, skin lesions, exophthalmia, and mortality in fish infected with *vibrio*. The published reports also mention some more symptoms, including tail rot, spleen enlargement, abdominal dropsy, intestinal inflammation, epidermal bleeding, and scale shedding. Conversely, studies on fish that have acquired vibriosis have revealed three stages of the disease: the first is the breakdown of tissue and host cells; the second is an escape that could cause injury to the host and ultimately result in death; and the third is an entry through the skin, fins, gills, and anus. Some names of *Vibrio* virulence agents related to this disease are: siderophores, hydrolytic enzymes, toxins, and extracellular products. Vibriosis resistance is determined through the pathogen, host, and environment interaction. It has been claimed that over 100% of fish mortality with specific *Vibrio* species has been documented in the published literature [87, 88]. These infections, which are spread from fish to people by the *Vibrio* species, can result in several illnesses, including tissue necrosis, erythema, septicemia, and lesions. Here, one of the primary causes of fish pathogen transfer into humans is the growing inclination of customers towards ready-to-eat seafood products (unprocessed fish flesh slices, for example, might result in illnesses related to seafood caused by *Vibrio parahaemolytica*). The important zoonotic pathogen *Vibrio vulnificus* is a public health concern since it has been shown to cause primary septicemia in humans who consume raw shellfish and secondary septicemia in those who get seawater in their wounds [89–91].

7.3.2.5 Aeromonadaceae

Another group of gram-positive bacteria called Aeromonas can infect fish and cause sickness. Until the fish experiences physical weakness and environmental stress, *Aeromonas* infections are asymptomatic. Fish are similarly impacted by the gram-positive (*Vibrio*) and gram-negative (*Aeromonas*) bacteria classes; freshwater is home to the majority of *Aeromonas* bacteria, whereas brackish, estuary and marine waters are home to most *Vibrio* bacteria. On the other hand, fish are fundamentally more effective at spreading pathogenic microorganisms to humans. *Aeromonas hydrophila* is the most prevalent pathogen among these species, which has been documented to have varying zoonotic potential. Other species include *Aeromonas caviae*, *Aeromonas jandaei*, *Aeromonas sorbia*, *Aeromonas salmonidae*, and *Aeromonas veroni* [69, 92, 93]. These days, several fish infections are widely recognised in fish and mimic the symptoms of *Aeromonas hydrophila*. These opportunistic illnesses affect weak fish as a secondary infection and have been linked to histopathological alterations in the liver, kidney, gills, stomach, and spleen, among other organs. Fish with *Aeromonas* infection may exhibit petechiae in the skin and fins, skin ulcers, arrhythmias, anorexia, exophthalmia, and stomach enlargement as clinical signs [94, 95]. *Aeromonas* pathogenicity is caused by a range of factors, including enzymes, enterotoxins, hemolysin, adhesin, flagella, lipopoly-saccharide, secretory systems, and quorum sensing. Although human infection is uncommon, these components can cause ulcers or ingestion in humans. Here, some of the clinical outcomes that impact an individual's muscles are discussed, including septicemia, cellulitis, and muscle necrosis. In humans, *Aeromonas* produce bacter-emia, a condition that can cause diarrhoea, gastroenteritis, sepsis, and urinary tract infections. *Aeromonas's* resistance to many antibiotics is a sign of a growing issue with human and aquatic health [69, 96–98].

7.3.2.6 Pseudomonadaceae

A Gram-negative bacillus called *Pseudomonas* is capable of both food poisoning and opportunistic diseases under stressful circumstances. It is a common component of the natural microbiota of fish and is very genetically flexible, allowing it to adapt to a wide range of animal and human situations. Owing to the virulence factors (enzymes, pili, flagella, LPS, and quorum sensing) of these motile bacteria, the inflammatory-invasive processes have been reinforced and have multiplied. *Pseudomonas fluorescent* has been identified as an opportunistic pathogen in the aquatic environment, the natural microbiota of the aquatic environment and fish, and the gut flora of healthy fish [99–101]. *Pseudomonas septicemia* has been observed to occur in freshwater, brackish water, and marine environments. Its agents in fish include *Pseudomonas aeruginosa*, *Pseudomonas anguilliseptica*, *Pseudomonas putida*, and *Pseudomonas fluorescent*. Several clinical signs have been observed, including ascites, ulceration, crowded gills, darkening of the skin, exophthalmia, eye cloudi-ness, and irregular body surface haemorrhages. These bacteria pose a risk to the general public's health since they are resistant to various medications and are spread through close human–animal interaction [102, 103].

7.3.2.7 Enterobacteriaceae

A family of bacteria known as *Enterobacteriaceae* is found in fish and is responsible for a wide range of human infections. These bacteria can cause quite serious diseases in humans. These bacterial species, which are referred to as fish zoonotic agents, include *Salmonella, Klebsiella*, and *Escherichia coli*. They are found in fish digestive tracts and aquatic settings, and they are members of the gram-negative family. In Iran, fish infections have been found to contain *Enterobacteriaceae* families, including *Salmonella, E. coli*, and *Klebsiella;* this suggests that these bacteria can spread to humans and cause infections [69, 104]. The most frequent ways to become infected with these bacteria are by opening wounds, touching fish, or scratches that result in infection and inflammation at the point of entry for the bacteria or systemic infections; however, food sources can also be a source of infection, as imported dried fish has been known to carry *Salmonella Typhimurium*. Different *E. coli* strains can be retained by fish as flora and moved to other water sources. Despite not being a part of a fish's natural microbiota, *E. coli* is frequently isolated from the fish's digestive system. Additionally, *E. coli* has been shown to infiltrate other fish tissues, including the kidney, bladder, muscle, and gills, in polluted situations [73, 105, 106]. These infections were also contingent on the season of the year, the individual's immunity, the polluted environment, and interaction with fish that were asymptomatic. Additionally, *E. coli* is the primary source of zoonotic infections—infections spread by fish and aquatic goods. Non-pathogenic strains can cause food sickness or diarrhoea in fish by releasing toxins, but some non-pathogenic strains can turn pathogenic when they spread from the gut to organs like the peritoneum or urinary system. In addition, publications from diverse nations indicate that different zoonotic agents, including enteropathogenic and enterotoxigenic, have been isolated from *E. coli* [107, 108]. A particular subspecies of *Salmonella enterica* has been discovered to be highly successful in causing digestive illnesses when it comes to fish, aquaculture products, and water. It is not a typical fish bacterium, and the aquatic environment and water quality affect its prevalence. Additionally, fish can serve as asymptomatic hosts for bacteria living in their guts or on their body's surface. *Salmonella Eastbourne, Salmonella Give, Salmonella Colindale, Salmonella Bredeney, Salmonella Poona, Salmonella Schwarzengrund,* and *Salmonella Llandoff* are some of the isolated species of fish and water that are included in this group. Humans contract salmonellosis from eating fish contaminated with species like *Salmonella typhimurium* and *Salmonella enteritidis.* They are a major contributor to environmental contamination and the spread of bacteria due to their tenacity in fish digestion and presence in excrement. Numerous factors contribute to virulence, such as secretory systems, tissue viability, intra-phagocyte proliferation, and intestinal lumen transfer [109]. *Salmonella-*contaminated fish can spread to humans through their skin, gills, and intestines and produce many clinical consequences, including sepsis, stomach discomfort, diarrhoea, and vomiting, along with symptoms including gastroenteritis, cramping in the abdomen, fever, and bacteremia. Like this, untreated water samples from dams, seawater, sediment, and the intestinal contents of shrimp and freshwater fish have been used to isolate and diagnose *Klebsiella pneumoniae* and *Klebsiella oxytoca* species. These samples reveal a variety of clinical hemorrhagic complications near the tail, as well as vacuolation and necrosis of hepatocytes in fish raised in India. These findings may be related to the multi-drug-resistant *Klebsiella* spp [110–114]. Once the *Klebsiella* species were isolated from the skin lesions of ornamental fish or carp, it became clear that the infectious process was caused by food processors' poor hygiene and that the fish's symptoms were caused by endotoxins acting directly on their bodies and by aberrant immune responses. Similarly, *Yersinia*, a different gram-negative bacterium, can infect both freshwater and marine fish. Additionally, *Y. ruckeri*, the causative agent of severe septicemia and enteric red mouth disease illness, has shown a sharp rise in cases in recent years. When fish experience the first signs of septicemia, which are caused by the bacteria *Aeromonas* and *Pseudomonas*, symptoms include a darker body, decreased appetite, exophthalmos, haemorrhages, mouth redness, enlarged kidney, and spleen. The secretory system, pili, enzymes, poisons, outer membrane proteins, flagella, iron acquisition system, heat sensitivity factor, and biofilm formation are some of the elements that determine a pathogen's pathogenicity. For a person who comes into contact with water and has the bacteria isolated from their wound, it is suggested that the bacterium is zoonotic, but more research is required to confirm this theory [115–118].

7.3.2.8 Hafniaceae

The family *Hafniaceae*, which includes the genera *Hafnia*, *Edwardsiella*, and *Obesumbacterium*, is a motile, anaerobic family of rod-shaped gram-negative bacteria in the order of Enterbacteriales. The pathogen *Edwardsiella* is harmful to aquatic life, particularly producing the fish sickness known as Edwardsiellosis, which has severely harmed the aquaculture sector's finances. The disease primarily affects fish in environments with higher ambient temperatures and higher levels of organic waste [119–122]. Before 1980, only one species the *Edwardsiella tarda* species was known to be the efficient cause of Edwardsiella. However, five well-known species have been identified as having this ability, including the three older species *E. tarda*, *Edwardsiella hoshnae*, and *Edwardsiella ictaturi* as well as the two most recent species *Edwardsiella piscicida* and *Edwardsiella anguillarum*. Except for *Edwardsiella hoshnae*, every species on the list is harmful to fish, and *E. tarda* is thought to be the primary source of infections in humans. However, according to the new categorization, *E. tarda* has not caused as much trouble in aquaculture as *E. piscicida*. More than 20 fish species in Europe and India are susceptible to this condition. Abnormal swimming, lateral movement, and spinning in the water column are some of the behavioural symptoms. Although extraintestinal Edwardsiellosis (*E. tarda*) can cause wound and liver infections, cholecystitis, peritonitis, meningitis, myonecrosis, osteomyelitis, sepsis, and bacteremia in humans, it is primarily an opportunistic pathogen that causes gastroenteritis [123–125]. Even though the *E. tarda* species of bacteria cause septicemia, it is a rare infection that affects fewer than 5% of people and is primarily caused by food and drink, yet it can still be fatal. These illnesses are particularly dangerous if someone has a weakened immune system or preexisting conditions like diabetes or hepatobiliary disease, which increases the risk of contracting an *E. tarda*-caused sickness. Numerous virulence mechanisms, including hemolysin EthA, translocation and assemble module (TAM), types III and VI secretion systems (T3SS and T6SS), and antibiotic resistance genes, are possessed by *Edwardsiella* species, particularly *E. tarda* and *E. piscicida* [126]. Through the plasmid, these bacteria can obtain mobile drug-resistant genes and subsequently transfer them into the microbiomes of humans, animals, and the environment. Additionally, eating raw fish, touching fish, swimming in tainted water, and immune system conditions can all transmit Edwardsiellosis. Furthermore, bacteria can enter human cells by attaching themselves to them, using hemolysin, and secreting systems. It grows in phagocytes and eventually spreads to neighbouring cells. The bacteria should be given more attention in the upcoming decades due to their significant contribution to the development of antibiotic resistance [125].

7.3.2.9 Other bacteria

The methicillin-resistant strain of *Staphylococcus aureus* (MRSA) in fish, is an infection either before or after harvest, or any other food-producing animal. Fish contamination can occur from food handlers who have *S. aureus* on their skin and mucous membranes. Other zoonotic bacteria that are linked to eating fish are *Campylobacter*, *Staphylococcus*, *Listeria*, and *Clostridium*. *Staphylococcus* research therefore appears as crucial for the fish food chain [73, 127, 128]. Since *S. aureus* enterotoxins are heat resistant, consuming fish or its byproducts can give humans gastroenteritis and raise health concerns for the public. Most of the research focuses on human infection from fish consumption, although *S. xylosus* has recently been identified as an emerging main pathogen that primarily kills fish. Fish death results from this new species' successful attack on fish immunity, which causes exophthalmos. There is a significant risk of illness transmission to humans when consuming these fish raw [73, 127, 129]. Although the bacterial TSST-1 toxin, which is thought to be a superantigen, enters the bloodstream and activates polyclonal T lymphocytes in peripheral blood, triggering a huge release of pro-inflammatory cytokines, these species can induce skin infections that can result in toxic shock syndrome. The most common type of bacteria found in fish and fisheries products is *Listeria monocytogenes*, according to statistics from the European Food Safety Authority (EFSA) in 2016. It was established that fish items included this infection [130, 131]. Gram-positive *Listeria monocytogenes* is a bacterium that can grow in several kinds of fresh and salty conditions and at temperatures as extreme as those found in refrigerators. Fish skin and faeces are the sources of disease transmission because the flora is native to the area and can be found on the water's surface, fish's exterior surface, mucus/mucosa, intestines, stomachs, and gills of infected fish. Nowadays, this bacterium infects people through food and raises problems for public health in relation to septicemia, meningitis, gastroenteritis, pneumonia, and abortion [132]. Human listeriosis is very common in older pregnant women and people with immunocompromised chronic illnesses. Anaerobic rod-shaped spore-forming bacteria (*Clostridium botulinum* and *Clostridium perfringens*) are other significant food-derived infections brought on by eating fish. The bacteria are linked to both fresh and canned fish and are found in soils, aquatic sediments, and natural anaerobic habitats [73, 133, 134]. *C. perfringens* produces enterotoxins (CPE) from the cpe gene, i.e. types A, C, and D, which cause gastroenteritis in humans, while *C. botulism* spores may persist in freshwater and marine sediments for decades. These toxins are absorbed from the intestine into the blood circulation and cause damage to tissues like the brain. In addition to being on the surface of healthy fish, they can also be found in their guts. The bacteria produce types A–H of botulinum toxins, which cause flaccid paralysis by preventing the release of acetylcholine from synaptic vesicles at neuromuscular junctions. Humans are poisonous to types A, B, E, and F [133, 135–139]. Fish intestines produce botulinum neurotoxins, which are heat-resistant and must be heated to a high temperature to eliminate their toxicity. As a result, eating food that has been improperly processed puts them at risk for botulism, which typically first appears as bloating, constipation, vomiting, dizziness, and diarrhoea. In a similar vein, Campylobacter is a frequent zoonotic agent bacterium that is present in the gastrointestinal tract of numerous animals. Although it is uncommon, these microorganisms can cause Campylobacteriosis when consumed. However, unclean water and the hands of food handlers are likely the sources of *Campylobacter jejuni* infection. The two most significant enteropathogens in this genus, *C. jejuni* and *C. coli*, cause campylobacteriosis, which presents as enteritis. They accomplish this by utilising bacterial motility, intestinal cell adhesion and invasion, disrupting intracellular signalling, killing cells, eluding the host immune system, and obtaining iron for growth and survival [140–142]. *Plesiomonas shigelloides* is a waterborne pathogen that causes disease in fish and is isolated from the infected fish; while *Legionella pneumophila*, which causes pneumonia and legionnaires' disease and is spread by water and aerosols, is isolated from patients who work at fish markets. Similarly, salmonids, eels, goldfish, sole, sturgeon, trout, carps, and turbot are susceptible to yersiniosis, also known as red mouth disease, which is caused by *Yersinia ruckeri*. Exophthalmos and blood spots in the eye are typical symptoms of the disease, and fish populations in Europe, North and South America, Australia, and New Zealand contain the bacterium that causes it [73, 143–146].

7.3.3 Parasites

Eating raw or undercooked fish or fish products is the main way that fish-derived parasites including tapeworms (like *Dibothriocephalus latum*), roundworms (like *Anisakis* spp.), and flukes (like *Metagonimus yokogawaii*) infect humans, causing illness rather than death. The importance of seafood in the worldwide diet and the growing health risks linked with parasite diseases and illnesses originating from seafood are topics covered in a wealth of research. It is well-recognised that many fish that are edible can have a variety of parasites that can infect humans. Some of these parasites, such as gnathostomiasis and anisakidosis, can pose a major threat to human health [49, 147–149]. Due to their ubiquity and high ranking on the list of food-borne illnesses, parasites are typically disregarded when talking about seafood safety. This is especially true with fish products. Therefore, parasites originating from fish are frequently overlooked and account for several newly discovered zoonotic illnesses. Standards for food inspection and the procedures for looking for disease agents differ greatly between nations and are sometimes insufficient and inconsistent [150–153]. It is possible to overlook import control for zoonotic parasite illnesses and food safety standards in food safety measures even in developed nations. The increased frequency, geographical range, and prevalence of health issues connected to zoonotic fish have been attributed to a growing taste for raw, undercooked, and foreign foods as well as climate changes. For instance, freshwater fish liver flukes are thought to presently afflict 45 million humans, and at least 680 million more are thought to be in danger of contracting them [154–158]. Of all the parasites discovered in seafood, helminthic is the most prevalent. It is diverse in aquatic environments and often spreads through fish. For example, a report from Vietnam makes it abundantly evident that 268 helminth species have been documented to harbour the parasite. This is one of the main causes of the most prevalent because of its trophic-oriented life cycle and needs on the food web for host transmission. Furthermore, helminthic parasites can use numerous kinds of edible teleost fish species as intermediate or even paratenic hosts. This means that the larger the host, the higher the risk of helminthic infection [159–163]. In general terms, it may also be claimed that as fish species diversify, so do the parasite organisms that host them. There are, nonetheless, many parasites that are less dangerous or even non-pathogenic; however, because of their small larval size and low-loaded sample, it is quite challenging to identify them. Shamsi claims that an investigation makes it abundantly evident that over 40 taxa of parasites are known to be harmful to humans, while other parasites are extremely rare and some are extremely pathogenic, posing a major risk to the general public's health. Based on this research, it can be projected that helminthic parasites have the potential to endanger the health of over 500 million people; however, they may also multiply if global warming continues to rise [164–168]. Both acute (constipation, stomach pain, and dysentery) and chronic (brain haemorrhage, hemiparesis, and cancer) diseases are linked to fish-derived helminthic illness. The presence of *Eustrongylides* sp., *Euclinostomum* sp. *from Channidae* fish and *Isoparorchis* sp. from Bagridae fish imported into Australia has been detected by several types of investigations undertaken to estimate the occurrence of zoonotic parasites. Routine surveillance is necessary to stop the importation of zoonotic parasites, even if freezing imported edible fish inactivates the parasites [169–172].

7.3.3.1 Trematodes (flukes)

Several trematode genera that induce derived zoonoses and are members of the Opisthorchiidae and Heterophyidae families are included below in this section. To prevent parasitic infections in farms, all these parasites are included in the project called Advanced Tools and Research Strategies for Parasite prevent in European Farmed Fish. Liver flukes, such as *Clonorchis sinensis, Opisthorchis viverrini*, and *Opisthorchis felineus*, and lung flukes, such as *Paragonimus westermani* and *Paragonimus heterotremus*, are examples of common flukes that infect fish and crabs. The high liver fluke is the source of inflammation and damage to the epithelial bile duct. The severity and duration of the infection determine the gastrointestinal distress and liver damage that result, as well as the possibility of serious clinical issues like pancreatitis, choledocholithiasis, cholangitis, and cholangiocarcinoma (CCA) [171, 173–177]. Eating freshwater crab or crayfish exposes a person to the fluke metacercariae that causes lung fluke illness, or paragonimiasis. Comparably, fish-derived trematodiosis is a leading cause of mortality in Southeast Asia and is quite common throughout Asian nations. Trematoda of zoonotic concern are found in marine, brackish, and freshwater fish, and these trematode infections typically result from eating raw fish or shellfish in freshwater. The causative agent of CCA, *Opisthorchis viverrini*, is extremely common in northeastern Thailand and Laos [49, 168, 171, 178–181]. A few additional species, including the *metacercariae of Clonorchis sinensis, Metagonimus* spp., *Centrocestus armatus, Echinostoma* spp., *Clinostomum complanatum, Opisthorchis viverrini*, and *Metorchis orientalis*, were also found in the freshwater fish sampled from the Republic of Korea. The infection of these viruses in aquaculture and their ability to spread pollutants across the ecosystem can put humans and other animals in danger. The following are the most often seen trematode species in fish that are transferred to people: *Haplorchis pumilio, Haplorchis yokokawi, Centrocestus formosanus*, and *Clonorchis sinensis* [182–184]. Hence with the goal of the circumstances, to simultaneously detect Opisthorchiid and Heterophyid metacercariae in fish or fish products, a quick and affordable multiplex PCR has been devised. The identification of the infectious stage of metacercariae in humans, which is challenging to spot visually in fish, would be greatly aided by all these developments. Digenetic trematodes, especially *Opisthorchis viverrini* and *Heterophyes* are primarily found in domestic dogs and cats, who serve as their reservoir hosts.

To maintain ongoing surveillance of fish zoonotic parasites in cats and dogs, more preventive and control measures must be implemented action [175, 185, 186].

7.3.3.2 Cestodes (tapeworms)

Another parasite in this instance is termed a cestode, more commonly referred to as a tapeworm. Unlike trematodes, cestodes are relatively big and can reach lengths of up to 20 metres. This group includes some of the most well-known parasites, including the Diphyllobothriidae order, which is known to cause the disease diphyllobothriosis. Of the approximately 50 species of the genus *Diphyllobothrium*, at least 14 are pathogenic to humans (the most pathogenic species being *Diplogonoporus balaenopterae, Adenocephalus pacificus, D. dendriticum*, and *D. nihonkaiense*) [187, 188]. Diphyllobothriosis is typically not fatal and is a minor illness. Most infected individuals do not show any symptoms, but others may have weight loss, diarrhoea, anaemia, abdominal pain, and vitamin B12 insufficiency. Up to 20 million individuals globally are thought to be affected. On the other hand, human tapeworm infection has decreased worldwide, apart from Japan and far Eastern Russia [187, 189, 190].

7.3.3.3 Nematodes (roundworms)

Human diseases that are caused by nematodes generated from fish are widespread and some have the potential to spread. Eating raw food can spread infections to humans, and some diseases are potentially fatal, especially for those who are afflicted. There is a lot of information available because most individuals do not understand these disorders. In South Africa, these species can be found in both fresh and saltwater. On the other hand, when these infections strike humans, they are reported by different nations worldwide, however, the frequency of occurrences differs among them. These illnesses are more common in areas where fish is the primary food source and energy source, especially on the western coast of South America [149, 191–194]. Popular table fish Chrysophrys auratus was sampled from Australian and New Zealand waters and it was found to contain zoonotic nematodes from the Anisakidae family. If Chrysophrys auratus is eaten raw as sashimi or in sushi, it poses a serious risk to human health. Furthermore, these nematodes exhibit limited host specificity in their larval stages, which are infectious for humans, and were found in the edible fish samples collected in Australia. Once the fish has died, the larval stage of the fish moves from the gastrointestinal system into the viscera and surrounding muscular tissues through the gastrointestinal mucosa. Thus, they may still be dangerous for people's health [61, 195–197]. *Anisakis* spp., *Pseudoterranova* spp., which cause anisakidosis, and members of the Gnathostomatidae family, which cause gnathostomiasis, are widespread fish nematodes that pose a health risk to humans. These species are all regarded as extremely significant on a global or regional scale. The most widespread fish nematodes in the Anisakidae family are found around the world in the genera *Anisakis, Pseudoterranova*, and *Contracaecum*. These larvae are among the most frequently recorded marine parasite larvae and have a significant zoonotic impact [166, 198–200]. Since its discovery in 1960, there has been a growing interest in the Anisakidae family, leading to several studies aimed at raising awareness, refining diagnostic methods, and understanding various facets of the species' pathogenicity and biology. The term 'anisakiasis' or 'anisakiosis' refers to the parasitic illness that nematodes of the genus Anisakis produce in humans. The parasite's third-stage larvae (L3) are the source of the infection. Anisakiasis is caused by members of the genus Anisakis, whereas it is caused by any member of the family Anisakidae. Anisakis simplex sensuality members are frequently the cause of anisakidosis. *A. physeteri, Pseudoterranova decipiens*, and *Contracaecum* spp. are some other nematodes [201–204]. Anisakis larvae have previously been found in fish products such as cod fillets, frozen fish fillets, fish fingers, and steaks. The nematodes of this parasite have also been studied in products made from farmed Atlantic salmon (*Salmo salar*) and smoked wild sockeye salmon (*Oncorhynchus nerka*). Although no parasites were found in the samples of farmed Atlantic salmon, suggesting minimal risk in farmed fish, the samples taken from smoked wild sockeye salmon tested positive in 10 of the 13 cases for Anisakis simplex larvae. A variety of tactics must be utilised to stop Anisakis and other parasites from getting into fish farms. These tactics include freezing waste fish that is used to feed farmed fish and fortifying water access points with nets to keep out wild fish [205, 206]. By heating the whole raw fish from above sixty degrees for more than a minute to below 20 degrees for a whole day before eating, the risk can be further decreased. The European Union (EU) mandated freezing treatment for fish products due to the possibility of zoonotic fish parasites spreading through raw or undercooked fish and fish products (regulation No. 1276/2011 amended the Annex III of Regulation (EC) No. 853/2004). Since human gastrointestinal tracts cause hypobiosis, or the stoppage of parasite development, humans are regarded as accidental hosts in the anisakid life cycle. Although dead anisakids can also spread disease, living anisakids are typically the cause of anisakidosis, and human gastroenteritis is the result of larvae entering the stomach or intestinal mucosa. Usually manifesting as gastrointestinal symptoms, anisakidosis frequently passes for food poisoning [148, 201, 206–210]. The duration of symptoms following a parasite infection can range from days to months, depending on the patient and the area. Additionally, their symptoms may be lessened by the body's natural regurgitation, elimination, or surgical removal. Anisakis-associated hypersensitivity is a serious issue that manifests as high sensitivity to death in extremely small quantities when a highly sensitive individual is exposed to it. When precisely prepared, a simplex substance might result in circumstances ranging from quickly fatal anaphylactic reactions to long-term incapacitating effects [155, 209, 211, 212]. Even though this condition has been documented globally and is more prevalent in Japan and Europe, it is thought to be grossly underreported and/or misdiagnosed because of patients' vague symptoms and the scarcity of diagnostic testing. For instance, research indicates that 60% of cases even in Japan, where the illness is well-known were incorrectly identified as diverticulitis, cholecystitis, appendicitis, ileitis, gastric and pancreatic cancer, and tuberculous peritonitis [213–216]. Similarly, consuming raw or undercooked foods like sushi and ceviche that contain fresh- and brackish-water fish species, as well as other freshwater creatures (amphibians, eels), can lead to infection from the highly significant nematode Gnathostomatidae. When infective larvae (L3) of the family

Gnathostomatidae, such as *Gnathostoma spinigerum*, *Gnathostoma doloresi*, *Gnathostoma hispidum*, *Gnathostoma binucleatum*, *Gnathostoma nipponicum*, and *Gnathostoma malaysiae*, as well as *Echinocephalus* sp., are consumed, the parasites cause gnathostomiasis [147, 217, 218]. The clinical symptoms of gnathostomiasis, which typically manifest 24–48 h after transmission and include nausea, vomiting, and abdominal discomfort, are like those of A. simplex except for hypoallergic reactions. As the parasite infective larva migrates through the subcutaneous tissues, it can cause typical inflammatory migratory swellings. It can also penetrate the skin, lungs, eyes, ears, gastrointestinal tract, and genitourinary system, and if it affects the nervous system, it can cause paresis, brain haemorrhage, or even death. Reports of the disease have come from Central and South America, Latin America, China, India, Japan, and Southeast Asia, namely from Thailand, Vietnam, Laos PDR, and Myanmar, as well as from travellers returning from these areas [149, 219, 220]. Gnathostoma spinigerum has been implicated in the majority of gnathostomiasis patients. Additional species of zoonotic significance include *G. hispidum*, *G. doloresi*, *G. binucleatum*, and *G. nipponicum* [218].

7.3.4 Viruses

Acute gastroenteritis can be brought on by Noroviruses (NoV) and is becoming more and more recognised as a food-borne illness that is becoming a global public health concern. This illness can arise from consuming ready-to-eat fisheries products and shellfish contaminated with faeces, either as isolated cases or as outbreaks [221–224]. The genus Norovirus contains only one species, the Norwalk virus. Noroviruses are nonenveloped positive-sense single-stranded RNA viruses that are members of the Caliciviridae family. It is further divided into seven geno groups (GI–GVII), the majority of which are pathogenic (GI–GII, for example). Gastroenteritis mediated by NoV can cause vomiting, nausea, watery diarrhoea, and abdominal discomfort as clinical symptoms. There may be symptoms such as low-grade fevers, headaches, lethargy, weakness, and loss of taste. Except for those with immunocompromised conditions who may develop a long-term infection with the virus-associated enteropathy and malabsorption, symptoms often appear 12–48 h after consuming the contaminated meal. Hepatitis A virus (HAV) infection can result from eating fresh or frozen foods, such as fish, bivalves, and water. Hepatitis is an inflammation of the liver that can cause fatigue, nausea, vomiting, diarrhoea, jaundice, dark urine, fever, abdominal pain, arthralgias, and myalgias. The illness can linger for a few weeks or several months. However, rare instances could result in liver failure or even death, especially in the elderly and those with long-term liver illness [225, 226].

7.3.5 Fungal zoonotic agent

Microorganisms classified as non-photosynthetic are called fungi. Usually, they are parasites of plants, animals, and people, or they exist as saprophytes in soil and dead organic debris. Just 300 of the 1.5 million fungus species that have been found are thought to be harmful to humans (Centres for Disease Control and Prevention (CDC) 2017). Fungal illnesses are frequently caused by common fungi found in the environment. Zoonotic fungi, which naturally spread from animals to people, can occasionally cause serious issues with public health. However, the development of preventative and control techniques has decreased because of international public health efforts giving zoonotic fungus little priority. Two categories of fish zoonotic fungus are listed below.

7.3.5.1 Basidiobolomycosis

Basidiobolomycosis caused by *Basidiobolus ranarum* is a rare fungal infection. This causative agent is found as a widespread environmental saprophyte isolated from putrefying plant materials, foodstuff, and leaves of deciduous trees, fruits, and soil. *B. ranarum* belongs to the class Zygomycetes, order Entomophthorales and the phylum Zygomycota). It has been hypothesized that the mode of acquisition of *B. ranarum* infection is through the skin following scratch, cut, or bite of insects. The other accessible sites of this fungal infection include the thigh, buttock, and perineum). It has also been reported in the gastrointestinal tracts of animals such as amphibians (e.g. toads and frogs), reptiles (e.g. geckos and garden lizards), and fish as well as mammals (e.g. insectivorous bats, dogs, horses, and humans) [227–235]. The illness typically manifests as a gastrointestinal and subcutaneous infection. Tropical regions, including those in Asia, Africa, Europe, South America, and the United States, have reported cases of the disease. After entering the body through a skin laceration, the spores of this fungus proliferate slowly and can result in an enlarged hard node beneath the skin, particularly in the arms and legs. A further method of spreading this zoonotic virus is via eating food contaminated with animal excrement or ingesting soil. It can infect vital organs like the brain and penetrate deeper tissues, which might result in the patient's death if treatment is not received. The first known instance of gastrointestinal basidiobolomycosis (GIB) in a young boy was documented in 1964 [227, 232, 236–239]. Furthermore, a three-year-old girl with a *B. ranarum* infection endured excruciating swelling and ulcerations on her right leg for one and six months, respectively. Dermal granulomatous inflammatory infiltrations with wide, septate fungal hyphae and yeast-like formations were seen in histological sectioning. Infants with painless leg swellings and ulcers that gradually spread into the underlying muscles have been documented in similar circumstances. Many more cases were subsequently recorded in several nations, including Kuwait, Saudi Arabia, the United States, and Iran. Following *B. ranarum* infection, it has been noted that levels of several cytokines, such as pro-inflammatory cytokine TNF-a and Th2-type cytokines (IL4, IL-10), are raised. The locally invasive fungus causes the production of IgM and IgG-specific antibodies, which may be used as a diagnosis [233, 240–242].

7.3.5.2 Sporotrichosis

The dimorphic fungus *Sporothrix schenckii*, which is especially common in tropical and subtropical parts of Mexico, Peru, Brazil, Uruguay, Japan, and India, is the source of the fungal disease known as sporotrichosis. The fungus is a saprophyte that feeds on living things (soil, plants, and animal waste) and puts them to rot. Therefore, soil, organic materials, and plants tainted by fungi are typically the source of infection. Activities such as farming, fishing, gardening, hunting, and other pursuits aid in the spread of the fungus [243–245]. Furthermore, accounts attest to the fungus's ability to spread by insect bites and animal scratches, including those from rats, dogs, horses, squirrels, cats, and birds. Additionally, *S. schenckii* has been isolated from aquatic creatures (mostly fish and dolphins) and insects that have come into close contact with the fungus. Sporotrichosis, which primarily affects farmers and woodcutters, has become an endemic illness in many rural areas. All ages and genders are at risk of contracting this fungal illness because the fungus is a free-living microbe in the environment [246–251]. A fisherman's finger injury from a fungus-infected dorsal fin spine (*Tilapia* sp.) resulted in ulceration, edoema, discomfort, and purulent discharge in the affected area, according to a report from a rural area of São Paulo state. About 98% of cases had localised versions of the disease. Fish in Guatemala were identified as an endemic focus for sporotrichosis in a report published in 1978. The isolation of the fungus from fish in this region supports the theory that the fish surface in this instance was most likely contaminated with *S. schenckii* inoculum after a hand-wound [251–253]. It appears less likely that *S. schenckii* was originally inoculated on the patient's hand surface area via other sources. Three primary clinical forms of sporotrichosis exist: There are three types of sporotrichosis: (1) lymphocutaneous; (2) fixed-cutaneous; and (3) multi-focal/disseminated-cutaneous. The hematogenous spread of the fungus from the main site of inoculation, lymph nodes, or patients with respiratory issues results in systemic sporotrichosis. The two most frequent types of sporotrichosis, cutaneous and subcutaneous, are easily treatable. For subcutaneous sporotrichosis, the Infectious Diseases Society of America advises using itraconazole oral medication as a first-line treatment [254–256].

7.4 Technique for the detection of pathogenic microbes in the marine ecosystem

Since a wide variety of bacteria or other pathogens are linked to the illness that affects marine fish and its byproducts, aquaculture is receiving more attention than the wild population. Many clinal symptoms, including erosions, swellings, ulcerations, and haemorrhagic septicemias, are present. In contrast to other veterinary and medical peers, diagnosis focuses more on pathogen identification than on the prompt management of the illness [257]. A team of medical professionals is needed to identify the pathogens, and based on the patient's obvious clinical symptoms, a diagnosis may be made and a treatment plan promptly put into place. Fish diagnosis has changed throughout time from the conventional histology and culture-dependent methods. An older procedure required obtaining cultures and spending a lot of time identifying them; nowadays, the focus is on culture-independent methods, particularly those that make use of advancements in molecular biology. These culture-independent methods have the advantages of speed and precision; bacteria can be identified and examined regardless of whether they can be grown in a lab. Additionally, there is a high degree of specificity, which is crucial for disease diagnosis [258]. There is currently a strong push for the study and development of biosensors, which use electronics to assess the functions of biological things.

Compounds may now be measured and detected even in complex environments due to advancements in electronics technology. Biosensors recognise target compounds by utilising molecular components of biological processes, such as enzymes and antibodies. The synthesis and consumption of chemical compounds by biocatalysts can result in small changes in parameters like current, resistance, and heat when they react with a target substance. By utilising signal conversion components like electrodes and optical devices to detect these minute changes and convert them into electric signals, biosensors can quickly and efficiently quantify target compounds [259–261]. Biosensors for assessing the health of aquatic organisms are currently being developed because of their excellent sensitivity and specificity. In the past few decades, scientists have created and utilised biosensor technology to assess fish health and design novel diagnostic techniques that have the potential to significantly enhance it. Many kinds of biosensors for fish health checks have been developed to increase the safety of cultured fish on the market. This research examines a few biosensor systems that measure blood components, estimate ovulation time, and identify harmful bacteria in fish—all of which are regarded as crucial indicators of fish health. Pathogens cause a wide range of infectious diseases in humans that ultimately result in death. These are the most common causes, which, with the right diagnosis, can be treated or prevented. Therefore, a variety of techniques that are scalable, dependable, affordable, and particularly effective in resource-constrained places are developed for the detection and diagnosis of infections. Two methods of detention are feasible. The first is the use of traditional techniques, such as culture methods, which are known as the gold standard for identifying harmful microorganisms and are highly helpful in diagnosis and identification. They can also be utilised for additional research and characterization [262–264]. These culturally specific methods are exceedingly labour-intensive, time-consuming, less sensitive, and ineffectual. To get around problems, biosensors are advised since they are more beneficial than traditional methods for identifying bacterial pathogens in real samples—such as food, water, and the environment—and have several advantages over them, including being quick, highly sensitive, selective, labour-saving, and very successful. The research that has been published indicates that there is a growing need for quick sensors in the control of food safety and water quality monitoring due to concerns about public health and safety. Due to this demand, the global biosensor market grew from $19.2 billion in 2019 to $31.5 billion in 2024 [262, 265]. The advanced and newest technology introduced is called a biosensor, which is made up of two parts: a biological element that interacts with the target analytes, such as cells, antibodies, nucleic acids, and enzymes, and a physicochemical transducer that converts the biological element's interaction with the analytes into an electrochemical, optical, or piezoelectric signal. The biosensor's biological component recognises the target analytes in the sample that was gathered by applying the concepts of the lock and key theory of enzyme complex. These procedures are the best, fastest, most sensitive, most specific, and most dependable when compared to the others. It is important to keep in consideration that a biosensor should be designed so that it is not affected by external factors like temperature or pH [264, 266]. The expense of commercialising biosensors for water quality and food safety is one of the biggest obstacles facing the technology, which limits its use in many fields. The biosensors present two challenges: stability and extreme heat sensitivity, which prevents heat sterilisation of the immobilised biosensors. In addition, the expense of research and development poses a significant barrier to the biosensors industry, making technology adoption challenging. Two essential characteristics that are consistently utilised to elucidate the biosensor's sensitivity are the limit of quantification (LOQ) and the limit of detection (LOD). The lowest concentration of sample needed to be detected by a fast sensor is indicated by its LOD, while the lowest concentration of sample that can be reliably and precisely detected and measured is indicated by its LOQ. Nonetheless, although these are crucial factors, researchers have accepted the LOD as the accepted benchmark for determining analytical sensitivity [265, 267]. Even though biosensors are more dependable and quicker in identifying the bacterial pathogen in the sample provided, because of the characteristics of the sample matrices they analyse, each biosensor has a different sensitivity. Because liquid samples are more homogeneous than slides, they can be examined more easily. When assessing solid food samples, the sample preparation process, which includes the culture of pathogens, validates the biosensor's sensitivity. A step that is required to ascertain the true sensitivity of these biosensors is the use of miniaturising devices, which do not require bacterial culture. While there are several fast sensors used to identify pathogens, biosensors and microfluidic sensors are the most often used fast sensors [268]. These rapid biosensors can directly identify bacterial pathogens without enrichment, even at very low concentrations, because of their extremely high sensitivity and specificity. Additionally, this technology may achieve high mobility and miniaturisation, which makes it possible to incorporate digital technologies like smartphones and makes it extremely important in environments with limited resources. These techniques are also manageable for technicians and non-experts, particularly in low-to-middle-income countries with high rates of infectious illness [266, 269, 270]. Thus, electrochemical biosensors are addressed here, which use an analyte and biorecognition agent to transform an interaction into an electrical signal related to the analyte's concentration. These biosensors fall into one of three categories: amperometric, which measure the change in current following the binding of analytes and biorecognition agents; potentiometric, which measure electric potential; or impedimetric, which measure impedance. Xu, Wang, and Li (2016) report that they designed an electrochemical biosensor with a limit of detection (LOD) of 102 CFU ml^{-1} that can identify *E. coli* O157:H7 in food, water, and ambient samples. Even though they are fast, electrochemical biosensors have several shortcomings. These include redox hindrances that can disrupt the electrochemical process and the saturation energy of enzyme-substrate reactions, which are particularly problematic when the biorecognition agent is an enzyme [264, 271–273]. Additional biorecognition components, such as nucleic acids and nanoparticles, may also be taken into consideration. Non-functionalized gold nanoparticles, for instance, have demonstrated remarkable potential in the identification of *E. coli* O157:H7. The uneven distribution of pathogens in food, drink, and environmental samples without sample pretreatment is causing issues for these sensors. Thus, Formisano *et al* have created an extended gate field-effect transistor functionalized with sugar-based biorecognition layers using a modular system that includes a signal reader that is unique to pathogenic *E. coli* and a disposable electrode. The physiochemical components of food and the viscosity of the liquid or solid samples may also have an impact on the biosensor's capacity to sense. To ensure rapid and effective findings, it is critical to concentrate on designing novel biosensors with enhanced analyte-biorecognition binding and amplified signal transduction designs. Because DNA-based nanosensors are extremely flexible and capable of adapting to any conformational change, unlike other biorecognition agents, they may be the best option for detecting binding activities during an analytical experiment [274–278].

7.5 Marine pollutants have ill effects on aquatic and human life

When humans are exposed to these contaminants directly or even indirectly—by breathing in or consuming tainted food or its byproducts—they become extremely dangerous. As a result, some of the most significant examples of contaminants that harm people, aquatic life, and other living things are given below. For example, an oil spill can be extremely dangerous to people both directly when it encounters them and indirectly when they breathe in the vapours of the pollutants or eat contaminated seafood. These can result in serious health consequences like nausea and dizziness, as well as certain types of cancer and problems with the central nervous system. Toxic hydrocarbons that can contaminate air quality include benzene, toluene, polyaromatic hydrocarbons, and oxygenated polycyclic aromatic hydrocarbons, which can be found in oil. The Kuwait Oil Fires, which occurred between January 16, 1991, and November 6, 1991, produced air pollution that resulted in respiratory distress, as reported in the literature [279–282]. The primary cause of water contamination is oil spills and its leakage into drinking water supplies. For instance, the 2013 oil spill in Miri, Malaysia, which poisoned the water supply for 300 000 locals, is a prime illustration of a lack of drinking water. Coral reefs serve as crucial nidification sites for fish, prawns, and other species and are essential parts of marine ecosystems. In addition to quickly deteriorating due to environmental and human stresses, the aquatic animals that reside within and around coral reefs are at risk of being exposed to the harmful compounds found in oil. As a result, they are experiencing notable global changes in 'habitat structure,' 'species diversity,' 'species abundance,' and 'species evenness'. Oil dispersants may be detrimental to aquatic life, such as coral reefs, and aquatic creatures that depend on the coral reefs as a nursery for their growth. [283, 284]. According to this documented literature, tests on coral nubbins discovered in reefs have revealed that oil dispersants and scattered oil are extremely detrimental to soft and hard coral in their early stages of development. Thus, physical contact—the primary method of exposure—causes the sea birds to experience the same consequences. For example, after the 2000 Treasure Oil disaster in South Africa, hundreds of African penguins (*Spheniscus demerus*) were covered in oil. Furred mammals are impacted by physical contact with oil because they depend on their outer coats for warmth and buoyancy. Thus, 'when oil flattens and adheres to the outer layer, these animals often succumb to hypothermia, drowning, and smothering' [285].

7.6 Conclusion and prospects

We have addressed the marine environment, a water-based system, in this chapter. Additionally, it contributes several kinds of raw materials to industries and bioactive phytochemicals that are highly beneficial for the treatment of a wide range of human disorders. In addition, this ecosystem supports a wide range of flora and wildlife and provides refuge for many species. As everyone is aware, it makes up 71% of the crust of the world and has a shortage of consumable fresh water. Even with this knowledge, people nevertheless often dump their waste sewage water into adjacent bodies of water. This ignorance and selfishness will have a disastrous effect on humanity because ecosystems are independent of humans and have the potential to wipe out human existence altogether. Humans are always building highways, canals, dams, factories next to rivers, and carry out deforestation to create more area for agriculture. All these activities seriously damage the ecosystems of the soil, water, and air. This study primarily concentrated on pathogenic microorganisms, including bacteria, viruses, protozoa, algae, and fungi, that are dispersed by human activity. Humans can encounter marine infections in a variety of ways, including through eating raw seafood, handling decorative aquatic animals, and getting cuts or wounds from pathogens. Numerous conventional methods, including chromatography and spectroscopy, are also given in the introduction section to identify and detect marine bacteria. While all of these methods are precise, they also have significant drawbacks, such as a small amount of harmful microorganisms, laborious sample preparation, and the need for trained professionals and experts to understand the data they provide. Biosensors are suggested here to address these kinds of environmental problems because there are a lot more problems that involve large procedures, necessitate certain pressure and temperature for their correct operation, and are not scalable. Scientists from all around the world have been very interested in biosensors, particularly in the detection and identification of microorganisms. This is because biosensors are compact, lightweight, portable, fast, efficient, highly sensitive, and have very high selectivity because they contain biological receptors. Biosensors are gaining popularity for many reasons, including their ease of use, lack of need for a skilled operator, and simplicity of result interpretation. If we devoted an entire day listing all the advantages over alternative methods, it could be ultimately concluded. However, similarly to how all methods have their limitations, so does this biosensor. For example, the biosensor uses heat-sensitive enzymes that need a certain temperature to function properly. Nevertheless, it is more advantageous than other methods because it produces results quickly and requires less processing time. A global team of scientists and researchers will keep working to reduce their costs and make them accessible to all those on a budget. Comparably, since marine diseases can infect humans and other animals, this work likewise concentrates on as well as focuses on them. Sometimes even mild ailments go undiagnosed and cannot be appropriately treated, which results in death for the patient. Recently, a very severe virus (Coronavirus) has forced all the world's most powerful nations to bow to its pressure. Thus, we have previously seen such a massive calamity that has demonstrated the ability to effectively eradicate human existence from the planet in just a couple of months. As a result, several initiatives should be started to raise public knowledge of marine pathogenic microorganisms, which are capable of infecting humans, animals, and aquatic life with all manner of illnesses. Although infections alter their genetic structures over time or in response to host conditions, even a seemingly minor symptom may not be neglected. While elderly people are the primary target of these kinds of covert intruders in the human body, it is also necessary to keep a closer watch on those with weakened immune systems. The last paragraph of the chapter discusses how biosensors are useful for detecting illnesses in fish or other household animals, in addition to helping to maintain a healthy lifestyle for human beings. These are adaptable technologies that support ecosystem maintenance, control, and regulation while also helping to maintain human standards. Because of this, scientists and researchers will soon suggest and vigorously campaign for this form of adaptable procedure-based study. They will have many options to continue developing the biosensor as more research into it is possible and it can be constructed for other purposes as well.

References

[1] Rocha-Santos T and Duarte A C 2014 Introduction to the analysis of bioactive compounds in marine samples *Comprehensive Analytical Chemistry* **vol 65** (Elsevier) ch 1

[2] Albaladejo C, Sánchez P, Iborra A, Soto F, López J A and Torres R 2010 Wireless sensor networks for oceanographic monitoring: a systematic review *Sensors* **10** 6948–68

[3] Justino C I L, Freitas A C, Duarte A C and Santos T A P R 2015 Sensors and biosensors for monitoring marine contaminants *Trends Environ. Anal. Chem.* **6–7** 21–30

[4] Ronan J M and McHugh B 2013 A sensitive liquid chromatography/tandem mass spectrometry method for the determination of natural and synthetic steroid estrogens in seawater and marine biota, with a focus on proposed water framework directive environmental quality standards *Rapid Commun. Mass Spectrom.* **27** 738–46

[5] Vidal-Dorsch D E, Bay S M, Maruya K, Snyder S A, Trenholm R A and Vanderford B J 2012 Contaminants of emerging concern in municipal wastewater effluents and marine receiving water *Environ. Toxicol. Chem.* **31** 2674–82

[6] Thévenot D R, Toth K, Durst R A and Wilson G S 2001 Electrochemical biosensors: recommended definitions and classification *Biosens. Bioelectron.* **34** 635–59

[7] Mills G and Fones G 2012 A review of *in situ*/IT methods and sensors for monitoring the marine environment *Sens. Rev* **32** 17–28

[8] Justino C I L, Rocha-Santos T A and Duarte A C 2010 Review of analytical figures of merit of sensors and biosensors in clinical applications *TrAC—Trends Anal. Chem.* **29** 1172–83

[9] Kröger S, Parker E R, Metcalfe J D, Greenwood N, Forster R M, Sivyer D B *et al* 2009 Sensors for observing ecosystem status *Ocean Sci.* **5** 523–35

[10] Yantasee W, Hongsirikarn K, Warner C L, Choi D, Sangvanich T, Toloczko M B *et al* 2008 Direct detection of Pb in urine and Cd, Pb, Cu, and Ag in natural waters using electrochemical sensors immobilized with DMSA functionalized magnetic nanoparticles *Analyst* **133** 348–55

[11] Barange M, Field J G, Steffen W, Harris R P, Hofmann E E, Perry R I *et al* 2010 *Marine Ecosystems and Global Change* (Oxford University Press)

[12] FAO 2013 *Guidelines to control water pollution from agriculture in China: Decoupling water pollution from agricultural production* 40 Food and Agriculture Organization

[13] Schwarzenbach R P, Egli T, Hofstetter T B, Von Gunten U and Wehrli B 2010 Global water pollution and human health *Annu. Rev. Environ. Resour.* **35** 109–36

[14] Bhat R A 2020 Biopesticides: The key component to remediate pesticide contamination in an ecosystem (unpublished)

[15] Bhat R A, Beigh B A, Mir S A, Dar S A, Dervash M A, Rashid A *et al* 2022 Biopesticide techniques to remediate pesticides in polluted ecosystems *Research Anthology on Emerging Techniques in Environmental Remediation* (IGI Global Scientific Publishing) ch 17

[16] Fao I 2017 *Water Pollution from Agriculture: A Global Review Executive Summary* (FAO IWMI)

[17] Selman M, Greenhalgh S, Diaz R and Sugg Z 2008 WRI Policy note 1: eutrophication and hypoxia in coastal areas: a global assessment of the state of knowledge *Water Quality: Eutrophication and Hypoxia* World Resources Institute

[18] FAO 2015 *Food and Agriculture Organization of the United Nations. FAOSTAT Statistical Database* (Rome: Encyclopedia of Food Health)

[19] Mateo-sagasta J 2010 *Agriculture and Water Quality Interactions: A Global Overview* (FAO)

[20] Lorenz J J 2014 A review of the effects of altered hydrology and salinity on vertebrate fauna and their habitats in Northeastern Florida Bay *Wetlands* **34** 189–200

[21] Herbert E R, Boon P, Burgin A J, Neubauer S C, Franklin R B, Ardon M *et al* 2015 A global perspective on wetland salinization: ecological consequences of a growing threat to freshwater wetlands *Ecosphere* **6** 1–43

[22] Boxall A 2012 New and emerging water pollutants arising from agriculture *Organization for Economic Co-Operation and Deveolpment Report* 1–48

[23] NORMAN 2022 Emerging substances *Network of Reference Laboratories, Research Centres and Related Organisations for Monitoring of Emerging Environmental Substances* https://www.norman-network.net/?q=node/19

[24] Thebo A L, Drechsel P, Lambin E F and Nelson K L 2017 A global, spatially-explicit assessment of irrigated croplands influenced by urban wastewater flows *Environ. Res. Lett.* **12** 074008

[25] Biney C, Amuzu A T, Calamari D, Kaba N, Mbome I L, Naeve H *et al* 1994 Review of heavy metals in the African aquatic environment *Ecotoxicol. Environ. Saf.* **28** 134–59

[26] Förstner U 2005 Metal speciation in solid wastes—factors affecting mobility *Speciation of Metals in Water, Sediment and Soil Systems* ed L Landner (Springer) pp 11–41

[27] Rashid A, Bhat R A, Qadri H, Mehmood M A and Shafiq-ur-Rehman 2019 Environmental and socioeconomic factors induced blood lead in children: an investigation from Kashmir, India *Environ. Monit. Assess.* **191** 76

[28] Alloway Brian J 2013 *Introduction Heavy Metals in Soils* (Dordrecht: Springer)

[29] Zhou Q, Zhang J, Fu J, Shi J and Jiang G 2008 Biomonitoring: an appealing tool for assessment of metal pollution in the aquatic ecosystem *Anal. Chim. Acta* **606** 135–50

[30] Harguinteguy C A, Cirelli A F and Pignata M L 2014 Heavy metal accumulation in leaves of aquatic plant Stuckenia filiformis and its relationship with sediment and water in the Suquía river (Argentina) *Microchem. J.* **114** 111–8

[31] Loska K and Wiechuła D 2003 Application of principal component analysis for the estimation of source of heavy metal contamination in surface sediments from the Rybnik Reservoir *Chemosphere* **51** 723–33

[32] Knapp S and Van De Velden M 2011 Global ship risk profiles: safety and the marine environment *Transp. Res. Part D Transp. Environ.* **16** 595–603

[33] Broekema W 2016 Crisis-induced learning and issue politicization in the eu: the braer, sea empress, erika, and prestige oil spill disasters *Public Adm.* **94** 381–98

[34] Alló M and Loureiro M L 2013 Estimating a meta-damage regression model for large accidental oil spills *Ecol. Econ.* **86** 165–75

[35] Bossert I, Kachel W M and Bartha R 1984 Fate of hydrocarbons during oily sludge disposal in soil *Appl. Environ. Microbiol.* **47** 763–7

[36] Margesin R and Schinner F 1997 Efficiency of indigenous and inoculated cold-adapted soil microorganisms for biodegradation of diesel oil in alpine soils *Appl. Environ. Microbiol.* **63** 2660–4

[37] Benko K L and Drewes J E 2008 Produced water in the Western United States: geographical distribution, occurrence, and composition *Environ. Eng. Sci.* **25** 26

[38] Das J and Acharya B C 2003 Hydrology and assessment of lotic water quality in Cuttack City, India *Water Air Soil Pollut.* **150** 163–75

[39] Vargas-González H H, Arreola-Lizárraga J A, Mendoza-Salgado R A, Méndez-Rodríguez L C, Lechuga-Deveze C H, Padilla-Arredondo G *et al* 2014 Effects of sewage discharge on trophic state and water quality in a coastal ecosystem of the Gulf of California *Sci. World J.* **2014** 618054

[40] Akpor O B and Muchie M 2011 Environmental and public health implications of wastewater quality *Afr. J. Biotechnol.* **10** 2379–87

[41] Pearce G R, Pearce G R, Ramzan Choudhury M and Ghulam S 1999 *A simple methodology for water quality monitoring* Report OD 142 Department for International Development

[42] Momba M N B, Osode A N and Sibewu M 2006 *The Impact of Inadequate Wastewater Treatment on the Receiving Water Bodies—Case Study: Buffalo City and Nkokonbe Municipalities of the Eastern Cape Province* (Water SA)

[43] Morrison G, Fatoki O S, Persson L and Ekberg A 2001 Assessment of the impact of point source pollution from the Keiskammahoek Sewage Treatment Plant on the Keiskamma River—pH, electrical conductivity, oxygen- demanding substance (COD) and nutrients *Water SA* **27** 475–80

[44] WHO (World Health Organization) 2006 Guidelines for the safe use of wastewater, excreta and greywater *Wastewater Use in Agriculture* (Geneva: World Health Organization)

[45] Chigor V N, Sibanda T and Okoh A I 2013 Studies on the bacteriological qualities of the Buffalo River and three source water dams along its course in the Eastern Cape Province of South Africa *Environ. Sci. Pollut. Res.* **20** 4125–36

[46] Genevieve M C and James P N 2008 *Water Quality for Ecosystem and Human Health* (United Nations Environments Progamme Global Monitoring System/Water Programme)

[47] CSIR 2010 A CSIR perspective on water in South Africa-2010 *CSIR Report* No. CSIR/NRE/PW/IR/2011/0012/AMarine Pollution Bulletin

[48] Christou L 2011 The global burden of bacterial and viral zoonotic infections *Clin. Microbiol. Infect.* **17** P326–30

[49] Shamsi S 2019 Seafood-borne parasitic diseases: a 'one-health' approach is needed *Fishes* **4** 9

[50] Han B A, Kramer A M and Drake J M 2016 Global patterns of zoonotic disease in mammals *Trends Parasitol.* **32** p565–77

[51] Gauthier D T 2015 Bacterial zoonoses of fishes: a review and appraisal of evidence for linkages between fish and human infections *Vet. J.* **203** 27–35

[52] Wolfe N D, Dunavan C P and Diamond J 2007 Origins of major human infectious diseases *Nature* **447** 279–83

[53] Barrett K A, Nakao J H, Taylor E V, Eggers C and Gould L H 2017 Fish-associated foodborne disease outbreaks: United States, 1998–2015 *Foodborne Pathogens Dis.* **14** 9

[54] Farzadnia A and Naeemipour M 2020 Molecular techniques for the detection of bacterial zoonotic pathogens in fish and humans *Aquac. Int.* **28** 309–20

[55] Tran A K T, Doan H T, Do A N, Nguyen V T, Hoang S X, Le H T T *et al* 2019 Prevalence, species distribution, and related factors of fish-borne trematode infection in Ninh Binh Province, Vietnam *BioMed. Res. Int.* **2019** 8581379

[56] Meurens F, Dunoyer C, Fourichon C, Gerdts V, Haddad N, Kortekaas J *et al* 2021 Animal board invited review: risks of zoonotic disease emergence at the interface of wildlife and livestock systems *Animal* **15** 100241

[57] Jones K E, Patel N G, Levy M A, Storeygard A, Balk D, Gittleman J L *et al* 2008 Global trends in emerging infectious diseases *Nature* **451** 990–3

[58] Aggarwal D and Ramachandran A 2020 One health approach to address zoonotic diseases *Indian J. Commun. Med.* **45** S6–8

[59] Golomazou E, Malandrakis E E, Panagiotaki P and Karanis P 2021 Cryptosporidium in fish: implications for aquaculture and beyond *Water Res.* **201** 117357

[60] Weir M, Rajić A, Dutil L, Cernicchiaro N, Uhland F C, Mercier B *et al* 2012 Zoonotic bacteria, antimicrobial use and antimicrobial resistance in ornamental fish: a systematic review of the existing research and survey of aquaculture-allied professionals *Epidemiol. Infect.* **140** 192–206

[61] Smith J W and Wootten R 1978 Anisakis and anisakiasis *Adv. Parasitol.* **16** 93–163

[62] Meron D, Davidovich N, Ofek-Lalzar M, Berzak R, Scheinin A, Regev Y *et al* 2020 Specific pathogens and microbial abundance within liver and kidney tissues of wild marine fish from the Eastern Mediterranean Sea *Microb. Biotechnol.* **13** 770–80

[63] Regev Y, Davidovich N, Berzak R, Lau S C K, Scheinin A P, Tchernov D *et al* 2020 Molecular identification and characterization of vibrio species and mycobacterium species in wild and cultured marine fish from the eastern mediterranean sea *Microorganisms* **8** 863

[64] Delghandi M R, El-Matbouli M and Menanteau-Ledouble S 2020 Mycobacteriosis and infections with non-tuberculous mycobacteria in aquatic organisms: a review *Microorganisms* **8** 1368

[65] Delghandi M R, Waldner K, El-Matbouli M and Menanteau-Ledouble S 2020 Identification mycobacterium spp. In the natural water of two austrian rivers *Microorganisms* **8** 1305

[66] Puk K and Guz L 2020 Occurrence of mycobacterium spp. In ornamental fish *Ann. Agric. Environ. Med.* **27** 535–9

[67] Hashish E, Merwad A, Elgaml S, Amer A, Kamal H, Elsadek A *et al* 2018 Mycobacterium marinum infection in fish and man: epidemiology, pathophysiology and management; a review *Vet. Quart.* **38** 35–46

[68] Gcebe N, Michel A L and Hlokwe T M 2018 Non-tuberculous mycobacterium species causing mycobacteriosis in farmed aquatic animals of South Africa *BMC Microbiol.* **18** 32

[69] Boylan S 2011 *Zoonoses Associated with Fish* (Veterinary Clinics of North America—Exotic Animal Practice)

[70] Bhambri S, Bhambri A and Del Rosso J Q 2009 Atypical mycobacterial cutaneous infections *Dermatol. Clin.* **27** 63–73

[71] Woo P T K and Leatherland J F 2011 *Fish Diseases and Disorders vol 3: Viral, Bacterial and fungal Infections* (CABI)

[72] Phillips Savage A C N, Blake L, Suepaul R, McHugh O, Rodgers R, Thomas C *et al* 2022 Piscine mycobacteriosis in the ornamental fish trade in Trinidad and Tobago *J. Fish Dis* **45** 547–60

[73] Novotny L, Dvorska L, Lorencova A, Beran V and Pavlik I 2004 Fish: a potential source of bacterial pathogens for human beings *Vet. Med.* **49** 343–58

[74] Iregui C A, Comas J, Vásquez G M and Verján N 2016 Experimental early pathogenesis of *Streptococcus agalactiae* infection in red tilapia Oreochromis spp *J. Fish. Dis.* **39** 205–15

[75] Haghighi Karsidani S, Soltani M, Nikbakhat-Brojeni G, Ghasemi M and Skall H F 2010 Molecular epidemiology of zoonotic streptococcosis/lactococcosis in rainbow trout (*Oncorhynchus mykiss*) aquaculture in Iran *Iran J. Microbiol.* **2** 198–209

[76] Leal C A G, Queiroz G A, Pereira F L, Tavares G C and Figueiredo H C P 2019 *Streptococcus agalactiae* sequence type 283 in farmed fish, Brazil *Emerg. Infect. Dis.* **25** 776–9

[77] Pradeep P J, Suebsing R, Sirthammajak S, Kampeera J, Jitrakorn S, Saksmerprome V *et al* 2016 Evidence of vertical transmission and tissue tropism of Streptococcosis from naturally infected red tilapia (Oreochromis spp *Aquac. Rep.* **3** 58–66

[78] Barkham T, Zadoks R N, Azmai M N A, Baker S, Bich V T N, Chalker V *et al* 2019 One hypervirulent clone, sequence type 283, accounts for a large proportion of invasive *Streptococcus agalactiae* isolated from humans and diseased tilapia in southeast asia *PLoS Negl. Trop. Dis.* **13** e0007421

[79] Baiano J C F and Barnes A C 2009 Towards control of *Streptococcus iniae Emerg. Infect. Dis.* **15** 1891–6

[80] Berzak R, Scheinin A, Davidovich N, Regev Y, Diga R, Tchernov D *et al* 2019 Prevalence of nervous necrosis virus (NNV) and *Streptococcus* species in wild marine fish and crustaceans from the Levantine Basin, Mediterranean Sea *Dis. Aquat. Organ.* **133** 7–17

[81] Haenen O L M, Evans J J and Berthe F 2013 Bacterial infections from aquatic species: potential for and prevention of contact zoonoses *OIE Rev. Sci. Tech.* **32** 497–507

[82] Pomaranski E K, Griffin M J, Camus A C, Armwood A R, Shelley J, Waldbieser G C *et al* 2020 Description of *Erysipelothrix piscisicarius* sp. Nov., an emergent fish pathogen, and assessment of virulence using a tiger barb (*Puntigrus tetrazona*) infection model *Int. J. Syst. Evol. Microbiol.* **70** 857–67

[83] Nielsen J J, Blomberg B, Gaïni S and Lundemoen S 2018 Aortic valve endocarditis with *Erysipelothrix rhusiopathiae*: a rare zoonosis *Infect. Dis. Rep.* **10** 7770

[84] Balootaki P A, Amin M, Haghparasti F and Rokhbakhsh-Zamin F 2017 Isolation and detection of *Erysipelothrix rhusiopathiae* and its distribution in humans and animals by phenotypical and molecular methods in Ahvaz-Iran in 2015 *Iran J. Med. Sci.* **42** 377–83

[85] Austin B 2010 Vibrios as causal agents of zoonoses *Vet. Microbiol.* **140** 310-17

[86] Helmi A M, Mukti A T, Soegianto A and Effendi M H 2020 A review of vibriosis in fisheries: public health importance *Syst. Rev. Pharm.* **11** 51–8

[87] Jun L and Woo N Y S 2003 Pathogenicity of vibrios in fish: an overview *J. Ocean Univ. Qingdao* **2** 117–28

[88] Huzmi H, I-SM Y, NFM I, Syukri F and Karim M 2019 Strategies of controlling vibriosis in fish *Asian J. Appl. Sci.* **7** 513–21

[89] Carmona-Salido H, Fouz B, Sanjuán E, Carda M, Delannoy C M J, García-González N *et al* 2021 The widespread presence of a family of fish virulence plasmids in Vibrio vulnificus stresses its relevance as a zoonotic pathogen linked to fish farms *Emerg. Microbes Infect.* **10** 2128–40

[90] Hernández-Cabanyero C and Amaro C 2020 Phylogeny and life cycle of the zoonotic pathogen Vibrio vulnificus *Environ. Microbiol.* **22** 4133–48

[91] You H J, Lee J H, Oh M, Hong S Y, Kim D, Noh J *et al* 2021 Tackling Vibrio parahaemolyticus in ready-to-eat raw fish flesh slices using lytic phage VPT02 isolated from market oyster *Food Res. Int.* **150** 110779

[92] Khardori N and Fainstein V 1988 *Aeromonas* and *Plesiomonas* as etiological agents *Annu. Rev. Microbiol.* **42** 395–419

[93] Abd-El-Malek A M 2017 Incidence and virulence characteristics of *Aeromonas* spp. in fish *Vet. World* **10** 34–7

[94] Agnew W and Barnes A C 2007 *Streptococcus iniae*: an aquatic pathogen of global veterinary significance and a challenging candidate for reliable vaccination *Vet. Microbiol.* **122** 1–15

[95] AlYahya S A, Ameen F, Al-Niaeem K S, Al-Sa'adi B A, Hadi S and Mostafa A A 2018 Histopathological studies of experimental *Aeromonas hydrophila* infection in blue tilapia, *Oreochromis aureus Saudi J. Biol. Sci.* **25** 182–5

[96] Odeyemi O A and Ahmad A 2017 Antibiotic resistance profiling and phenotyping of *Aeromonas* species isolated from aquatic sources *Saudi J. Biol. Sci.* **24** 65–70

[97] Jin L, Chen Y, Yang W, Qiao Z and Zhang X 2020 Complete genome sequence of fish-pathogenic *Aeromonas hydrophila* HX-3 and a comparative analysis: insights into virulence factors and quorum sensing *Sci. Rep.* **10** 15479

[98] Volpe E, Mandrioli L, Errani F, Serratore P, Zavatta E, Rigillo A *et al* 2019 Evidence of fish and human pathogens associated with doctor fish (Garra rufa, Heckel, 1843) used for cosmetic treatment *J. Fish Dis* **42** 1637–44

[99] Algammal A M, Mabrok M, Sivaramasamy E, Youssef F M, Atwa M H, El-kholy A W *et al* 2020 Emerging MDR-*Pseudomonas aeruginosa* in fish commonly harbor oprL and toxA virulence genes and bla TEM, bla CTX-M, and tetA antibiotic-resistance genes *Sci Rep.* **10** 15961

[100] Yagoub S O 2009 Isolation of *Enterobacteriaceae* and Pseudomonas spp. from raw fish sold in fish market in Khartoum state *J. Bacteriol. Res.* **1** 085–8

[101] Benie C K D, Dadié A, Guessennd N, N'gbesso-Kouadio N A, Kouame N D, N'golo D C *et al* 2017 Characterization of virulence potential of *Pseudomonas aeruginosa* isolated from bovine meat, fresh fish, and smoked fish *Eur. J. Microbiol. Immunol.* **7** 55–64

[102] Fernandes M R, Sellera F P, Moura Q, Carvalho M P N, Rosato P N, Cerdeira L *et al* 2018 Zooanthroponotic transmission of drug-resistant pseudomonas aeruginosa, Brazil *Emerg. Infect. Dis.* **24** 1160–2

[103] I. E, Ismail M, El Lamei M and A. H 2017 Studies on pseudomonas septicemia in some Tilapia in Ismailia *Suez Canal Vet. Med. J. SCVMJ* **22** 107–17

[104] Pelli A 2017 Disease infection by *Enterobacteriaceae* family in fishes: a review *J. Microbiol. Exp.* **4** 00128

[105] Martins M, Barbosa C, Pinto F D R, Ribeiro L F, Sobue C, Guriz L *et al* 2014 Serology and patterns of antimicrobial susceptibility in *Escherichia coli* isolates from pay-to-fish ponds *Arq. Inst. Biológico.* **81** 43–8

[106] Hansen D L, Clark J J, Ishii S, Sadowsky M J and Hicks R E 2008 Sources and sinks of *Escherichia coli* in benthic and pelagic fish *J. Great Lakes Res.* **34** 228–34

[107] Adanech B H and Temesgen K G 2018 Isolation and identification of *Escherichia coli* and *Edwardsiella tarda* from fish harvested for human consumption from Zeway Lake, Ethiopia *Afr. J. Microbiol. Res.* **12** 476-80

[108] Cardozo M V, Borges C A, Beraldo L G, Maluta R P, Pollo A S, Borzi M M *et al* 2018 Shigatoxigenic and atypical enteropathogenic *Escherichia coli* in fish for human consumption *Brazil. J. Microbiol.* **49** 936–41

[109] Traoré O, Nyholm O, Siitonen A, Bonkoungou I J O, Traoré A S, Barro N *et al* 2015 Prevalence and diversity of Salmonella enterica in water, fish and lettuce in Ouagadougou, Burkina Faso *BMC Microbiol.* **15** 151

[110] Das A, Acharya S, Behera B K, Paria P, Bhowmick S, Parida P K *et al* 2018 Isolation, identification and characterization of *Klebsiella pneumoniae* from infected farmed Indian Major Carp Labeo rohita (Hamilton 1822) in West Bengal, India *Aquaculture* **482** 111–6

[111] Bibi F, Qaisrani S N, Ahmad A N, Akhtar M, Khan B N and Ali Z 2015 Occurrence of salmonella in freshwater fishes: a review *J. Anim. Plant. Sci.* **25** 303–10

[112] Lehane L and Rawlin G T 2000 Topically acquired bacterial zoonoses from fish: a review *Med. J. Aust.* **173** 256–9

[113] Gundogan N 2014 Klebsiella *Encyclopedia of Food Microbiology* 2nd edn ed C A Batt (Academic)

[114] Gopi M, Thankappanpillai Ajith Kumar T, Prakash S and Mohan Gopi C 2016 Opportunistic pathogen Klebsiella pneumoniae isolated from Maldive's clown fish Amphiprion nigripes with hemorrhages at Agatti Island, Lakshadweep archipelago *Int. J. Fish Aquat. Stud.* **4** 464–7

[115] de Keukeleire S, de Bel A, Jansen Y, Janssens M, Wauters G and Piérard D 2014 Yersinia ruckeri, an unusual microorganism isolated from a human wound infection *New Microbes New Infect* **2** 134–5

[116] Wrobel A, Leo J C and Linke D 2019 Overcoming fish defences: the virulence factors of Yersinia ruckeri *Genes* **10** 700

[117] Oliveira R V, Peixoto P G, Ribeiro D D C, Araujo M C, do Santos C T B, Hayashi C *et al* 2014 Klebsiella pneumoniae as a main cause of infection in nishikigoi Cyprinus carpio (carp) by inadequate handling *Brazil. J. Vet. Pathol.* **7** 86–8

[118] Diana T C and Manjulatha C 2012 Incidence and identification of *Klebsiella pneumoniae* in mucosal buccal polyp of *Nemipterus japonicus* of Visakhapatnam Coast, India *J. Fish. Aquat. Sci.* **7** 454–60

[119] Miniero Davies Y, Xavier de Oliveira M G, Paulo Vieira Cunha M G, Soares Franco L, Pulecio Santos S L, Zanolli Moreno L *et al* 2018 *Edwardsiella tarda* outbreak affecting fishes and aquatic birds in Brazil *Vet. Q* **38** 99–105

[120] Adeolu M, Alnajar S, Naushad S and Gupta R S 2016 Genome-based phylogeny and taxonomy of the 'Enterobacteriales': proposal for Enterobacterales ord. nov. divided into the families *Enterobacteriaceae, Erwiniaceae Int. J. Syst. Evol. Microbiol.* **66** 5575–99

[121] Park S B, Aoki T and Jung T S 2012 Pathogenesis of and strategies for preventing *Edwardsiella tarda* infection in fish *Vet. Res.* **43** 67

[122] Yu J E, Cho M Y, Woo K J and Kang H Y 2012 Large antibiotic-resistance plasmid of *Edwardsiella tarda* contributes to virulence in fish *Microb. Pathog* **52** 259–66

[123] Kerie Y, Nuru A and Abayneh T 2019 *Edwardsiella* species infection in fish population and its status in Ethiopia *Fish Aquac. J* **10** 100266

[124] Buján N, Toranzo A E and Magariños B 2018 *Edwardsiella piscicida*: a significant bacterial pathogen of cultured fish *Dis. Aquat. Organ.* **131** 59–71

[125] Leung K Y, Wang Q, Yang Z and Siame B A 2019 *Edwardsiella piscicida*: a versatile emerging pathogen of fish *Virulence* **10** 555–67

[126] Wimalasena S H M P, Pathirana H N K S, De Silva B C J, Hossain S, Sugaya E, Nakai T *et al* 2019 Antibiotic resistance and virulence-associated gene profile of *Edwardsiella tarda* isolated from cultured fish in Japan *Turkish J. Fish. Aquat. Sci.* **19** 141–8

[127] Obaidat M M, Salman A E B and Lafi S Q 2015 Prevalence of *Staphylococcus aureus* in imported fish and correlations between antibiotic resistance and enterotoxigenicity *J. Food. Prot.* **78** 1999–2005

[128] Vaiyapuri M, Joseph T C, Rao B M, Lalitha K V and Prasad M M 2019 Methicillin-resistant *Staphylococcus aureus* in seafood: prevalence, laboratory detection, clonal nature, and control in seafood chain *J. Food Sci.* **84** 3341–51

[129] Oh W T, Jun J W, Giri S S, Yun S, Kim H J, Kim S G *et al* 2019 *Staphylococcus xylosus* infection in rainbow trout (*Oncorhynchus mykiss*) as a primary pathogenic cause of eye protrusion and mortality *Microorganisms* **7** 330

[130] Skowron K, Wiktorczyk N, Grudlewska K, Wałecka-Zacharska E, Paluszak Z, Kruszewski S *et al* 2019 Phenotypic and genotypic evaluation of *Listeria monocytogenes* strains isolated from fish and fish processing plants *Ann. Microbiol.* **69** 469–82

[131] Rukkawattanakul T, Sookrung N, Seesuay W, Onlamoon N, Diraphat P, Chaicumpa W *et al* 2017 Human scFvs that counteract bioactivities of *Staphylococcus aureus* TSST-1 *Toxins (Basel)* **9** 50

[132] Jami M, Ghanbari M, Zunabovic M, Domig K J and Kneifel W 2014 *Listeria monocytogenes* in aquatic food products-a review *Compr. Rev. Food Sci. Food Saf.* **13** 798–813

[133] Sabry M, El-Moein K A, Hamza E and Kader F A 2016 Occurrence of *Clostridium perfringens* types A, E, and C in fresh fish and its public health significance *J. Food Prot.* **79** 994–1000

[134] Gillesberg Lassen S, Ethelberg S, Björkman J T, Jensen T, Sørensen G, Kvistholm Jensen A *et al* 2016 Two listeria outbreaks caused by smoked fish consumption—using whole-genome sequencing for outbreak investigations *Clin. Microbiol. Infect.* **22** P620–4

[135] Collins M D and East A K 1998 Phylogeny and taxonomy of the food-borne pathogen *Clostridium botulinum* and its neurotoxins *J. Appl. Microbiol.* **84** 5–17

[136] Barash J R and Arnon S S 2014 A novel strain of *Clostridium botulinum* that produces type B and type H botulinum toxins *J. Infect. Dis.* **209** 183–91

[137] Freedman J C, Shrestha A and McClane B A 2016 *Clostridium perfringens* enterotoxin: action, genetics, and translational applications *Toxins* **8** 73

[138] Uzal F A, Freedman J C, Shrestha A, Theoret J R, Garcia J, Awad M M *et al* 2014 Towards an understanding of the role of *Clostridium perfringens* toxins in human and animal disease *Fut. Microbiol.* **9** 361–77

[139] Espelund M and Klaveness D 2014 Botulism outbreaks in natural environments—an update *Front. Microbiol.* **5** 287

[140] Epps S V R, Harvey R B, Hume M E, Phillips T D, Anderson R C and Nisbet D J 2013 Foodborne campylobacter: infections, metabolism, pathogenesis and reservoirs *Int. J. Environ. Res. Public Health* **10** 6292–304

[141] Facciolà A, Riso R, Avventuroso E, Visalli G, Delia S A and Laganà P 2017 Campylobacter: from microbiology to prevention *J. Prev. Med. Hygiene* **58** E79–92

[142] Rasetti-Escargueil C, Lemichez E and Popoff M 2019 Public health risk associated with botulism as foodborne zoonoses *Toxins* **12** 17

[143] Tobback E, Decostere A, Hermans K, Haesebrouck F and Chiers K 2007 Yersinia ruckeri infections in salmonid fish *J. Fish Dis.* **30** 257–68

[144] Kumar G, Menanteau-Ledouble S, Saleh M and El-Matbouli M 2015 Yersinia ruckeri, the causative agent of enteric redmouth disease in fish *Vet. Res.* **46** 103

[145] Carson J and Wilson T 2019 Yersiniosis in fish *The Australian and New Zealand standard diagnostic procedure (ANZSDP)* Department of Agriculture, Fisheries and Forestry https://www.agriculture.gov.au/agriculture-land/animal/health/laboratories/procedures/anzsdp/yersiniosis-in-fish

[146] Nakajima H, Inoue M and Mori T 1991 Isolation of Yersinia, campylobacter, *Plesiomonas* and *Aeromonas* from environmental water and fresh water fishes *[Nippon kōshūeisei zasshi] Japanese J. Public Heal* **38** 815–20

[147] Daengsvang S 1981 Gnathostomiasis in southeast Asia *Southeast Asian J. Trop. Med. Public Health* **12** 319–32

[148] Audicana M T, Ansotegui I J, De Corres L F and Kennedy M W 2002 Anisakis simplex: dangerous—dead and alive? *Trends Parasitol.* **18** P20–5

[149] Herman J S and Chiodini P L 2009 Gnathostomiasis, another emerging imported disease *Clin. Microbiol. Rev.* **22** 484–92

[150] Shamsi S 2020 Seafood-borne parasites in Australia: human health risks, fact or fiction? *Microbiol. Aust* **41** 33–7

[151] Williams M, Hernandez-Jover M and Shamsi S 2020 A critical appraisal of global testing protocols for zoonotic parasites in imported seafood applied to seafood safety in Australia *Foods* **9** 448

[152] Huss H H, Reilly A and Embarek K B P 2000 Prevention and control of hazards in seafood *Food Control* **11** 149–56

[153] Dorny P, Praet N, Deckers N and Gabriel S 2009 Emerging food-borne parasites *Vet. Parasitol* **163** 196–206

[154] Saijuntha W, Sithithaworn P, Petney T N and Andrews R H 2021 Foodborne zoonotic parasites of the family Opisthorchiidae *Res. Vet. Sci.* **135** 404–11

[155] Chai J Y, Murrell K D and Lymbery A J 2005 Fish-borne parasitic zoonoses: status and issues *Int. J. Parasitol.* **35** 1233–54

[156] Lõhmus M and Björklund M 2015 Climate change: what will it do to fish-parasite interactions? *Biol. J. Linn. Soc.* **116** 397–411

[157] Shamsi S 2016 Seafood-borne parasitic diseases in Australia: how much do we know about them? *Microbiol. Aust* **37** 27–9

[158] Shamsi S and Sheorey H 2018 Seafood-borne parasitic diseases in Australia: are they rare or underdiagnosed? *Intern. Med. J* **48** 591–6

[159] Marcogliese D J 2003 Food webs and biodiversity: are parasites the missing link? *J. Parasitol* **89** 106–13

[160] Chai J Y, Shin E H, Lee S H and Rim H J 2009 Foodborne intestinal flukes in Southeast Asia *Korean J. Parasitol.* **47** S69–102

[161] Ogbeibu A E, Okaka C E and Oribhabor B J 2014 Gastrointestinal Helminth Parasites Community of Fish Species in a Niger Delta Tidal Creek, Nigeria *J. Ecosyst.* **2014** 246283

[162] Nguyen T H, Dorny P, Nguyen T T G and Dermauw V 2021 Helminth infections in fish in Vietnam: a systematic review *Int. J. Parasitol.: Parasites Wildlife* **14** 13–32

[163] Polley L and Thompson R C A 2009 Parasite zoonoses and climate change: molecular tools for tracking shifting boundaries *Trends Parasitol.* **25** P285–91

[164] Fiorenza E A, Wendt C A, Dobkowski K A, King T L, Pappaionou M, Rabinowitz P *et al* 2020 It's a wormy world: meta-analysis reveals several decades of change in the global abundance of the parasitic nematodes Anisakis spp. and Pseudoterranova spp. in marine fishes and invertebrates *Glob. Chang. Biol.* **26** 2854–66

[165] Lowry T and Smith S A 2007 Aquatic zoonoses associated with food, bait, ornamental, and tropical fish *J. Am. Vet. Med. Assoc.* **231** 876–80

[166] Shamsi S and Suthar J 2016 A revised method of examining fish for infection with zoonotic nematode larvae *Int. J. Food Microbiol.* **227** 13–16

[167] Deardorff T L 1991 Epidemiology of marine fish-borne parasitic zoonoses *Southeast Asian J. Trop. Med. Public Health* **22** 146–9

[168] Lima dos Santos C A M and Howgate P 2011 Fishborne zoonotic parasites and aquaculture: a review *Aquaculture* **318** 253–61

[169] Williams M, Hernandez-Jover M and Shamsi S 2022 Parasites of zoonotic interest in selected edible freshwater fish imported to Australia *Food Waterborne Parasitol* **26** e00138

[170] Germann R, Schächtele M, Neßler G, Seitz U and Kniehl E 2003 Cerebral gnathostomiasis as a cause of an extended intracranial bleeding *Klin. Padiatr* **215** 223–5

[171] Sripa B, Bethony J M, Sithithaworn P, Kaewkes S, Mairiang E, Loukas A *et al* 2011 Opisthorchiasis and *Opisthorchis*-associated cholangiocarcinoma in Thailand and Laos *Acta Trop* **120** S158–68

[172] Cong W and Elsheikha H M 2021 Biology, epidemiology, clinical features, diagnosis, and treatment of selected fish-borne parasitic zoonoses *Yale J. Biol. Med.* **94** 297–309

[173] Choi B I, Han J K, Hong S T and Lee K H 2004 Clonorchiasis and cholangiocarcinoma: etiologic relationship and imaging diagnosis *Clin. Microbiol. Rev.* **17** 540–52

[174] Boerlage A S, Graat E A M, Verreth J A and de Jong M C M 2013 Effect of control strategies on the persistence of fish-borne zoonotic trematodes: a modelling approach *Aquaculture* **408–409** 106–12

[175] Caffara M, Gustinelli A, Mazzone A and Fioravanti M L 2020 Multiplex PCR for simultaneous identification of the most common European Opisthorchiid and Heterophyid in fish or fish products *Food Waterborne Parasitol* **19** e00081

[176] Lin R, Li X, Lan C, Yu S and Kawanaka M 2005 Investigation on the epidemiological factors of *Clonorchis sinensis* infection in an area of south China *Southeast Asian J. Trop. Med. Public Health* **36** 1114–7

[177] Hung N M, Dung D T, Lan Anh N T, Van P T, Thanh B N, Van Ha N *et al* 2015 Current status of fish-borne zoonotic trematode infections in Gia Vien district, Ninh Binh province, Vietnam *Parasites Vectors* **8** 21

[178] Maleewong W, Wongkham C, Intapan P, Pariyananda S and Morakote N 1992 Excretory-secretory antigenic components of *Paragonimus heterotremus* recognized by infected human sera *J. Clin. Microbiol.* **30** 2077–9

[179] Tantrawatpan C, Intapan P M, Janwan P, Sanpool O, Lulitanond V, Srichantaratsamee C *et al* 2013 Molecular identification of *Paragonimus* species by DNA pyrosequencing technology *Parasitol Int* **62** 341–5

[180] Fürst T, Keiser J and Utzinger J 2012 Global burden of human food-borne trematodiasis: a systematic review and meta-analysis *Lancet Infect. Dis* **12** 210–21

[181] Prueksapanich P, Piyachaturawat P, Aumpansub P, Ridtitid W, Chaiteerakij R and Rerknimitr R 2018 Liver fluke-associated biliary tract cancer *Gut Liver* **12** 236–45

[182] Clausen J H, Madsen H, Murrell K D, Van P T, Thu H N T, Do D T *et al* 2012 Prevention and control of fish-borne zoonotic trematodes in fish nurseries, Vietnam *Emerg. Infect. Dis* **18** 1438–45

[183] Sohn W M, Na B K, Cho S H, Ju J W, Kim C H, Hwang M A *et al* 2021 Prevalence and infection intensity of zoonotic trematode metacercariae in fish from soyang-cheon (Stream), in Wanju-gun, Jeollabuk-do, Korea *Korean J. Parasitol.* **59** 265–71

[184] Manivong K, Komalamisra C and Radomyos P 2009 *Opisthorchis viverrini* metacercariae in cyprinoid fish from three rivers in Khammouane Province, Lao PDR *J. Trop. Med. Parasitol.* **32** 23–9

[185] Enes J E, Wages A J, Malone J B and Tesana S 2010 Prevalence of *Opisthorchis viverrini* infection in the canine and feline hosts in three villages, Khon Kaen Province, northeastern Thailand *Southeast Asian J. Trop. Med. Public Health* **41** 36–42

[186] El-Seify M A, Sultan K, Elhawary N M, Satour N S and Marey N M 2021 Prevalence of heterophyid infection in tilapia fish 'Orechromas niloticus' with emphasize of cats role as neglected reservoir for zoonotic *Heterophyes heterophyes* in Egypt *J. Parasit. Dis.* **45** 35–42

[187] Scholz T and Kuchta R 2016 Fish-borne, zoonotic cestodes (*Diphyllobothrium* and relatives) in cold climates: a never-ending story of neglected and (re)-emergent parasites *Food Waterborne Parasitol.* **4** 23–38

[188] Jones S R M 2015 Transmission dynamics of foodborne parasites in fish and shellfish *Foodborne Parasites in the Food Supply Web: Occurrence and Control* (Elsevier) pp 293–315

[189] Dick T A 2007 Diphyllobothriasis: the *Diphyllobothrium latum* human infection conundrum and reconciliation with a worldwide zoonosis *Food-Borne Parasitic Zoonoses* (Springer) pp 151–84

[190] McConnaughey M 2014 Life cycle of parasites *xPharm: The Comprehensive Pharmacology Reference* Reference Module in Biomedical Sciences S J Enna and Bylund D B (Elsevier)

[191] Eiras J C, Pavanelli G C, Takemoto R M and Nawa Y 2018 Fish-borne nematodiases in South America: neglected emerging diseases *J. Helminthol.* **92** 649–54

[192] Steffen R, DeBernardis C and Baños A 2003 Travel epidemiology—a global perspective *Int. J. Antimicrob. Agents* **21** 89–95

[193] Butt A A, Aldridge K E and Sanders C V 2004 Infections related to the ingestion of seafood part I: viral and bacterial infections *Lancet Infect. Dis.* **4** 201–12

[194] Nawa Y and Nakamura-Uchiyama F 2004 An overview of gnathostomiasis in the world *Southeast Asian J. Trop. Med. Public Health* **35** 87–91

[195] Hossen M S, Wassens S and Shamsi S 2021 Occurrence and abundance of zoonotic nematodes in snapper *Chrysophrys auratus*, a popular table fish from Australian and New Zealand waters *Food Waterborne Parasitol.* **23** e00120

[196] Salikin N H, Nappi J, Majzoub M E and Egan S 2020 Combating parasitic nematode infections, newly discovered antinematode compounds from marine epiphytic bacteria *Microorganisms* **8** 1963

[197] Suthar J and Shamsi S 2021 The occurrence and abundance of infective stages of zoonotic nematodes in selected edible fish sold in Australian fish markets *Microb. Pathogen.* **154** 104833

[198] Borges J N, Cunha L F G, Santos H L C, Monteiro-Neto C and Santos C P 2012 Morphological and molecular diagnosis of anisakid nematode larvae from cutlassfish (*Trichiurus lepturus*) off the coast of Rio de Janeiro, Brazil *PLoS One* **7** e40447

[199] Bao M, Pierce G J, Strachan N J C, Pascual S, González-Muñoz M and Levsen A 2019 Human health, legislative and socioeconomic issues caused by the fish-borne zoonotic parasite Anisakis: challenges in risk assessment *Trends Food Sci. Technol.* **86** 298–310

[200] Safonova A E, Voronova A N and Vainutis K S 2021 First report on molecular identification of Anisakis simplex in *Oncorhynchus nerka* from the fish market, with taxonomical issues within Anisakidae *J. Nematol.* **53** e2021–23

[201] Audicana M T and Kennedy M W 2008 Anisakis simplex: from obscure infectious worm to inducer of immune hypersensitivity *Clin. Microbiol. Rev.* **21** 360–79

[202] Eiras J C, Pavanelli G C, Takemoto R M and Nawa Y 2018 An overview of fish-borne nematodiases among returned travelers for recent 25 years—unexpected diseases sometimes far away from the origin *Korean J. Parasitol.* **56** 215–27

[203] Shamsi S 2014 Recent advances in our knowledge of Australian anisakid nematodes *Int. J. Parasitol.: Parasites and Wildlife* **3** 178–87

[204] Adroher-Auroux F J and Benítez-Rodríguez R 2020 Anisakiasis and Anisakis: an under-diagnosed emerging disease and its main etiological agents *Res. Vet. Sci.* **132** 535–45

[205] Ramos P 2020 Parasites in fishery products—laboratorial and educational strategies to control *Exp. Parasitol.* **211** 107865

[206] González MÁP, Cavazza G, Gustinelli A, Caffara M and Fioravanti M 2020 Absence of anisakis nematodes in smoked farmed Atlantic salmon (*Salmo salar*) products on sale in European countries *Ital. J. Food Saf.* **9** 8615

[207] Fioravanti M L, Gustinelli A, Rigos G, Buchmann K, Caffara M, Pascual S *et al* 2021 Negligible risk of zoonotic anisakid nematodes in farmed fish from European mariculture, 2016 to 2018 *Eurosurveillance* **26** 1900717

[208] Anderson R C 2000 *Nematode Parasites of Vertebrates: Their Development and Transmission* (CABI)

[209] Aibinu I E, Smooker P M and Lopata A L 2019 Anisakis nematodes in fish and shellfish-from infection to allergies *Int. J. Parasitol.: Parasites Wildlife* **9** 384–93

[210] Ramanan P, Blumberg A K, Mathison B and Pritt B S 2013 Parametrial anisakidosis *J. Clin. Microbiol.* **51** 3430–4

[211] Shamsi S and Butcher A R 2011 First report of human anisakidosis in Australia *Med. J. Aust* **194** 199–200

[212] Shimamura Y, Muwanwella N, Chandran S, Kandel G and Marcon N 2016 Common symptoms from an uncommon infection: gastrointestinal Anisakiasis *Can. J. Gastroenterol. Hepatol.* **2016** 5176502

[213] Yokogawa M and Yoshimura H 1967 Clinicopathologic studies on larval anisakiasis in Japan *Am. J. Trop. Med. Hyg.* **16** 723–8

[214] Valle J, Lopera E, Sánchez M E, Lerma R and Ruiz J L 2012 Spontaneous splenic rupture and Anisakis appendicitis presenting as abdominal pain: a case report *J. Med. Case Rep.* **6** 114

[215] Nieuwenhuizen N E 2016 Anisakis—immunology of a foodborne parasitosis *Parasite Immunol.* **38** 548–57

[216] Rahmati A R, Kiani B, Afshari A, Moghaddas E, Williams M and Shamsi S 2020 World-wide prevalence of Anisakis larvae in fish and its relationship to human allergic anisakiasis: a systematic review *Parasitol. Res.* **119** 3585–94

[217] Pinheiro R H da S, Santana R L S, Melo F T V, Santos dos J N and Giese E G 2017 Gnathostomatidae nematode parasite of *Colomesus psittacus* (osteichthyes, tetraodonti-formes) in the Ilha de Marajó, Brazilian amazon *Rev. Bras. Parasitol. Vet.* **26** 340–7

[218] Shamsi S, Steller E and Zhu X 2021 The occurrence and clinical importance of infectious stage of *Echinocephalus* (Nematoda: Gnathostomidae) larvae in selected Australian edible fish *Parasitol. Int* **83** 102333

[219] Liu G H, Sun M M, Elsheikha H M, Fu Y T, Sugiyama H, Ando K *et al* 2020 Human gnathostomiasis: a neglected food-borne zoonosis *Parasit. Vectors* **13** 616

[220] Sawadpanich K, Chansuk N, Boonroumkaew P, Sadaow L, Rodpai R, Sanpool O *et al* 2021 An unusual case of gastric gnathostomiasis caused by gnathostoma spinigerum confirmed by video gastroscopy and morphological and molecular identification *Am. J. Trop. Med. Hyg.* **104** 2050–4

[221] Pavoni E, Consoli M, Suffredini E, Arcangeli G, Serracca L, Battistini R *et al* 2013 Noroviruses in seafood: a 9-year monitoring in Italy *Foodborne Pathogens Disease* **10** 1399

[222] Li D, Stals A, Tang Q J and Uyttendaele M 2014 Detection of noroviruses in shellfish and semiprocessed fishery products from a Belgian seafood company *J. Food Prot.* **77** 1342–7

[223] Kittigul L, Thamjaroen A, Chiawchan S, Chavalitshewinkoon-Petmitr P, Pombubpa K and Diraphat P 2016 Prevalence and molecular genotyping of noroviruses in market oysters, mussels, and cockles in Bangkok, Thailand *Food Environ. Virol.* **8** 133-40

[224] Marsh Z, Shah M P, Wikswo M E, Barclay L, Kisselburgh H, Kambhampati A *et al* 2018 Epidemiology of foodborne norovirus outbreaks—United States, 2009–2015 *Food Saf* **6** 58–66

[225] Atmar R L, Baehner F, Cramer J P, Lloyd E, Sherwood J, Borkowski A *et al* 2020 Persistence of antibodies to 2 virus-like particle norovirus vaccine candidate formulations in healthy adults: 1-year follow-up with memory probe vaccination *J. Infect. Dis.* **220** 603–14

[226] Vinjé J, Green J, Lewis D C, Gallimore C I, Brown D W G and Koopmans M P G 2000 Genetic polymorphism across regions of the three open reading frames of 'Norwalk-like viruses *Arch. Virol* **145** 223–41

[227] Okafor J I, Testrake D, Mushinsky H R and Yangco B G 1984 A *Basidiobolus* sp. and its association with reptiles and amphibians in southern Florida *Med. Mycol* **22** 47–51

[228] Zahari P, Hirst R G, Shipton W A and Campbell R S F 1990 The origin and pathogenicity of *Basidiobolus* species in northern Australia *Med. Mycol.* **28** 461–8

[229] Gugnani H C 1999 A review of zygomycosis due to *Basidiobolus ranarum Eur. J. Epidemiol.* **15** 923–9

[230] Bigliazzi C, Poletti V, Dell' A D, Saragoni L and Colby T V 2004 Disseminated basidiobolomycosis in an immunocompetent woman *J. Clin. Microbiol.* **42** 1367–9

[231] El-Shabrawi M H F and Kamal N M 2011 Gastrointestinal basidiobolomycosis in children: an overlooked emerging infection? *J. Med. Microbiol.* **60** 871–80

[232] Shreef K, Saleem M, Saeedd M A and Eissa M 2018 Gastrointestinal basidiobolomycosis: an emerging, and a confusing, disease in children (a multicenter experience) *Eur. J. Pediatr. Surg.* **28** 194–9

[233] Anaparthy U and Deepika G 2014 A case of subcutaneous zygomycosis *Indian Dermatol. Online J.* **5** 51–4

[234] Mugerwa J W 1976 Subcutaneous phycomycosis in Uganda *Br. J. Dermatol.* **94** 539–44

[235] Singh R, Xess I, Ramavat A and Arora R 2008 Basidiobolomycosis: a rare case report *Indian J. Med. Microbiol.* **26** 265–7

[236] Rabie M E, El Hakeem I, Al-Shraim M, Al Skini M S and Jamil S 2011 Basidiobolomycosis of the colon masquerading as stenotic colon cancer *Case Rep. Surg.* **32011** 685460

[237] Mantadakis E and Samonis G 2009 Clinical presentation of zygomycosis *Clin. Microbiol. Infect.* **15(supp. 5)** 15–20

[238] Ageel H I, Arishi H M, Kamli A A, Hussein A M, Bhavanarushi S and Arishi M 2017 Unusual presentation of gastrointestinal basidiobolomycosis in a 7-year-old child—case report *Am. J. Med. Case Rep.* **5** 131–4

[239] Bittencourt A L, Londero A T, Araujo M D G S, Mendonça N and Bastos J L A 1979 Occurrence of subcutaneous zygomycosis caused by *Basidiobolus haptosporus* in Brazil *Mycopathologia* **68** 101–4

[240] Khan Z U, Khoursheed M, Makar R, Al-Waheeb S, Al-Bader I, Al-Muzaini A *et al* 2001 *Basidiobolus ranarum* as an etiologic agent of gastrointestinal zygomycosis *J. Clin. Microbiol.* **39** 2360–3

[241] Sackey A, Ghartey N and Gyasi R 2017 Subcutaneous basidiobolomycosis: a case report *Ghana Med. J.* **51** 43–6

[242] Mendiratta V, Karmakar S, Jain A and Jabeen M 2012 Severe cutaneous zygomycosis due to *Basidiobolus ranarum* in a young infant *Pediatr. Dermatol.* **29** 121–3

[243] Roberts G D 1988 Medical mycology: the pathogenic fungi and the pathogenic actino-mycetes *Mayo Clin. Proc.* **63** P1061–2

[244] de Lima Barros B, de Almeida Paes R and Schubach A O 2011 *Sporothrix schenckii* and sporotrichosis *Clin. Microbiol. Rev.* **24** 633–54

[245] Kenyon E M, Russell L H and McMurray D N 1984 Isolation of *Sporothrix schenckii* from potting soil *Mycopathologia* **87** 128

[246] Sharma N L, Sharma R C, Gupta M L, Singh P and Gupta N 1990 Sporotrichosis-study of 22 cases from Himachal Pradesh *Indian J. Dermatol. Venereol. Leprol.* **56** 296–8

[247] Mahajan V K, Sharma N L, Sharma R C, Gupta M L, Garg G and Kanga A K 2005 Cutaneous sporotrichosis in Himachal Pradesh, India *Mycoses* **48** 25–31

[248] Saravanakumar P S, Eslami P and Zar F A 1996 Lymphocutaneous sporotrichosis associated with a squirrel bite: case report and review *Clin. Infect. Dis.* **23** 647–8

[249] Fleury R N, Taborda P R, Gupta A K, Fujita M S, Rosa P S, Weckwerth A C *et al* 2001 Zoonotic sporotrichosis. Transmission to humans by infected domestic cat scratching: report of four cases in São Paulo, Brazil *Int. J. Dermatol.*

[250] Migaki G, Font R L, Kaplan W and Asper E D 1978 Sporotrichosis in a Pacific white-sided dolphin (*Lagenorhynchus obliquidens*) *Am. J. Vet. Res.* **39** 1916–9

[251] Jr V H, Miot H A, Bartoli L D, Cardoso A D C and Pires De Camargo R M 2002 Localized lymphatic sporotrichosis after fish-induced injury (*Tilapia* sp.) *Med. Mycol.* **40** 425–7

[252] Mayorga R, Cáceres A, Toriello C, Gutiérrez G, Alvarez O, Ramirez M E *et al* 1978 An endemic area of sporotrichosis in Guatemala *Sabouraudia* **16** 185–98

[253] Bargman H 1983 Sporotrichosis of the skin with spontaneous cure—report of a second case *J. Am. Acad. Dermatol.* **8** 261–2

[254] Mahajan V K 2014 Sporotrichosis: an overview and therapeutic options *Dermatol. Res. Pract.* **2014** 272376

[255] Itoh M, Okamoto S and Kariya H 1986 Survey of 200 cases of sporotrichosis *Dermatology* **172** 209–13

[256] Bonifaz A, Saúl A, Paredes-Solis V, Fierro L, Rosales A, Palacios C *et al* 2007 Sporotrichosis in childhood: clinical and therapeutic experience in 25 patients *Pediatr. Dermatol.* **24** 369–72

[257] Austin B and Austin D A 2012 *Bacterial Fish Pathogens: Disease of Farmed and Wild Fish* (Taylor & Francis)

[258] Austin B 2017 The value of cultures to modern microbiology *Anton. Leeuw. Int. J. Gen. Mol. Microbiol.* **110** 1247–56

[259] Grieshaber D, MacKenzie R, Vörös J and Reimhult E 2008 Electrochemical biosensors—sensor principles and architectures *Sensors* **8** 1400–58

[260] Yogeswaran U and Chen S M 2008 A review on the electrochemical sensors and biosensors composed of nanowires as sensing material *Sensors* **8** 290–313

[261] Windmiller J R and Wang J 2013 Wearable electrochemical sensors and biosensors: a review *Electroanalysis* **25** 29–46

[262] Paul R, Saville A C, Hansel J C, Ye Y, Ball C, Williams A *et al* 2019 Extraction of plant DNA by microneedle patch for rapid detection of plant diseases *ACS Nano* **13** 6540–9

[263] Mao K, Min X, Zhang H, Zhang K, Cao H, Guo Y *et al* 2020 Paper-based microfluidics for rapid diagnostics and drug delivery *J. Control. Release* **322** 187–99

[264] Hameed S, Xie L and Ying Y 2018 Conventional and emerging detection techniques for pathogenic bacteria in food science: a review *Trends Food Sci. Technol.* **81** 61–73

[265] Chocarro-Ruiz B, Fernández-Gavela A, Herranz S and Lechuga L M 2017 Nanophotonic label-free biosensors for environmental monitoring *Curr. Opin. Biotechnol.* **45** 175–83

[266] Mehrotra P 2016 Biosensors and their applications—a review *J. Oral Biol. Craniofac. Res.* **6** 153–9

[267] Forootan A, Sjöback R, Björkman J, Sjögreen B, Linz L and Kubista M 2017 Methods to determine limit of detection and limit of quantification in quantitative real-time PCR (qPCR) *Biomol. Detect. Quantif.* **12** 1–6

[268] Li F, Hu Y, Li Z, Liu J, Guo L and He J 2019 Three-dimensional microfluidic paper-based device for multiplexed colorimetric detection of six metal ions combined with use of a smartphone *Anal. Bioanal. Chem.* **411** 6497–508

[269] Poghossian A, Geissler H and Schöning M J 2019 Rapid methods and sensors for milk quality monitoring and spoilage detection *Biosens. Bioelectron* **140** 111272

[270] Kruk M E, Gage A D, Joseph N T, Danaei G, García-Saisó S and Salomon J A 2018 Mortality due to low-quality health systems in the universal health coverage era: a systematic analysis of amenable deaths in 137 countries *Lancet* **392** P2203–12

[271] Wang Y and Salazar J K 2016 Culture-independent rapid detection methods for bacterial pathogens and toxins in food matrices *Compr. Rev. Food Sci. Food Saf* **15** 183–205

[272] Zhang Z, Zhou J and Du X 2019 Electrochemical biosensors for detection of foodborne pathogens *Micromachines* **10** 222

[273] Xu M, Wang R and Li Y 2016 An electrochemical biosensor for rapid detection of: *E. coli* O157:H7 with highly efficient bi-functional glucose oxidase-polydopamine nanocomposites and Prussian blue modified screen-printed interdigitated electrodes *Analyst* **141** 5441–9

[274] Soukarié D, Ecochard V and Salomé L 2020 DNA-based nanobiosensors for monitoring of water quality *Int. J. Hyg. Environ. Health* **226** 113485

[275] Xu M, Wang R and Li Y 2017 Electrochemical biosensors for rapid detection of *Escherichia coli* O157:H7 *Talanta* **162** 511–22

[276] Cesewski E and Johnson B N 2020 Electrochemical biosensors for pathogen detection *Biosens. Bioelectron* **159** 112214

[277] Wan J, Ai J, Zhang Y, Geng X, Gao Q and Cheng Z 2016 Signal-off impedimetric immunosensor for the detection of *Escherichia coli* O157:H7 *Sci. Rep.* **6** 19806

[278] Formisano N, Bhalla N, Heeran M, Reyes Martinez J, Sarkar A, Laabei M *et al* 2016 Inexpensive and fast pathogenic bacteria screening using field-effect transistors *Biosens. Bioelectron.* **85** 103–9

[279] Petruccelli B P, Goldenbaum M, Scott B, Lachiver R, Kanjarpane D, Elliott E *et al* 1999 Health effects of the 1991 Kuwait oil fires: a survey of US army troops *J. Occup. Environ. Med.* **41** 433–9

[280] Aguilera F, Méndez J, Pásaroa E and Laffona B 2010 Review on the effects of exposure to spilled oils on human health *J. Appl. Toxicol* **30** 291–301

[281] Major D N and Wang H 2012 How public health impact is addressed: a retrospective view on three different oil spills *Toxicol. Environ. Chem.* **94** 442–67

[282] Tidwell L G, Allan S E, O'Connell S G, Hobbie K A, Smith B W and Anderson K A 2015 Polycyclic aromatic hydrocarbon (PAH) and oxygenated PAH (OPAH) air–water exchange during the deepwater horizon oil spill *Environ. Sci. Technol.* **49** 141–9

[283] Shafir S, Van Rijn J and Rinkevich B 2007 Short and long term toxicity of crude oil and oil dispersants to two representative coral species *Environ. Sci. Technol.* **41** 5571–4

[284] Perkol-Finkel S and Benayahu Y 2007 Differential recruitment of benthic communities on neighboring artificial and natural reefs *J. Exp. Mar. Bio. Ecol.* **340** 25–39

[285] Lin C Y and Tjeerdema R S 2008 Crude oil, oil, gasoline and petrol *Encyclopedia of Ecology, Five-Volume Set* ed B D Fath (Elsevier)

IOP Publishing

Sensors for Marine Biosciences
Next-generation sensing approaches
Shyam S Pandey, Rout George Kerry and Kshitij RB Singh

Chapter 8

Advancement of sensors in preservation and packaging of marine products

Ekta Poonia, Rohit Ranga, Heena Dahiya, Vinita Bhankar and Krishan Kumar

Modern market trends and scientific advancements, along with customer demands that are increasingly high, push manufacturers and scientists to produce novel solutions across a range of industries, including the packaging and preservation sector. The development of innovative techniques for clever preservation and packaging has made it possible to increase the quality of marine products and guarantee that eating marine food won't be harmful to consumers' health. The condition of marine food may deteriorate as a result of chemical, physical or biological reasons. The cause of the degradation could be related to storage, transportation, or other logistical and sales operations that compromise the preservation and packaging, which is the marine food items' protective aspect. Because the majority of marine food products are extremely perishable and maintaining a food product's quality is the most crucial concern along the whole supply chain, this is especially crucial. This chapter's primary goal is to give a broad overview of the use of sensors in intelligent marine product preservation and packaging, given the significance of the subject.

8.1 Introduction

The modern market's advancement and worldwide development, coupled with consumers' high standards and preferences for product quality and safety (both environmental and health-related), as well as the tighter consumer laws, have intensified efforts to develop novel preservation and packaging concepts and technologies. This makes the safety of marine products which is defined as having the right physical and sensory qualities, being microbiologically safe, having a suitable chemical composition, and having nutritional value, seem to be especially crucial. It is well known that raw materials go through a variety of processes that alter their chemical makeup and physical characteristics [1, 2].

doi:10.1088/978-0-7503-5999-3ch8

It is important to keep in mind that the consumer bases their decisions on expectations and perception from their senses. Therefore, a possibility remains that the consumer may reject the product if it does not meet their expectations in terms of freshness, smell, look, or other sensory attributes. It is well recognized that a variety of biological, chemical, or physical processes that many marine products go through can eventually cause the product to spoil [3]. Environmental factors are equally significant because contemporary customers are becoming more conscious, particularly when it comes to protecting the environment and cutting down on the waste of marine items.

The degradation of marine food quality might happen gradually during storage, transportation, or other sales and logistical activities. This degradation can be caused by microbes, chemical or physical changes in the marine food, or both. Therefore, it would seem to be of utmost importance to continuously check a few quality attributes of the product as it is being transported and stored throughout the whole production and logistics processes. Because the majority of maritime items are extremely perishable and maintaining a marine product's quality is the most crucial problem in the whole supply chain, this is especially crucial [4]. It should be highlighted that customers find it challenging to estimate the aforementioned changes, particularly when it comes to packaged marine food. Furthermore, if a product deviates even slightly from the standard, for example, in terms of color or consistency, this results in discarding the product.

Traditional techniques of preserving marine products include chilling, freezing, drying, pickling, curing, smoking, and canning. In order to reduce unfavourable changes during processing and produce nutrient-dense foods, fresh-tasting, without the use of heat or chemical preservatives, a variety of innovative processing technologies are being investigated and put into practice. Advancements in recent times have yielded better methods for product production, treatment, stuffing, preservation, and stowage.

Based on the use of heat to preserve food, thermal and nonthermal technologies are also used for the preservation of marine products. On the other hand, microwave processing, high pressure processing, ohmic heating, irradiation, high pressure freezing and thawing, pulsed electric field, pressure-shift freezing, high intensity pulsed light technology, etc are emerging technologies with broader applications in food processing in the future, as listed in figure 8.1.

Food products can be preserved using a new nonthermal cold pasteurization method called high pressuring processing, also known as ultra-high pressure (UHL) or high hydrostatic pressure (HHP). High pressure equipment, such as temperature control units, pressure generator systems, high pressure vessels and associated closures, and material handling systems, uses high level hydrostatic pressure, which ranges from 100 to 600 MPa (87 000 psi). HPP is performed at cooled or mild process temperatures (<45 °C). It has been demonstrated that the items to be preserved have negligible influence on their nutritional content, texture, flavour, or appearance. Figure 8.2 explains the advantages of HPP.

The shellfish business, which processes shrimp, oysters, crab, clams, mussels, lobsters, and other ready-to-eat seafood meals, is the primary application of HPP in

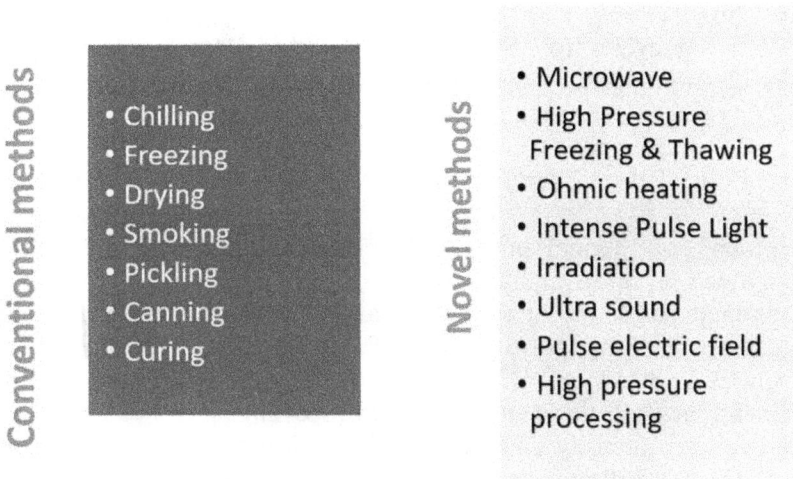

Figure 8.1. Conventional and novel methods for processing and preservation of marine products.

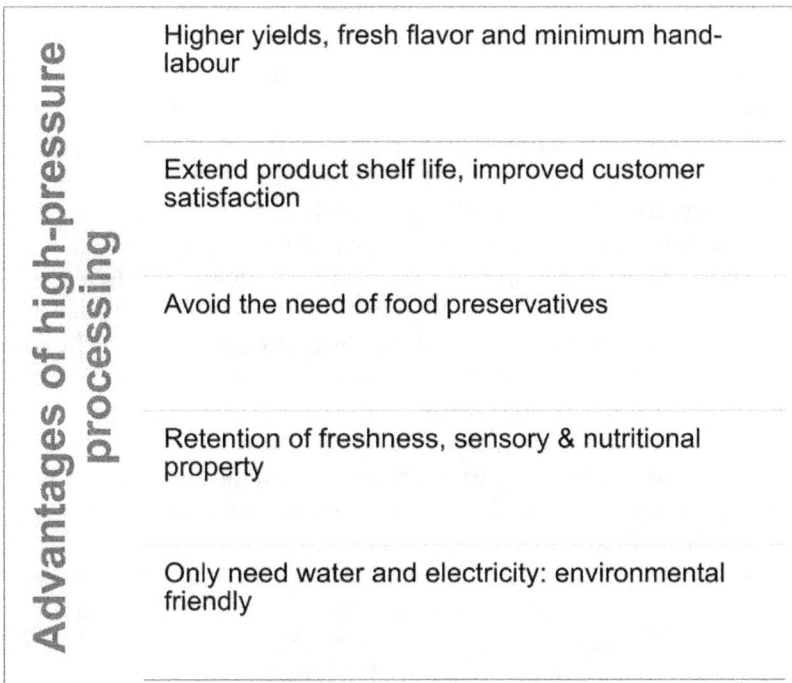

Figure 8.2. High-pressure processing (HPP) advantages.

the processing sector. Samples are cooled to −20 °C under 200 MPa in high pressure aided freezing without ice forming. Water can stay liquid under high pressures of up to 210 MPa and temperatures as low as −22 °C. This makes quick freezing and thawing possible. Salmon items are frozen using an impact pressure shift technique.

Pressure shift freezing, with pressure release in 18 min., improved reduction to 2 and 2.5 log cycles for *Listeria innocua* and *Micrococcus luteus*. Asian seabass samples that were freeze-dried quickly at temperatures between $-15\,°C$ and $-25\,°C$ revealed shrunken and gray muscle tissue along with white ice crystals, while samples that were stored at a similar temperature after pressure shift freezing revealed intact muscle tissue with no deformation and smaller, less icy crystals.

Another new non-thermal processing approach is pulsed electric field (PEF) processing, which produces high-quality meals with good sensory, nutritional, and shell life characteristics. In a treatment chamber, a high voltage pulse generator produces a high intensity pulsed electric field (HIPEF), with voltages typically ranging from 30 to 80 $kV\,cm^{-1}$. The fixed pulse frequency time ranges from 10 to 10 000 μs, contingent on the items requiring processing.

Both the voltage intensity and frequency duration are maintained by a control mechanism. This procedure is used to prepare marine items such as marinated fish fillets, dried and salted marine products, and fresh fish fillets. Inactivating pathogenic strains of microbes including *Clostridium welchii*, *Bacillus cereus*, *Escherichia coli*, *Staphylococcus aureus*, and others is possible with a high voltage electric field (up to 25 kV). PEF demonstrated enhanced water binding and better brine microdiffusion in marinades as a result of the interaction between phosphates, salt, and protein.

Foods can be preserved without heat treatment thanks to a novel technological idea called high intensity pulsed light technology (PLT). This makes use of brief, strong bursts of broad spectrum 'white light,' such as pulsed UV-light, which has an intensity 20 000 times greater than that of sunshine at sea level and lasts only a fraction of a second. PLT has been demonstrated to lower levels of spoilage and pathogenic bacteria and serves as a sterilizing technique for pharmaceuticals, packaging materials, and products with uniform surfaces. This technique aids in preserving the nutritional and sensory qualities while lowering the requirement for chemical preservatives and disinfectants.

UV, visible, and infrared (IR) light (200–1000 nm) are the primary light sources used. PLT's efficient penetration to food depths extends the shelf life of mildly preserved, fish products (cold-smoked). Additionally, this technique is employed as a means of managing microbes on food surfaces.

A relatively new technique for preserving food goods is ultrasound assisted technology, which helps cut down on the amount of water needed for processing, thawing, and rehydration times. Here, mechanical waves of a specific frequency are used in ultrasound technology; the frequency range is fixed based on the energy output. This method is frequently used to suppress a variety of microorganisms, ensure effective pasteurization, and preserve flavour and color.

Utilizing technology, dried marine goods are being used to revive fish and echinoderms, such as sea cucumbers. When ultrasonic power was increased from 100 to 300 W @ (28, 35, and 45 kHz) up to 12 times, water holding capacity and rehydration ratio were shown to increase. Due to the need to protect product quality, technology is developing quickly in the EU and the USA. It has been shown to be very advantageous for businesses that process seafood, as it increases production capacity by roughly 35%, earnings by 50%, and margin by 2%.

Ultrasound has a variety of chemical and biological effects, such as sterilization of processing equipment, enhanced bactericidal action, changes in enzyme activity, etc. Likewise, mechanical effects also apply to homogenization, tenderization, and flavour extraction.

Ohmic heating is the process of directly heating food through resistance while an electric current passes through it, producing heat internally. Ohmic heating aids in the quick generation of heat during the cooking process, which speeds up the cooking of the product. This technique is also used to defrost frozen fisheries products more quickly, which lowers the growth of microorganisms, stops nutrients from leaching out, and conserves more water. In addition, ohmic heating is utilized as a substitute for vacuum evaporation to achieve homogeneous heating through electroconductive blanching. Ohmic cooking is used to prepare surimi items from Pacific whiting, catfish, pollock, etc, and it has greatly increased the gel qualities and color of these goods.

A relatively new method called 'food irradiation' uses x-rays, electron beam and gamma rays to treat food. It is mostly employed on fisheries goods such as raw, cooked or ready-to-cook mussels and oysters, frozen, partially cooked, cooked, shelled, or dried crustaceans, etc. Not only can radiation destroy bacteria and insects without producing heat in crabs, but it also significantly extends the shelf life of meat when it comes to harmful strains of *E. coli*, *Vibrio*, and *Listeria* that do not form spores.

Radiation-treated samples also showed improvements in the sanitary quality of frozen goods for export, including finfish, fillets, cuttlefish, squid, frozen shrimp, and other marine products. It has been shown that the product quality of chilled freshwater and marine fisheries goods increases two to three times, whereas radiation processes completely eliminate insect eggs and larvae in dried products. To keep seafood and other food goods from spoiling, proper packaging is crucial.

Consequently, producers of packaging have started working on developing what is known as 'sustainable packaging,' which is a product that is safe to use for the duration of its life cycle and is produced, recycled, or biodegraded in a way that is good for the environment. To be more precise, packaging should promote sustainable marine product management, which addresses the issue of marine food waste by maintaining food items' quality and safety (minimizing the usage of artificial additives and preventing food-borne illnesses) [5, 6]. To help accomplish this, sophisticated packaging systems, such as intelligent packaging, are used. The capacity to track and display product condition without needing to adhere to approximate expiration dates sets them apart from conventional packaging. It is a really useful option when it comes to food, medications, or hazardous materials.

The notion of intelligent packaging has been devised in relation to the marine industry with the aim of mitigating the loss of marine product samples utilized for chemical and microbiological investigations, as well as expanding the scope of analysis [7]. Conversely, deconsumption (i.e., balancing quality over quantity

Table 8.1. Intelligent packaging classification based on mode of operation.

Component	Sensor	Indicators
Base	Fluorescence-based oxygen sensor, gas sensor, biosensors	Time–temperature indicators, gas leakage concentration indicators, microbial growth and freshness indicators

consumed), changing lifestyles, the ongoing pursuit of better living standards, and the desire to preserve the environment are the driving forces behind the increasing significance of intelligent packaging in consumer choices. The physical character-istics and enhanced aspect of food packaging confinement are expectations that impact product sales and consumer sentiments, as demonstrated by Tu *et al*'s 2015 research [8]. Some even go so far as to say that the development of 'smart' packaging is a direct result of intelligent technology becoming prevalent in most areas of our lives: today, the majority of the products in our environment such as refrigerators, phones, televisions, cars, and so forth are smart or soon will be. On the other hand, the other kind of intelligent packaging is distinguished by the inclusion of sensors or indications that enable the package to monitor its surroundings as explained in table 8.1 [7]. The following are the primary categories of tools used in intelligent packaging.

8.1.1 Sensors

1. Gas sensors: Instruments that track alterations in the gases' composition within the package.
2. Fluorescence oxygen sensors: Identifying food deterioration and evaluating product freshness.
3. Biosensors.

8.1.2 Indicators

Interactive indicators can be found inside or outside of the packaging, and they respond to changes in the internal environment by disclosing dye dispersion or altering in color and intensity. In this approach, they provide the consumer with semi-qualitative or quantitative information on the freshness, quality and other aspects of the packed product [9, 10].

1. Devices that take the shape of different kinds of labels or stickers and are affixed to the exterior of group or individual packaging, capturing even brief temperature variations in the surroundings (an increase or fall in temperature that is higher or lower than what is suitable for a specific product) while food is being stored, transported, and distributed. Time-temperature indicators (TTIs) can be designed to influence biological activities (enzymes, spores, enzymes) or chemical reactions (polymerization, melting etc) or physical changes. These kinds of indications are usually utilized in the packing of

semi-finished goods meant to be prepared in ovens or microwaves, as well as in the case of chilled and frozen food [9, 11].

2. Concentration/gas leakage indicators: Devices that track the amount of CO_2 in packaging can adjust how tight it is without sacrificing the integrity of the material. The gas leak could foster the growth of microorganisms in the packaging. Microbial growth indicators that interact with metabolites produced by microorganisms to aid in the observation of potential microbial growth within the packed product.

3. Freshness indicators: Assessing freshness through color changes or variations in color. They offer precise details on the quality of the product as a result of chemical alterations or microbial development in the food item. The interactions between the metabolites of microbial growth, such as organic acids, biogenic ammines, glucose, sulfur compounds, CO_2, volatile nitrogen, and the indicators included inside the package can be used to assess the microbiological growth [12, 13]. But all of these packaging techniques aim to give the business and the client the greatest experience possible and complete control over the product's quality, regardless of the technology employed.

Since the sensors can track a wide variety of indicators, both consumers and businesses may use this crucial information to make well-informed decisions. It is important to note that this type of packaging has the potential to be a solution in the industrial internet of things space since it can utilize software and sensors to communicate data about an article's state with other devices connected to a network.

8.2 Intelligent packaging based on sensors

A sensor is a device that sends out signals to measure chemical or physical parameters after detecting, locating, or identifying events or changes in the nearby environment. Typically, sensors consist of a transmitter that transforms the energy into an optical, thermal, electrical, analytical and chemical signal and a receptor that transforms chemical or physical information into a form of energy [14, 15]. Conversely, smart or intelligent sensors include specialized signal processing features that allow for more design freedom when creating sensor devices and the addition of additional detection capabilities [16]. The current state of science has led to the discovery of numerous structures that have significant potential for application in intelligent packaging and are necessary for sensors to work. Chemical sensor-based devices function by converting chemical information from the surrounding environment into a useful measurement signal. One feature that sets chemical sensors apart is their ability to selectively capture an ion or molecule. The real analyte is adsorbed in the recognition layer, forming the analytical signal. When analyte detection and energy conversion are combined, the receptor's pH, light level, temperature or redox potential change. Furthermore, a reversible chemical interaction between the chosen analyte and chemical sensors allows for the qualitative and quantitative detection of the analyte. They are therefore the greatest substitute for labor-intensive analytical tools such as gas chromatography–mass spectrometry (GC–MS), which can only be employed at the expense of the food package's integrity [13]. The sensor action has two common chemical components. One of these, the receptor (the sensing portion), is responsible for precisely detecting a particular component in the region where a chemical reaction occurs, and the other is the converter (the transduction portion). The converter facilitates the transformation of one energy type into another and the transfer of information into an analytical signal. A signal that appears as a heat release, fluorescence, color shift or shift in the crystal oscillator's frequency is produced by the developing chemical reaction [13, 17]. The authors Hanrahan *et al* [18] propose that the optimal sensor should possess five key characteristics: (1) sensitivity to variations in target species concentrations; (2) specificity or selectivity to the target species; (3) fast reaction time; (4) extended shelf life, spanning several months; and (5) small size, allowing for low-cost production. Different parameters are tested by different kinds of sensors. In figure 8.1, one of their divisions is displayed. Furthermore, in order to guarantee the maximum level of food safety, particular consideration should be given to sensors that measure bacteria, pH changes, humidity, time, temperature, and specific chemicals like total volatile basic nitrogen (TVB-N). Figure 8.3 lists the many kinds of sensors that are employed in smart packaging [14]. Furthermore, in order to guarantee the maximum level of food safety, particular consideration should be given to sensors that measure bacteria, pH changes, humidity, time, temperature, and specific chemicals like TVB-N. One important factor indicating the food's quality and edibility is its physical and nutritional state, which is represented by the indicators [19].

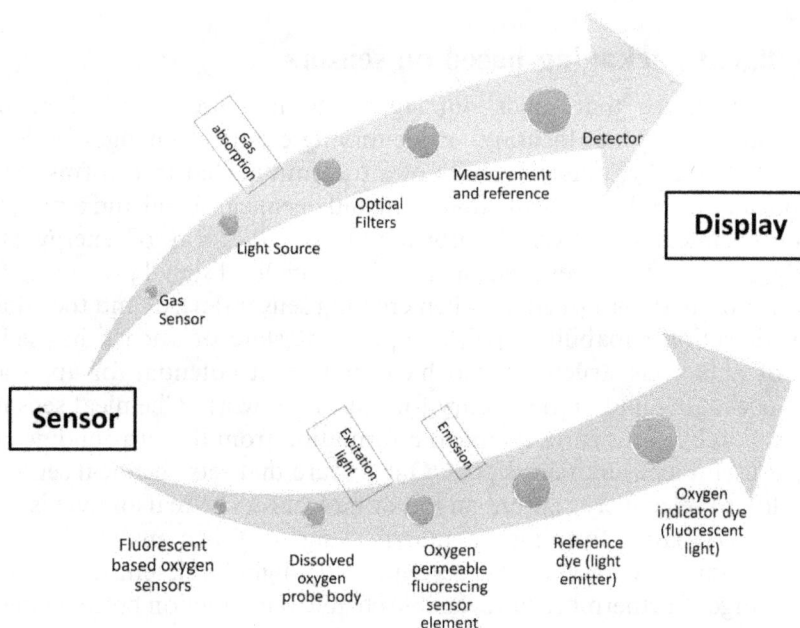

Figure 8.3. Various sensors used for intelligent packaging.

The following describes a number of advanced packaging metrics that are used to assess if a product has maintained its level of safety and quality attributes.

8.2.1 Oxygen- and carbon dioxide-based sensors

Gas sensors are most widely used solutions in the field of sensor-based intelligent packaging systems, as food product deterioration is detected by the release of gases like H_2S or CO_2. Gas sensors track alterations in a package brought on by a variety of external inputs, such as weather. These instruments can frequently ascertain the potency of the active components called O_2 and CO_2 scavengers in the container [20]. Gas sensors exhibit quantitative responses. Changes in the physical properties of the sensor can indicate the presence of gas because of their transient nature [10]. Because carbon dioxide has an antibacterial effect on specific germs, it is frequently utilized in packaging that has a changed atmosphere. Utilizing such a solution contributes to the product's longer shelf life [21]. It is also important to remember that some substances dissolve CO_2 because they are more permeable to it than to oxygen. The carbon-dioxide generator, which helps to regulate the concentration of this molecule and stops the package from deforming, is therefore a crucial tool to stop these undesirable effects [22]. Non-dispersive infrared technology is the foundation of CO_2 sensors. But they can also be sensors that work on the basis of chemical changes. Crucially, non-dispersive infrared sensors are spectroscopic sensors since they measure the CO_2 level by measuring the gas's absorption at a certain wavelength. Solid or polymeric electrolytes are used by chemical sensors to measure CO_2 levels. On the other hand, O_2 is detected by infrared sensors and devices that rely on electrochemical, laser, or ultrasonic transformations [23, 24]. Redox-based colorimetric indicators are the quickest and most straightforward approach to see changes in packaging. These indicators are made of potassium or sodium hydroxide compounds. Utilizing a colorimetric indicator that combines methylene blue with a reducing agent (such as glucose), it is possible to identify the presence of O_2, which is the cause of food degradation caused by microorganisms [25, 26]. Organic conductive polymers, semiconductor metal oxide transistors, piezoelectric crystal sensors, amperometric O_2 sensors and potentiometric CO_2 sensors are examples of gas sensors [12]. But keep in mind that the reducing agent is reactive to O_2, thus these indicators need to be maintained in anaerobic environments to avoid unfavorable reactions that could lead to the indicator malfunctioning. Another significant issue would be the dubious safety of a food item contained in a container with this kind of sensor. The unintentional contamination of food with a hazardous synthetic sensor component needs special consideration [27].

8.2.2 Sensors based on pH changes and specific chemicals

The growth of pathogenic bacteria that cause food deterioration is linked to pH changes and the creation of particular chemical compounds that are frequently toxic to humans. The method of regulating pH variations and the synthesis of particular chemical compounds is the same as that of O_2 and CO_2, i.e., colorimetric testing. These tests are typically integrated within the package to enable the user to quickly and easily see the changes. Kuswandi *et al*'s polyaniline foil indicators for perishable fish appear to be an intriguing approach [28]. Scientists have produced packaging that reacts with color when food quality deteriorates and amine chemicals start to develop. Electrochemical sensors, whose operation is dependent on the redox--reaction on the electrode surface, are widely employed to detect changes in pH. It emits an electrical signal with the ability to identify volatile amine chemicals that are created when food oxidizes. Meat and fish products, in particular, benefit greatly from this [29, 30]. Bhadra *et al*'s construction of a passive pH sensor wrapped in a unique hydrogel [31] made it easier to measure the concentration of volatile substances. The entire apparatus is based on the resonance frequency effect, which varies with the number of volatile amines present in a certain environment. By adjusting the impedance of the outer coil connected to the sensor, any variations in the aforementioned parameter can be found. The quickness and sensitivity of decay process detection is one of these sensors' primary benefits. In addition, these sensors offer crucial information about the concentration of the hazardous substance being monitored. They are also more sophisticated than traditional colorimetric testing. Since this technique is far more complex than color measurement, some drawbacks are already apparent when it is used in packaging systems [13].

8.2.3 Sensors based on humidity

Food moisture can frequently pose a serious threat to maintaining food safety and quality along the whole supply chain. Due to their high-water activity, products including meat, fish, fruit, and vegetables frequently lose moisture and become dry [32]. However, low water activity items typically take in moisture from the air, which degrades their quality, including their organoleptic qualities [30]. Therefore, it is best to adjust the package's humidity level in real time. This parameter is now being monitored using a variety of techniques, including as measurements of changes in crystal resonant frequency, capacitance, and dielectric characteristics. Intelligent packaging that senses humidity frequently uses protein, which is sensitive to humidity. Nanocomposite foil packaging is one such example, which has a zinc oxide layer that is meant to detect the signal in addition to a layer based on gelatin because of its dielectric qualities [29, 33]. Wheat gluten was utilized by Bibi *et al* [34] to identify variations in package relative humidity. Due to gluten's dielectric qualities, a method of operation that is extremely similar to that of the nano-composite foil was employed. Additionally, sensors inserted inside the materials used to produce the package are frequently used in intelligent packaging. The biodegradable package's interlayer-electrode (IDE) serves as an induction layer. The capacitance between electrodes is defined by the atmospheric relative humidity function. It should be noted that as moisture from the atmosphere is absorbed, the dielectric constant increases [35]. In addition, the increased humidity inside the container causes the water vapor to be absorbed by a specially developed layer, which is often inductive. The capacitor's changed capacitance and resonant frequency are the end products of this operation. This kind of sensor is made of a mixture of starch and polypropylene (PP), polylactic acid (PLA), polyvinyl alcohol, glossy photo paper and planarized polyethylene terephthalate (PET) [36]. There are several uses for intelligent packaging, but it is particularly useful in the case of products that are prone to moisture. They are particularly helpful for dried goods like cereals, flakes, and powders, as well as sweet and salty snacks, as the water activity of these goods is significantly higher and could cause the quality attributes to degrade.

8.2.4 Sensors based on time and temperature (TTI)

Two crucial elements that affect food quality, safety, and freshness are temperature and time. The majority of food items have recommended use-by dates and storage temperatures (room temperature or refrigerator). Determining whether a particular product has been stored correctly is likewise quite challenging [37]. Since temperature has a major influence on almost all quality metrics, it is critical to monitor it and take appropriate action when a crisis arises. Throughout the production chain, time–temperature sensors enable real-time batch control of consumer goods [38].

Time–temperature sensors, for instance, are available for purchase from the US-based company Evigence Sensors Inc. The company makes sure that their TTIs are inexpensive and activated automatically, therefore, that they can be used on production lines; additionally, they feature a digital and visual indicator that makes it easier to read (e.g., with a monitoring device such as smartphone); and lastly, but perhaps most importantly, they are not degradable when exposed to UV or sun light. The TTI sensor's operating approach is predicated on reactions occurring between two or more substrates, wherein the dominant reaction gives rise to a dominating hue in the indicator [39]. These days, TTI can be categorized into multiple types based on the type of response.

Among these are chemical TTIs, which use high temperatures to polymerize 1,4-additive monomers. Polydiacetylene compounds (PDAs) are created in this manner. It is crucial that this reaction be irreversible and that its rate increase in tandem with the ensuing rise in temperature. Furthermore, the idea of thermally induced decay in the reverse reaction of photoluminescent compounds serves as their foundation for photoluminescence. Increases in temperature and time will affect the reaction process, which will affect how much the indication fades and becomes visible on the packaging [38]. Physical TTIs: electrons, diffusion, and nanoparticles form the basis of their working principle. According to the temperature of the reaction or the transition from solid to liquid, relevant changes are generated. The indicator's color changes as a result of these dependencies. The surface of metal NPs (Ag/Au NPs) is altered by heat-absorbing nanoparticles with thermochromic characteristics; this causes the wavenumber to shift to the visible surface and alters color [40]. Enzymatic TTIs rely on the indicator's substrate's hydrolysis reaction with the enzyme to produce the desired hue. The enzymatic reaction intensifies with time and temperature, revealing a hue that corresponds to the degree of exposure to these two variables. Using these indications, different food kinds can be managed based on the chosen substrates, activators, inhibitors, and/or enzymes [41]. Microbial TTIs modify the indicator element based on metabolites generated by microorganisms under suitable environmental conditions (temperature and time). These days, the most common microorganisms employed for anaerobic acid generation are yeast or lactic acid bacteria. The indicator changes color as a result [38]. An increasing corpus of research on TTI indicators made of materials that generate themselves is available. Choi *et al* [42], for instance, used aromatic disulfide as a basis for thermoplastic polyurethane (TPU) to generate an elastic fabric. When temperature changes and time passes, this opaque nanofiber turns transparent. The integrated monitor also makes temperature adjustment easier. This material also resists mechanical effects quite well.

8.2.5 Characteristics of optical sensors and others

The enormous potential that optical sensors have for application in food packaging is making them a more and more common study topic. It has been demonstrated that these sensors are less expensive than others. Most importantly, they can operate properly without the need for extra devices or electronic connections. Also, inexperienced operators can use them. Their basic working principle is derived from the following systems: plasmonic particles and photonic structures, fluorogenic receptors, or just a colorimetric test [43]. The sensors that Baleizao *et al* [44] have already discussed are a few examples of those that are employed in intelligent packaging systems. In order to detect oxygen at various temperatures, the study team created a dual optical sensor that is extremely sensitive. It is made up of two light-emitting substances, one of which is capable of detecting oxygen and the other of which can measure temperature. The scientists' findings verified that the dual sensor could identify variations in temperature within the range of 0 °C–120 °C. New concepts for this kind of sensor, which may be applied to food product packaging, were developed in the ensuing years. An optochemical sensor, for instance, was created by Borchert, Kerry, and Papkovsky to measure CO_2 in food kept in packaging with a changed atmosphere. It was discovered that the sensor could detect CO_2 at 4 °C for nearly three weeks [45]. Heising *et al* have created a sensor that uses volatile molecules emitted from the fish during storage to determine the freshness of packaged cod fillets [46]. Moreover, luminescence sensors are available that measure the chemiluminescence or fluorescence signals that are released once the analyte has been kept in the proper carrier. Then solid phase luminescence (SPL) or its counterpart (SML) starts to react. This reaction facilitates the identification of the intended food ingredients and confirms the presence of impurities in a particular product [47]. The most promising indicators are those that can detect changes in pH, such as acid or alkaline gasses, brought on by the deterioration of food products [48]. Fiber optic O_2 sensors based on fluorescence should also be taken into account. They are also useful for measuring liquids and gases, do not require oxygen, and are immune to electromagnetic interference, in contrast to traditional gas sensors. O_2xyDot, manufactured by OxySense in Delware, USA, is the most widely used fluorescence-based oxygen sensor available. Its primary benefit is that it may be read without damaging it by reading it from the outside of the packaging [9]. It should be noted, therefore, that the application of this technology is limited to transparent packaging only. The great measurement sensitivity and precision of the defined sensors are their key features. Furthermore, they offer numerical data regarding the package's temperature, humidity, and gas concentration. The sensors can also be connected to other intellectual property systems, like radio frequency identification (RFID). As a result, the user is provided with more precise information [2].

8.3 Conclusion and prospects

Modern methods of preservation and packaging help to increase the safety of marine goods and prolong the shelf life of packaged foods. But in the seafood industry, technology is changing, and many of these systems are still in the early stages of development. Further advancements in the quality, safety, and stability of marine food are anticipated as a result of ongoing developments in active and sophisticated preservation and packaging.

Manufacturers and researchers are interested in intelligent packaging solutions because of the many advantages they provide. Intelligent packaging not only offers consumer convenience and greater food safety and quality, but it also gives management over the entire manufacturing and logistical chain. Though there are still some gaps in the research, it appears that optometric intelligent packaging for food will be a viable venture in the future.

As a result, these would be a standard component of food packaging. It is necessary to convey both the technological and economic aspects, as they are closely linked to the viability of different innovative systems. From the standpoint of the manufacturers implementing these technologies widely, this is crucial. With the help of businesses, research facilities, and governmental organizations, this is possible. Researchers, producers, marketers, and consumers can all benefit from the thorough explanation of the applied technologies in this chapter as they work to customize integrated packaging systems for specific uses. Remarkably, the food business currently uses intelligent packaging the most, with other industries using it less frequently. It is worthwhile to take into account additional life sciences sectors, such as the pharmaceutical or cosmetics industries, where intelligent packaging may show to be beneficial. Cosmetics and medications should have the same level of safety as goods meant for ingestion. Barcodes would enhance the traceability or temperature control of such products, just like in the food business. As previously stated, concerns regarding the cost of intelligently packed goods must be taken into consideration. Such products' greater pricing may successfully turn away potential customers. It makes sense that customers will not want to pay extra for their preferred goods just because they arrive in different packaging. Therefore, it is important to share the benefits of these platforms. It is crucial to establish consumer trust about the safety of intelligent packaging. It is necessary to promote and educate the public about such systems extensively. Finally, because food products may be controlled in real-time, using intelligent packaging can provide a major competitive edge. Producers and consumers will benefit from intelligent packaging in a number of ways if all efforts are made to bring such solutions to market. Given the many advantages intelligent packaging can provide, it is crucial to keep researching this field.

References

[1] Wikstrom F, Verghese K, Auras R, Olsson A, Williams H, Wever R, Gronman K, Kvalvag Pettersen M, Moller H and Soukka R 2019 Packaging strategies that save food: a research agenda for 2030 *J. Ind. Ecol.* **23** 532–40

[2] Fuertes G, Soto I, Carrasco R, Vargas M, Sabattin J and Lagos C 2016 Intelligent packaging systems: sensors and nanosensors to monitor food quality and safety *J. Sens.* **2016** 4046061

[3] Fung F, Wang H S and Menon S 2018 Food safety in the 21st century *Biomed. J.* **41** 88–95

[4] Lorite G S, Selkala T, Sipola T, Palenzuela J, Jubete E, Vinuales A, Cabanero G, Grande H J, Tuominen J and Uusitalo S 2017 Novel, smart and RFID assisted critical temperature indicator for supply chain monitoring *J. Food Eng.* **193** 20–8

[5] Tichoniuk M, Bieganska M and Cierpiszewski R 2021 Intelligent packaging: sustainable food processing and engineering challenges *Sustainable Food Processing and Engineering Challenges* (Cambridge, MA: Academic) pp 279–313

[6] Molina-Besch K and Olsson A 2022 Innovations in food packaging-sustainability challenges and future scenarios *Future Foods* (Cambridge, MA: Academic) pp 375–92

[7] Bagchi A and He T 2012 Intelligent sensing and packaging of foods for enhancement of shelf life: concepts and applications *Int. J. Sci. Eng. Res.* **3** 1–13

[8] Tu Y, Yang Z and Ma C 2015 Touching tastes: the haptic perception transfer of liquid food packaging materials *Food Qual. Prefer.* **39** 124–30

[9] Mohebi E and Marquez L 2015 Intelligent packaging in meat industry: an overview of existing solutions *J. Food Sci. Technol.* **52** 3947–64

[10] Kuswandi B, Wicaksono Y, Abdullah A, Heng L Y and Ahmad M 2011 Smart packaging: sensors for monitoring of food quality and safety *Sens. Instrum. Food Qual. Saf.* **5** 137–46

[11] Fang Z, Zhao Y, Warner R D and Johnson S K 2017 Active and intelligent packaging in meat industry *Trends Food Sci. Technol.* **61** 60–71

[12] Chowdhury E U and Morey A 2019 Intelligent packaging for poultry industry *J. Appl. Poult. Res.* **28** 791–800

[13] Ghaani M, Cozzolino C A, Castelli G and Farris S 2016 An overview of the intelligent packaging technologies in the food sector *Trends Food Sci. Technol.* **51** 1–11

[14] Lee S Y, Lee S J, Choi D S and Hur S J 2015 Current topics in active and intelligent food packaging for preservation of fresh foods *J. Sci. Food Agric.* **95** 2799–810

[15] Mlalila N, Kadam D M, Swai H and Hilonga A 2016 Transformation of food packaging from passive to innovative via nanotechnology: concepts and critiques *J. Food Sci. Technol.* **53** 3395–407

[16] Yamasaki H 1996 What are the intelligent sensors *Handbook of Sensors and Actuators* **3** (Amsterdam: Elsevier) pp 1–17

[17] Cattrall R W 1997 *Chemical Sensors* (Oxford: Oxford University Press)

[18] Hanrahan G, Patil D G and Wang J 2004 Electrochemical sensors for environmental monitoring: design, development and applications *J. Environ. Monit.* **6** 657–64

[19] Yousefi H, Su H M, Imani S M, Alkhaldi K, Filipe M C D and Didar T F 2019 Intelligent food packaging:a review of smart sensing technologies for monitoring food quality *ACS Sens.* **4** 808–21

[20] Pavelkova A 2013 Time temperature indicators as devices intelligent packaging *Acta Univ. Agric. Silvic. Mendel. Brun.* **61** 245–51

[21] Drago E, Campardelli R, Pettinato M and Perego P 2020 Innovations in smart packaging concepts for food: an extensive review *Foods* **9** 1628

[22] Han J W, Ruiz Garcia L, Qian J P and Yang X T 2018 Food packaging: a comprehensive review and future trends *Compr. Rev. Food Sci. Food Saf.* **17** 860–77

[23] Pandey S K and Kim K H 2007 The relative performance of NDIR-based sensors in the near real-time analysis of CO_2 in air *Sensors* **7** 1683–96

[24] Park Y W, Kim S M, Lee J Y and Jang W 2015 Application of biosensors in smart packaging *Mol. Cell. Toxicol.* **11** 277–85

[25] Putra B T W and Kuswandi B 2022 Smart food sensing and IoT technologies *Bio- and Nano-Sensing Technologies for Food Processing and Packaging* (London: Royal Society of Chemistry) pp 129–50

[26] Won S and Won K 2021 Self-powered flexible oxygen sensors for intelligent food packaging *Food Packag. Shelf Life* **29** 100713

[27] Mirza Alizadeh A, Masoomian M, Shakooie M, Zabihzadeh Khajavi M and Farhoodi M 2021 Trends and applications of intelligent packaging in dairy products: a review *Crit. Rev. Food Sci. Nutr.* **62** 383–97

[28] Kuswandi B, Jayus Restyana A, Abdullah A, Heng L Y and Ahmad M 2012 A novel colorimetric food package label for fish spoilage based on polyaniline film *Food Control* **25** 184

[29] Pereira P F, Picciani P H, Calado V and Tonon R V 2020 Gelatin-based nanobiocomposite films as sensitive layers for monitoring relative humidity in food packaging *Food Bioprocess Technol.* **13** 1063–73

[30] Gaikwad K K, Singh S and Ajji A 2019 Moisture absorbers for food packaging applications *Environ. Chem. Lett.* **17** 609–28

[31] Bhadra S, Narvaez C, Thomson D J and Bridges G E 2015 Non-destructive detection of fish spoilage using a wireless basic volatile sensor *Talanta* **134** 718–23

[32] Das R, Bej S, Hirani H and Banerjee P 2021 Trace-level humidity sensing from commercial organic solvents and food products by an AIE/ESIPT-triggered piezochromic luminogen and ppb-level 'OFF–ON–OFF' sensing of Cu^{2+}: a combined experimental and theoretical outcome *ACS Omega* **6** 14104–21

[33] Tan E L, Ng W N, Shao R, Pereles B D and Ong K G 2007 A wireless, passive sensor for quantifying packaged food quality *Sensors* **7** 1747–56

[34] Bibi F, Guillaume C, Vena A, Gontard N and Sorli B 2016 Wheat gluten, a bio-polymer layer to monitor relative humidity in food packaging: electric and dielectric characterization *Sens. Actuators* A **247** 355–67

[35] Wawrzynek E, Baumbauer C and Arias A C 2021 Characterization and comparison of biodegradable printed capacitive humidity sensors *Sensors* **21** 6557

[36] Raju R and Bridges G E 2021 Radar cross section-based chipless tag with built-in reference for relative humidity monitoring of packaged food commodities *IEEE Sens. J.* **21** 18773–80

[37] Cheng H, Xu H, McClements D J, Chen L, Jiao A, Tian Y, Miao M and Jin Z 2022 Recent advances in intelligent food packaging materials: principles, preparation and applications *Food Chem.* **375** 131738

[38] Wang S, Liu X, Yang M, Zhang Y, Xiang K and Tang R 2015 Review of time temperature indicators as quality monitors in food packaging *Packag. Technol. Sci.* **28** 839–67

[39] Firouz M S, Mohi-Alden K and Omid M 2021 A critical review on intelligent and active packaging in the food industry: research and development *Food Res. Int.* **141** 110113

[40] Lim S, Gunasekaran S and Imm J Y 2012 Gelatin-templated gold nanoparticles as novel time-temperature indicator *J. Food Sci.* **77** N45–9

[41] Siddiqui J, Taheri M, Alam A U and Deen M J 2022 Nanomaterials in smart packaging applications: a review *Small* **18** 2101171

[42] Choi S, Eom Y, Kim S M, Jeong D W, Han J, Koo J M, Hwang S Y, Park J and Oh D X 2020 A self-healing nanofiber-based self-responsive time-temperature indicator for securing a cold-supply chain *Adv. Mater.* **32** 1907064

[43] Dodero A, Escher A, Bertucci S, Castellano M and Lova P 2021 Intelligent packaging for real-time monitoring of food-quality: current and future developments *Appl. Sci.* **11** 3532

[44] Baleizao C, Nagl S, Schaferling M, Berberan Santos M N and Wolfbeis O S 2008 Dual fluorescence sensor for trace oxygen and temperature with unmatched range and sensitivity *Anal. Chem.* **80** 6449–57

[45] Borchert N B, Kerry J P and Papkovsky D B 2013 A CO_2 sensor based on PT-porphyrin dye and fret scheme for food packaging applications *Sens. Actuators* B **176** 157–65

[46] Heising J K, Dekker M, Bartels P V and Van Boekel M 2014 Monitoring the quality of perishable foods: opportunities for intelligent packaging *Crit. Rev. Food Sci. Nutr.* **54** 645–54

[47] Ibanez G A and Escandar G M 2011 Luminescence sensors applied to water analysis of organic pollutants-an update *Sensors* **11** 11081–102

[48] Dalmoro V, Dos Santos J H Z, Pires M, Simanke A, Baldino G B and Oliveira L 2017 Encapsulation of sensors for intelligent packaging *Food Packaging* (Amsterdam: Elsevier) pp 111–45

IOP Publishing

Sensors for Marine Biosciences
Next-generation sensing approaches
Shyam S Pandey, Rout George Kerry and Kshitij RB Singh

Chapter 9

Nucleic acid-based biosensor for detection of infectious pathogens in marine products

Kanishk Singh, Parshant Kumar Sharma, Yi-Hsiang Huang, Sucharita Khuntia, Getaneh Berie Tarekegn, Wei-Chen Huag and Li-Chia Tai

The growing demand for marine products as a key source of nutrition and economic growth in recent decades has underlined the critical need for effective pathogen detection tools to assure consumer safety. Traditional methods for detecting infectious microorganisms in marine products are sometimes time-consuming and labor-intensive, resulting in delays in identifying contaminated batches and executing essential recall actions. In order to address these challenges, nucleic acid-based biosensors have become more widely used as a promising technology for prompt and sensitive pathogen detection in marine products. This chapter of the book explores the development and use of biosensors based on nucleic acids as cutting-edge instruments for the identification of infectious microorganisms in marine products. It begins by presenting an overview of infectious pathogens prevalent in marine ecosystems and their potential threats to human health. Subsequently, it explores the principles and mechanisms behind nucleic acid-based biosensors, discussing various detection strategies such as optical, electrochemical, and piezoelectric techniques. Furthermore, the chapter discusses the challenges and advancements in designing biosensors for marine product safety. Recent uses of nucleic acid-based biosensors for identifying specific diseases in marine products are thoroughly reviewed. Finally, this comprehensive book chapter provides useful insights into the development of cutting-edge biosensing technologies that can revolutionize the detection of infectious diseases in marine products, providing a safer and more sustainable seafood business in the future.

9.1 Introduction

Foodborne ailments stand as a paramount concern for public health on a global scale. Among the diverse range of food options available, seafood has notably piqued the interest of consumers across various regions. A multitude of seafood products are prepared with minimal processing, making them conveniently ready to eat. However, due to their inherently short shelf life, traditional methods employed for detecting pathogens encounter limitations in ensuring their safety [1]. These processed seafood items have been connected to various instances of foodborne outbreaks caused by harmful microorganisms, prompting recalls of contaminated batches. These recalls cause major financial losses globally, especially in the seafood sector, in addition to serious health hazards to the general population [2]. The occurrence of such outbreaks emphasizes the pressing need for enhanced measures to safeguard the production and distribution of seafood products, ensuring their safety for consumers and preventing adverse health impacts and financial repercussions. Maintaining stringent measures and implementing effective strategies within the seafood industry is essential to uphold food quality standards. A paramount focus on seafood safety serves as a crucial deterrent against pathogen contamination and subsequent transmission to consumers. Adopting an integrated approach and utilizing various tools across the entire food chain are instrumental in achieving these safety objectives [3]. For this reason, achieving the principal objective of assuring seafood safety depends on the detection, verification, and constant surveillance of viruses inside seafood.

Foodborne pathogens encompass biological agents such as viruses, bacteria, and parasites, capable of inducing infections through contaminated food consumption. A foodborne disease epidemic is defined as the occurrence of multiple instances of the same illness following eating the same food. Initially, in 1986, pathogens were divided into categories according to the type of disease they produced. The International Commission on Microbiological Specifications for Foods, however, updated this categorization in 2000 and provided a more thorough framework. This revised categorization offers a more nuanced understanding of these pathogens and their respective impacts on health [4]. In certain instances, detecting specific foodborne pathogens can be challenging. This is where 'indicator organisms' play a crucial role. These organisms, primarily associated with intestinal sources, serve as indicators that can reveal the potential presence of challenging-to-detect foodborne pathogens. They act as markers or signals, aiding in the assessment of food safety by indicating the possible contamination or presence of harmful pathogens that might otherwise be difficult to directly detect or identify. In food safety assessments, commonly used indicator organisms include *Enterobacteriaceae, coliforms, fecal streptococci, and Escherichia coli* [5]. These microorganisms serve as reliable indicators due to their distinctive characteristics and their ability to reflect potential contamination or the presence of harmful pathogens within food samples. In essence, the use of index organisms or microbial indicators represents a valuable strategy for monitoring and ensuring food safety standards. Their adherence to specified criteria enables more effective surveillance, early detection, and mitigation of potential health risks associated with foodborne pathogens.

9.2 Critical role of marine products in foodborne poisoning

Seafood is a highly nutritious food source rich in proteins, vitamin D, selenium, iodine, and unsaturated fatty acids. Its consumption holds significant importance for health, particularly during pregnancy and the early stages of growth [6]. Additionally, seafood has proven beneficial for individuals with cardiac conditions [7]. Over the years, there has been a consistent increase in seafood consumption in the all over the world. Despite its nutritional benefits, seafood serves as a primary carrier for various bacterial diseases, contributing to food poisoning in humans. Pathogenic bacteria commonly transmitted through finfish, shellfish, and other seafood products, are responsible for causing illnesses in individuals.

Fish are susceptible to infections caused by various pathogens, including *Listeria monocytogenes, E. coli, Yersinia enterocolitica, Vibrio parahaemolyticus, Vibrio cholera, Staphylococcus aureus*, Salmonella species, and *Campylobacter jejuni* [8–10]. Coastal areas, where seafood is primarily sourced, serve as significant hubs for pathogenic microorganisms due to their proximity to densely populated regions. In seafood, the primary concerns include pathogens like Norovirus, Salmonella, and Vibrio. Secondary concerns involve Shigella, hepatitis A virus (HAV), *L. monocytogenes, Clostridium botulinum*, and microbial toxins such as *Staphylococcal enterotoxins A* (SEA), and C (SEC) [11, 12]. These agents, though termed minor, can still lead to significant adverse outcomes. These pathogens and their effects are discussed below.

9.3 Pathogens of major concern

9.3.1 Norovirus

Numerous norovirus outbreaks have been documented worldwide, stemming from the ingestion of tainted seafood. Norovirus, a highly transmissible RNA calicivirus, is responsible for infecting humans and manifests through various clinical symptoms such as headaches, fever, diarrhea, cramping in the abdomen, nausea, and vomiting. Remarkably, eating infected shellfish is linked to over half of all human norovirus cases [13]. Norovirus is the leading cause of gastroenteritis not caused by bacteria. People often contract it by eating undercooked or raw shellfish, especially oysters, or by coming into contact with contaminated water. The risk of human infection increases significantly when fish and shellfish are farmed near waters polluted with sewage [14]. While norovirus can withstand freezing, it is relatively vulnerable to disinfection with free chlorine. Its ability to resist various commercial disinfectants depends on the specific product used. Autoclaving can affect the virus's sensitivity, but it remains stable in both aquatic environments and shellfish. Even in purified shellfish, norovirus can persist for up to a week.

9.3.2 *Vibrio* spp.

Vibrio spp. is a type of Gram-negative bacteria. They have a rod-shaped and curved structure, thrive in salty environments (halophilic), can survive with or without oxygen (facultative anaerobic), do not produce spores, and can move using polar flagella and sheath. These bacteria test positive for oxidase. They are found abundantly in sediments, water, plankton, and various kinds of plants and animals in coastal locations, where they are found in estuaries and coastal regions by nature. The three main Vibrio species that are harmful to human health are *V. cholera*,

Vibrio vulnificus, and *V. parahaemolyticus* [15]. When individuals eat raw or undercooked contaminated seafood, these bacteria can cause outbreaks and occasionally food-borne disease, which presents an extreme risk to human health.

V. parahaemolyticus is a significant contributor to sporadic infections and gastroenteritis outbreaks. The pathogenicity of *V. parahaemolyticus* is often linked to the presence of two crucial virulence genes: thermostable direct hemolysin (tdh) and TDH-related hemolysin (trh). Consumption of raw or undercooked oysters can commonly lead to illness when these virulence factors are present [16].

V. vulnificus biotype 1 is known to pose a threat to humans through two main avenues: consumption of contaminated seafood and entry via open wounds. Conversely, biotypes 2 and 3 have been associated with a lower incidence of direct wound infections. The pathogenicity of *V. vulnificus* stems from multiple virulence factors, including its iron acquisition capability, the presence of a capsule in encapsulated phase variants, particular proteins found in type II and type IV pilus, zinc metalloprotease, the transmembrane regulatory protein ToxR, and hemolysin encoded by vvh (sometimes referred to as cytotoxin-hemolysin). Of note, the metalloprotease, referred to as *V. vulnificus* protease (VVP) contributes a significant role in the development of skin infection resulting from *V. vulnificus* infections. Individuals with liver diseases, hemochromatosis, or compromised immune systems are at heightened risk of experiencing severe illness due to *V. vulnificus* [17, 18].

Both *V. cholera* and *Vibrio mimicus* are closely associated, and live in freshwater and brackish habitats. Severe diarrhea, electrolyte imbalances, and dehydration are hallmarks of this bacterial infection. Their pathogenicity is associated with several virulent chemicals, the most important of which are the cholera toxin and toxins-coregulated pilus (TCP), which are known to cause epidemic cholera.

9.3.3 Salmonella

Salmonellosis is the second most common cause of food-borne disease in the USA and a worldwide health problem. Salmonellae are facultative anaerobes that are tiny, rod-shaped, Gram-negative bacteria that typically migrate with peritrichous flagella [19]. They test positive for catalase and negative for oxidase, and they produce gas from glucose. Salmonella does not ferment lactose. These bacteria generate enterotoxins, prompting an inflammatory response and leading to diarrhea. Symptoms usually manifest 12–72 h after consuming contaminated food. Acute symptoms may last for a couple of days or longer, contingent upon the specific host, the quantity ingested, and the specific bacterial strains implicated. Salmonella virulence is reliant on various determinants, including sodC1, which aids in evading macrophage attacks, mgtBC facilitating intracellular nourishment, spvB disrupting polymerization of actin, and altering vesicular trafficking by SPI-2 [20]. Raw or undercooked fish and crustaceans are the main source of salmonella illnesses from eating seafood items. Salmonella infection of seafood can happen during preparation and storage.

9.4 Pathogens of minor concern

9.4.1 Hepatitis A virus

Within the Picornaviridae family, Hepatovirus A, often referred to as hepatitis A virus (HAV), is a non-enveloped virus belonging to the genus Hepatovirus. With about 1.4 million cases reported annually, hepatitis A is considered a global epidemic. The onset of symptoms for this disease typically ranges from 2–3 weeks according to the World Health Organization, but it can extend up to 45 days [21]. While most people have moderate symptoms, children often lack symptoms at all. Nonetheless, people who are elderly or have compromised immune systems may experience more severe effects. This illness frequently presents as diarrhea, jaundice, a high body temperature stomach discomfort, and lack of appetite. Though it has a low death rate, the hepatitis A virus (HAV) is a dangerous infection. Remarkably, Asia accounts for half of the cases that were recorded.

9.4.2 Listeria monocytogenes

L. monocytogenes is, non-spore-forming, rod-shaped, Gram-positive bacteria which is classified as a facultative food-borne pathogen affecting both humans and animals. It exhibits resilience in refrigerated temperatures, low pH conditions, and high-salinity environments [22]. This bacterium is capable of producing lactic acid. Being ubiquitous, *L. monocytogenes* thrives in various environments including soil, water, and among animals and plants due to its adaptability to a broad limit of temperatures and pH levels. *Listeria* spp. is responsible for a food-borne illness known as Listeriosis, often associated with seafood consumption [23]. Consuming packaged seafood items can expose individuals to this sickness. Symptoms of Listeriosis include chills, nausea, fever, and gastroenteritis, which can progress to highly intense conditions such as meningitis, septicemia, encephalitis, abortion, and in extreme cases, leads to death. Pregnant women, immunocompromised individuals, and the elderly are particularly susceptible to contracting listeriosis. Less than one colony forming unit (CFU) per serving is present in 93% of raw seafood, according to the study.

9.4.3 *Shigella* spp.

Shigella species are non-motile, oxidase-negative rods that are Gram-negative. The four groups that these species divide into, A, B, C and D subclasses, are represented as *Shigella dysenteriae*, *Shigella flexneri*, *Shigella boydii*, and *Shigella sonnei* respectively, according to serological classification based on their somatic antigen O [24, 25]. Shigellosis can cause mild symptoms like loose, watery stools, or more severe ones like a high temperature, tenesmus, stomach discomfort, and bloody diarrhea. Serious side effects in adolescents involve toxic megacolon, Reiter's disease, and hemolytic uremic syndrome. Three toxins produced by Shigella bacteria cause bloody diarrhea in those who are afflicted. Watery diarrhea can be caused by *Shigella enterotoxins* 1 and 2. These infections have a variety of virulence characteristics that allow them to attach to and penetrate intestinal cells, tolerate stomach acidity, overcome the body's defenses, and exude poisons.

9.4.4 *Clostridium botulinum*

C. botulinum is a spore-forming, anaerobic, Gram-positive bacillus bacterium known for producing a potent neurotoxin that can contaminate food. Human botulism, caused by this bacterium, has two transmission types: through consumption of contaminated fish, via wounds, affecting infants, adults, and individuals with weak immunity. Botulism symptoms exhibit weakness in the muscles, exhaustion, double vision, trouble swallowing, and difficulty speaking. The ingestion of seafood products that have been infected with bacteria that produce neurotoxins is a common cause of this ailment. Many fish species, including, cod, flounder, whitefish, rockfish, and smoked fish, are frequently discovered to have *C. botulinum* spores and toxins [26, 27].

9.4.5 *Staphylococcus aureus*

S. aureus is a Gram-positive and catalase-positive bacterium, known for being a chemotrophic bacterium. It is recognized for causing Staphylococcal food poisoning, presenting as a food-borne intoxication syndrome. However, it is generally not considered a major concern for raw seafood [28]. The pathogenicity and virulence of *S. aureus* are attributed to *S. aureus* enterotoxins (SEs). From SEA to SEE and SEG to SEJ, nine serological kinds of SEs are all recognized for their emetic properties. Of these, food poisoning linked to Staphylococcus is known to be mostly caused by SEA.

9.5 Traditional method for detection of pathogens in marine products

For many years, conventional methods reliant on culture have been the primary approach for identifying pathogens in food products. These methods, though relatively straightforward and cost-effective, suffer from time constraints as they rely on the growth of the specific pathogen. Detecting the presence of the target pathogen alone can take anywhere from 2 to 3 days, with confirmation requiring up to 7 days. Moreover, culture-based techniques have limitations—they may fail to detect pathogens that do not grow in culture conditions, are unable to distinguish between cells that are viable for culture and cells that are viable but not culturable (VBNC), and cannot differentiate among various strains [29]. This approach's drawbacks underscore the need for alternative methods that offer faster and more comprehensive pathogen detection and identification in food products.

Over the last decade, a range of analytical tools has emerged for detecting chemical contaminants in seafood products, focusing on both qualitative and quantitative assessment. Numerous analytical methodologies are commonly employed in detecting contaminants in seafood. These include traditional methods like gas chromatography–mass spectrometry (GC–MS), high-performance liquid chromatography (HPLC), UV–Vis spectrophotometry (UV–Vis), inductively coupled plasma mass spectroscopy (ICP-MS), atomic absorption spectroscopy (AAS) and capillary electrophoresis (CE). More advanced tandem chromatographic techniques such as GC–MS–MS, LC–MS–MS, and GC infrared (GC-IR), as well as liquid chromatography nuclear magnetic resonance spectroscopy (LC-NMR), have also been developed for this purpose [30–32].

However, there are also significant drawbacks to these methods, including lengthy extraction processes requiring many steps, sluggish reaction times, and expensive equipment acquisition and operation expenses [33]. In addition, there are major obstacles due to the complex infrastructure needed and the requirement for specific technical competence for analysis.

9.6 Nucleic acid-based method of detection of seafood pathogens

Nucleic acid-based techniques operate by detecting particular DNA or RNA sequences within the pathogen of interest. This approach relies on the hybridization of the target nucleic acid with a synthetic oligonucleotide, such as probes or primers, designed to be complementary to the specific sequence of the target. Several bacterial pathogens, including, *S. aureus*, *L. monocytogenes*, *C. botulinum*, and *V. cholerae*, produce toxins responsible for causing marine product foodborne diseases. Nucleic acid-based techniques can be used to identify and detect the genes linked to these toxins in these infections.

Nucleic acid-based methods are adept at identifying specific genes within the target pathogens, ensuring more precise and accurate results while minimizing ambiguity or misinterpretation. The following are some of the most recent nucleic acid-based techniques:

(a) **Polymerase chain reaction (PCR):** A popular technique for amplifying a particular DNA region is PCR, allowing for the detection and analysis of targeted genes.

(b) **Multiplex polymerase chain reaction (mPCR):** mPCR amplifies multiple DNA sequences in a single reaction, enabling the simultaneous detection of various target genes from the same sample.

(c) **Quantitative polymerase chain reaction (qPCR):** Quantitative PCR, or qPCR, is a real-time technique that amplifies DNA while simultaneously determining its amount in a sample. It provides information on the initial DNA quantity, enabling accurate quantification.

(d) **Nucleic acid sequence-based amplification (NASBA):** RNA sequences are amplified precisely using the isothermal nucleic acid amplification method known as NASBA. It operates under constant temperature conditions without the need for a thermal cycler.

(e) **Loop-mediated isothermal amplification (LAMP):** Another isothermal amplification method called LAMP replicates DNA quickly and with great specificity in a constant temperature environment. It is known for its simplicity and robustness.

(f) **Microarray technology:** Microarrays enable the simultaneous detection of multiple genes or DNA sequences within a sample by immobilizing probes on a solid surface and detecting hybridization with target nucleic acids.

These cutting-edge nucleic acid-based techniques provide a variety of ways to identify and examine certain genes in target pathogens, each with benefits for application in various laboratory settings and advantages in terms of speed, accuracy, specificity, and flexibility.

9.7 Drawbacks of using traditional nucleic acid-based detection method

Nucleic acid-based methods such as real-time/quantitative PCR (qPCR), multiplex PCR (mPCR), PCR, NASBA, microarray technology, and LAMP are widely employed, offer powerful tools for detecting specific genes in pathogens, but are not without their limitations. The possibility of false positives or false negatives is one of the main disadvantages. Contamination or nonspecific amplification can produce false positives, which result in the target gene being incorrectly identified [34]. Conversely, false negatives may arise from insufficient sensitivity, causing the method to overlook the presence of the gene of interest, especially when the amount of the target gene is low or the assay is not optimized properly. Additionally, these methods often require specialized equipment, skilled personnel, and a controlled laboratory setting, making them less accessible and more expensive for resource-limited settings or smaller laboratories. The need for proper sample preparation and enrichment, including DNA or RNA extraction, can also introduce variability and potential errors, as shown in figure 9.1. Moreover, some techniques like PCR and microarrays may have limitations in multiplexing, either due to primer interactions in PCR or limited probe capacity in microarrays, restricting the simultaneous detection of numerous genes. Lastly, the development and validation of primers/probes for specific gene targets demand comprehensive knowledge of the target sequences and design considerations,

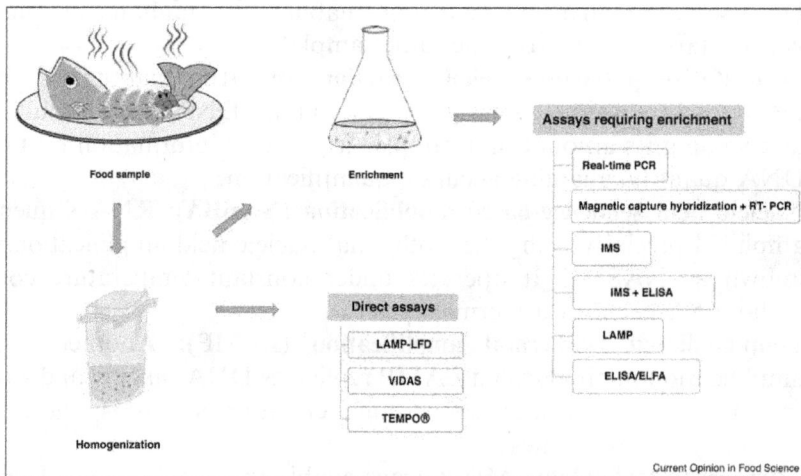

Figure 9.1. Various methods for detection of foodborne pathogens, some using enrichment to increase pathogen levels in the sample. Reprinted from [1], Copyright (2024), with permission from Elseiver.

which can be challenging and time-consuming. Overall, while these methods are highly sensitive and specific, their limitations encompass technical complexity, potential for errors, and cost, which must be considered when employing them for pathogen detection and gene identification.

9.8 Nucleic acid-based biosensor for seafood pathogen detection

A biosensor is a tool that detects and measures biomolecules. It has three main parts: a recognition element (or bio-receptor), a transducer, and a signal processor. When a sample is put into the biosensor, the recognition element grabs onto specific target molecules. The transducer then turns this interaction into signals that can be measured and analyzed by the detector [35]. This overall concept is depicted in figure 9.2. Three primary types of bio-receptors exist that are responsible for identifying the target analyte:

- Biological material: This includes antibodies, enzymes, nucleic acids, and cell-binding receptors.
- Biologically obtained material: Such as aptamer and synthetic antibodies.
- Biomimicry: This class includes artificial catalysts that replicate biological recognition mechanisms and imprint polymeric materials.

Among the different kinds of biosensors, the nucleic acid-based biosensor has gained considerable attention from the research community. Biosensors that utilize nucleic acids as their recognition component fall under the category of affinity sensors.

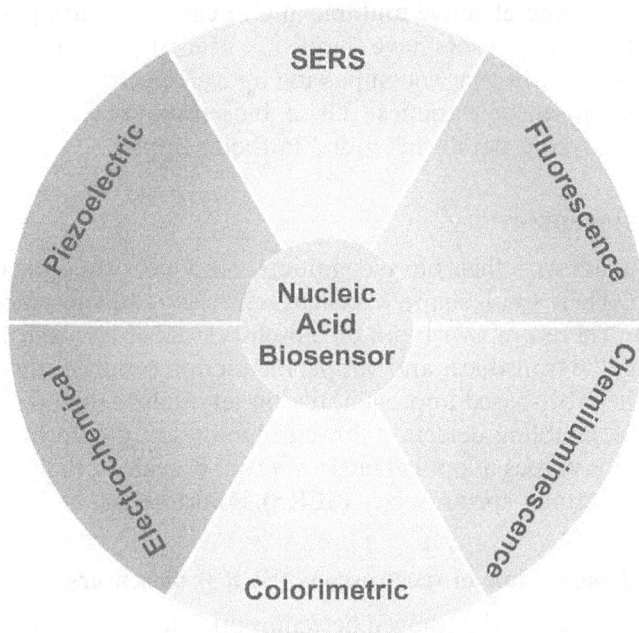

Figure 9.2. The representing set of nucleic-based biosensors based on different transducer.

These sensors detect binding reactions which lead to the physicochemical changes occurring between the nucleic acid sequence and the target bioanalyte [36]. This capacity enables nucleic acid-based biosensors to be widely employed for quantification of various targets, such as viruses and bacteria. Nucleic acid biosensors have a distinct advantage over traditional molecular biology methods and immunological methods due to their ability to bypass sample pre-enrichment requirements [37, 38]. Unlike the conventional approaches that necessitate sample pre-enrichment to concentrate pathogens before detection, nucleic acid-based biosensors do not require this preliminary step. This characteristic allows them to directly detect and analyze target pathogens without the need for prior concentration or enrichment of the sample, streamlining the detection process and potentially reducing the time and complexity involved in pathogen identification.

However, due to the susceptibility of nucleic acids to environmental factors such as temperature and pH, along with their vulnerability to cleavage by restriction enzymes, the applications of nucleic acid-based biosensors have been somewhat restricted. However, among these biosensors, those based on aptamer have garnered significant attention due to their exceptional thermal stability and low immunogenicity. Aptamers are distinct oligonucleotide sequences that are chosen from a variety of pools and have the capacity to bind to certain target biomolecules with an excellent level of affinity. The capacity of the single-stranded nucleic acids to hybridize is promoted by hydrophobic bonds, electrostatic interaction, or complementary structures. It is possible to develop the aptamer to bind with a wide range of targets, including microbes and very small compounds, thereby offering diverse applications. Additionally, their small size makes them well-suited for high-density immobilization, enabling effective multiplexing in various assays [39–41].

Nucleic acid-based biosensors have frequently been employed in recent times for detecting food-borne pathogens encompassing optical, electrochemical, mass-based, and colorimetric biosensor varieties. These biosensors play a common role in identifying and analyzing pathogens found in food samples.

9.9 Optical biosensor

Optical fields interact with their bio-recognition components in optical biosensors to enable detection. Their small shape, excellent sensitivity, and specificity make them commonly used. There are two types of biosensors: label-free and label-based. In label-free sensing, the transducer and sample interaction produces the signal directly. Conversely, in the label-based approach, the target analyte requires tagging with a reporter molecule, enabling detection through luminescent, colorimetric signals, or fluorescence. Various types of optical biosensors are prevalent, including fluorescent, surface-enhanced Raman spectroscopy (SERS), colorimetric, and fluorescence.

9.9.1 Surface-enhanced Raman spectroscopy (SERS) biosensors

SERS biosensors for seafood pathogen detection operate on the principle of leveraging nanostructured surfaces or nanoparticles to greatly amplify Raman scattering signals. These nanostructures possess unique properties that intensify Raman signals when exposed to incident light. Functionalized with specific biomolecular recognition elements—such as DNA probes or aptamer sequences, these SERS-active substrates or nanoparticles selectively bind to target seafood pathogens present in a sample. Upon interaction with laser light, the captured pathogens on the SERS-active surface generate highly amplified Raman scattering signals, producing characteristic molecular fingerprints. Analysis of these unique spectral patterns enables the identification and quantification of the pathogens within the seafood sample, facilitating rapid and precise detection. SERS-based biosensors offer exceptional sensitivity, selectivity, and the potential for multiplexed detection, presenting promising applications in ensuring seafood safety and quality through efficient pathogen monitoring. Ongoing advancements aim to further refine these biosensors, enhancing their sensitivity, selectivity, and suitability for real-time and quick detection in the seafood industry [42].

As depicted in figure 9.3(a), A SERS-based aptasensor technique for the quantitative identification of infectious microorganisms was presented by Duan *et al.* To capture the target molecules, they used a SERS substrate with Au@Ag core/shell nanoparticles (NPs) functionalized with aptamer 1 (apt 1). They used *X*-rhodamine (ROX)-modified aptamer 2 (apt 2) as the Raman reporter and recognition element. Specifically, *Salmonella typhimurium* reacted with the aptamer to create complexes that resembled sandwiches: Target-apt 1-Au@Ag-apt 2-ROX. This novel aptasensor structure facilitated the determination of *S. typhimurium* concentration, yielding a calibration curve ranging from $15–1.5 \times 10^6$ CFU ml^{-1}, with a limit of detection (LOD) of 15 CFU ml^{-1} [43]. The researchers used this method to authenticate food samples, demonstrating consistent results comparable to those obtained using traditional plate counting methods. Another study presents a highly sensitive SERS-based aptasensor for detecting *V. parahaemolyticus* in food. Fe_3O_4@Au nanoparticles enclosed with graphene oxide served as both SERS substrates and distinct tools [44]. Aptamer-coated nanostructures captured the pathogen, while a Raman reporter-modified aptamer facilitated detection. The formed sandwich complex was magnetically separated and quantified by SERS intensity. As shown in figure 9.3(b), the assay detected *V. parahaemolyticus* from 1.4×10^2 to 1.4×10^6 CFU ml^{-1}. The SERS assay's relative intensity ($I–I_0$) displayed a direct proportionality to the concentration of the target within a spectrum from $1.4 \times 10^2–1.4 \times 10^6$ CFU ml^{-1}. The process for identifying *V. parahaemolyticus* using the SERS aptasensor, which relies on SiO_2@Au surface, is delineated in figure 9.3(c). The SiO_2@Au core/shell nanoparticles were first synthesized as a substrate for SERS detection. Thiolated aptamers were then coated onto the SiO_2@Au surface via Au–S bonding. When *V. parahaemolyticus* is present, aptamer 1 on the nanoparticle surface binds to the target. Adding Cy3-labeled aptamer 2 results in the formation of SiO_2@Au-apt 1-*V. parahaemolyticus*-Cy3-apt 2 sandwiches, causing nanoparticle aggregation and enhancing Raman scattering.

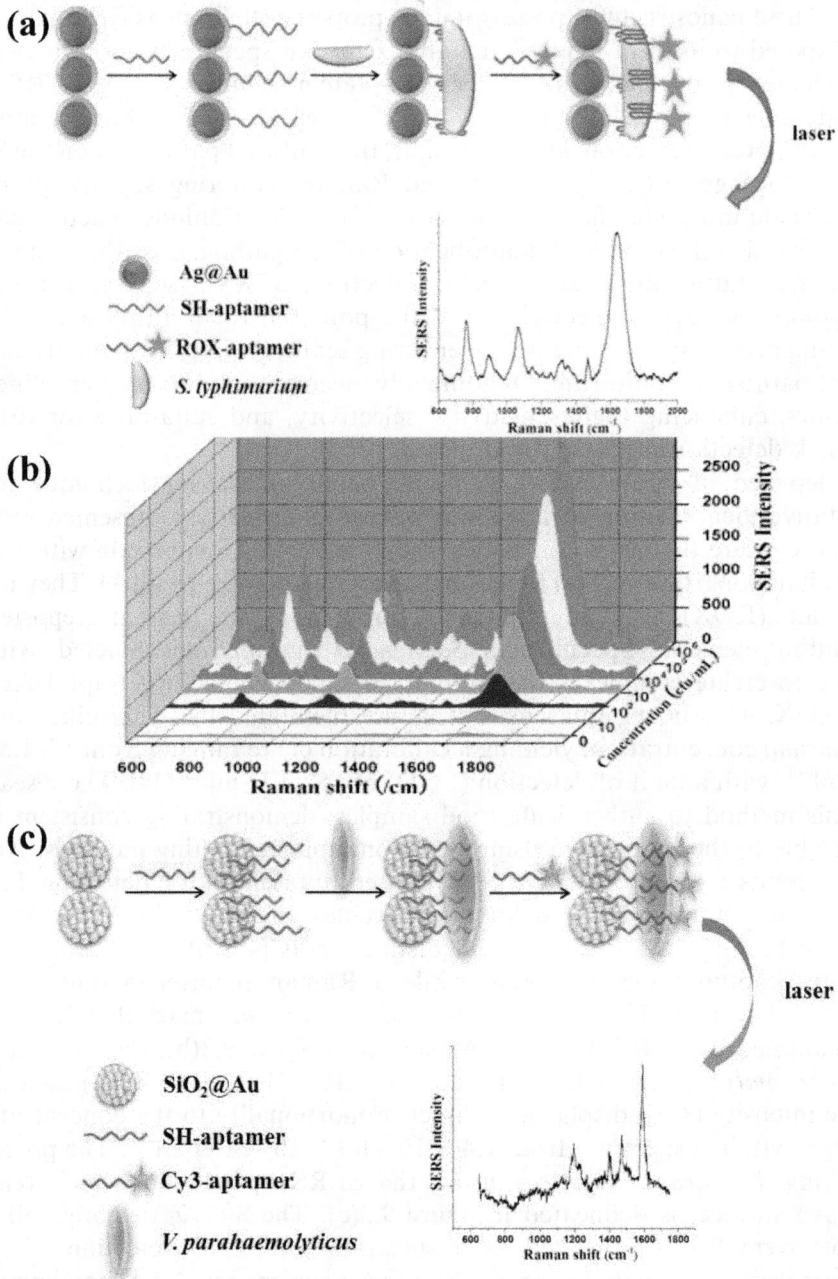

Figure 9.3. (a) SERS aptasensor for *S. typhimurium* quantification. (b) The SERS response spectra of the aptasensor were recorded across *S. typhimurium* concentrations ranging from 1.4×10^2–1.4×10^6 CFU ml^{-1}. (c) Diagrammatic representation of the SiO$_2$@Au core/shell NPs-based SERS-based aptasensor for *V. parahaemolyticus* detection. Reprinted from [43, 44], Copyright (2016), with permission from Elseiver.

9.9.2 Fluorescence-based biosensor

Fluorescence-based biosensors for seafood pathogen detection via nucleic acids function by employing specific genetic recognition principles. First, a special sequence of nucleic acids called a probe is made to exactly match the genetic material (DNA or RNA) of the pathogen that is being targeted. This probe is then labeled with a fluorescent molecule or fluorophore. When introduced to a sample suspected of containing the pathogen, the labeled probe selectively binds to the complementary sequence of the pathogen's genetic material through a process called hybridization. This binding event alters the fluorescent properties of the fluorophore, leading to changes in its emission characteristics, such as intensity or wavelength. The detection equipment, like fluorescence spectrometers or microplate readers, then captures and quantifies these changes in fluorescence, indicating the existance and sometimes the amount of the pathogen in the sample. These biosensors offer high specificity and sensitivity, enabling quick and reliable identification of seafood pathogens, important for ensuring food safety and preventing potential health risks associated with contaminated seafood consumption [45, 46].

In this context, Fan *et al* fabricated a DNAzyme-based fluorescent sensor to quantify *Vibrio vulnificus* in marine products. Following screening and mutation, the DNAzyme 'RFDVV-M2' displayed exceptional activity, specificity, and sensitivity [47]. This sensor achieved an LOD of 2.2×10^3 CFU ml^{-1} in time span of 5–10 min. The study also determined that a protein with a molecular weight of 50–100 kDa is DNAzyme RFD-VV-M2's target.

Ren *et al* developed a new aptamer-based paper sensor for the fast, accurate, lightweight, and precise detection of *V. parahaemolyticus* in seafood by utilizing the fluorescence resonance energy transfer (FRET) principle [48]. As shown in figure 9.4 efficient and straightforward FRET biosensor is used with a smartphone.

Initially, the aptamer was combined with AuNPs-attached probes, creating a ds-DNA structure. The aptamer was particularly associated with *V. parahaemolyticus* when it was present, which resulted in the release of AuNPs-probes. Subsequently, these released AuNPs attached probes were gently mixed and incubated with CDs-cDNA. The obtained mixture was then carefully applied drop by drop onto filter paper. The combination of AuNPs' large extinction coefficient and wide energy absorption band in the visible range with CDs' emission spectrum led to fluorescence quenching, which turned the vivid blue paper strip colorless. These distinctive characteristics facilitated the development of nanoprobes capable of fluorescence response to *V. parahaemolyticus*. When the integrated 365 nm UV lamp was turned on, integration with a smartphone-based fluorescent device allowed the creation of visual fluorescence pictures showing *V. parahaemolyticus* at different concentrations. These fluorescent photos were transformed into digital RGB signals using the smartphone's color selection application, allowing for a quantitative examination of *V. parahaemolyticus* using B/R values. High sensitivity, specificity, and exceptional durability against interference were demonstrated by this method, which made it easier to conduct accurate and trustworthy tests for the presence of *V. parahaemolyticus*.

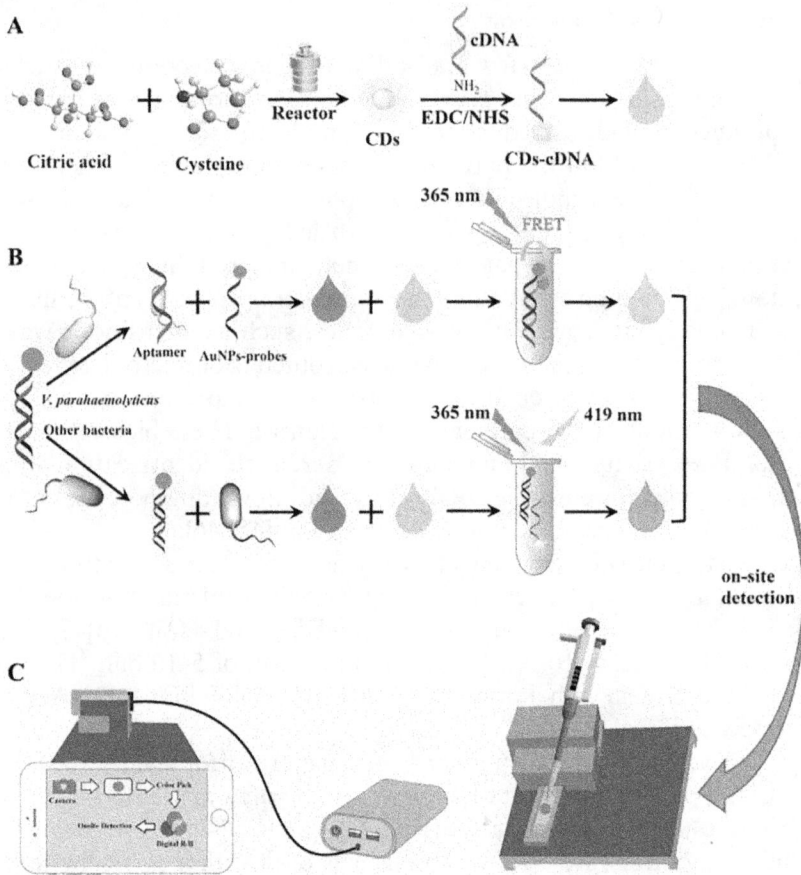

Figure 9.4. Real-time detection scheme: FRET biosensor utilizing smartphone. Reprinted from [48], Copyright (2023), with permission from Elseiver.

9.9.3 Chemiluminescence-based sensor

The principle of chemiluminescence-based sensors for seafood pathogen detection involves the use of bioluminescent or chemiluminescent reactions to identify the existance of pathogens. These sensors utilize the emission of light by living micro-organisms or the light produced during a chemical reaction to identify specific microorganisms and microbial toxins. The use of biosensors, including biolumines-cence-based biosensors, is considered a novel and rapid approach for the identi-fication of food-borne pathogens in seafood and its other associated products, providing results in less time. The specific detection principle applied in these sensors depends on light emission by the target microorganisms or the chemical reaction with the pathogens, allowing for the quick and highly sensitive identification of seafood pathogens [49].

Sun *et al* performed the development of a unique microfluidic chemiluminescence biosensor for highly sensitive and rapid detection of the food-borne pathogen *E. coli*

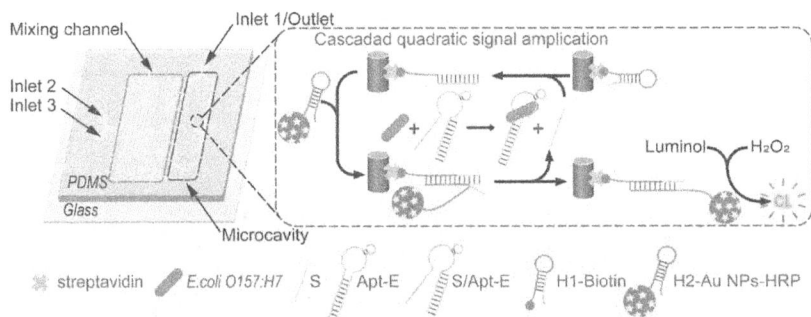

Figure 9.5. Chemiluminescence biosensor microfluidic chip with multi-signal amplification achieved through CHA and H$_2$–Au NPs-HRP for *E. coli* O157:H7 detection. Reprinted from [60], Copyright (2022), with permission from Elseiver.

O157:H7 [50]. As shown in figure 9.5, two functional units are integrated into a microfluidic chip by the biosensing platform—an efficient micromixer and a detection microchamber embedded with micropillars. The underlying sensing mechanism relies on a dual signal amplification strategy coupling catalytic hairpin assembly and horseradish peroxidase-gold nanoparticle-mediated chemiluminescence. Specifically, target bacteria competitively bind to an aptamer, releasing an initiator strand to trigger toehold-mediated enzyme-free DNA strand displacement and assembly of hairpin probes. This generates numerous labeled hairpin duplexes, immobilizing horseradish peroxidase-gold nanoparticles to catalyze the chemiluminescent reaction between luminol and hydrogen peroxide. By converting pathogen detection to a DNA-based signal readout, this bimolecular recognition approach allows sensitive quantification over a wide dynamic range. Under optimal conditions, the microfluidic biosensor can detect 130 CFU ml^{-1} of *E. coli* O157:H7 within 1.5 h using just 10 μl of sample.

9.10 Colorimetric based biosensor

Colorimetric-based biosensors are a type of biosensor that detects the presence of a seafood pathogen by measuring a color change. These biosensors use chromogenic substrates that change color in response to the availbility of the target pathogen. The catalytic interaction between the target pathogen and the chromogenic substrate is responsible for the color shift. The colorimetric biosensors are easy, quick, and inexpensive, making them ideal for on-site quantification of seafood pathogens. Numerous colorimetric biosensors have been created to identify infections found in seafood, including *S. typhimurium* and *V. parahaemolyticus* [50].

Wang *et al* devised a colorimetric detection method using metal nanoparticles and aptamer loaded into ZIF-8 via biomimetic mineralization. Their approach, outlined in figure 9.6, involves mixing Fe$_3$O$_4$-Ap@ZIF-8 and Pt-Ap@ZIF-8 with the target, forming the Fe$_3$O$_4$4-Ap@ZIF-8-VP-Pt-Ap@ZIF-8 immune complex [51]. After magnetic separation to remove unbound components, treatment with mono-acid releases Pt from Pt-Ap@ZIF-8. The addition of ABTS and H$_2$O$_2$ triggers a color

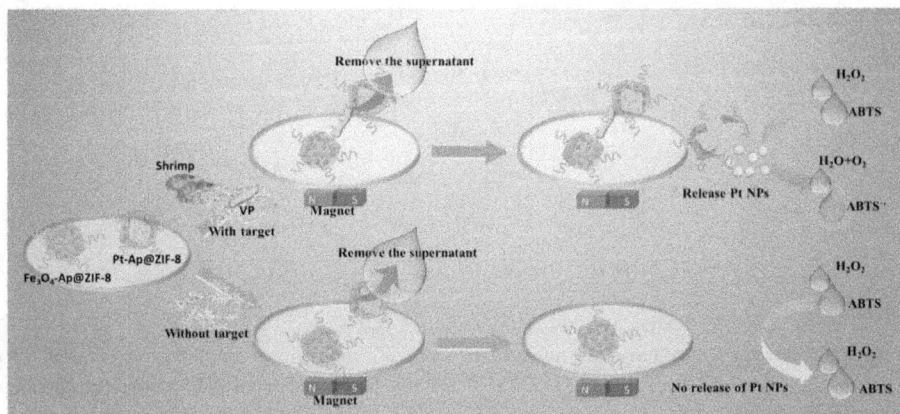

Figure 9.6. Sketch diagram depicting the visual sensing approach for VP utilizing Fe_3O_4-Ap@ZIF-8 and Pt-Ap@ZIF-8. Reprinted from [51], Copyright (2022), with permission from Elseiver.

reaction for signal output, allowing qualitative or quantitative detection of *V. parahaemolyticus* by either naked-eye observation or UV–vis spectrophotometry. This method offers a straightforward, universal approach with potential applications in rapid seafood pollutant detection, early disease screening, and environmental monitoring.

The colorimetric detection strategy for *V. parahaemolyticus* is depicted in figure 9.7. First, a specially designed forward primer includes three regions: a sequence complementary to the target DNA of *V. parahaemolyticus*, a poly A linker, and an anti-HRPzyme sequence for signal reporting. Using this forward primer alongside a conventional reverse primer, the presence of *V. parahaemolyticus* DNA leads to the production of amplified double-stranded PCR products containing the HRPzyme sequence. These products catalyze the oxidation of TMB by forming a G-quadruplex/hemin complex, resulting in a distinct blue color. The optical intensity of this reaction can be observed visually or quantified using a UV–vis spectrometer.

9.11 Electrochemical biosensor

Nucleic acid-based electrochemical biosensors are advanced tools created to identify specific DNA or RNA sequences, facilitating the detection of pathogens in seafood samples. These biosensors operate on the principle of molecular recognition and electrochemical signal transduction. They involve specialized DNA or RNA probes that are meticulously designed to hybridize exclusively to the unique genetic sequences of the target pathogen. These probes are often labeled with electroactive molecules or nanoparticles that facilitate signal generation upon binding to the complementary target sequences present in the sample. When the seafood sample containing the suspected pathogen is introduced to the biosensor, if the target genetic material is present, the probes bind to it, resulting in a change in the electrical properties of the biosensor's surface. An electrochemical signal corresponding to the pathogen concentration is produced by this binding event, which is then detected by

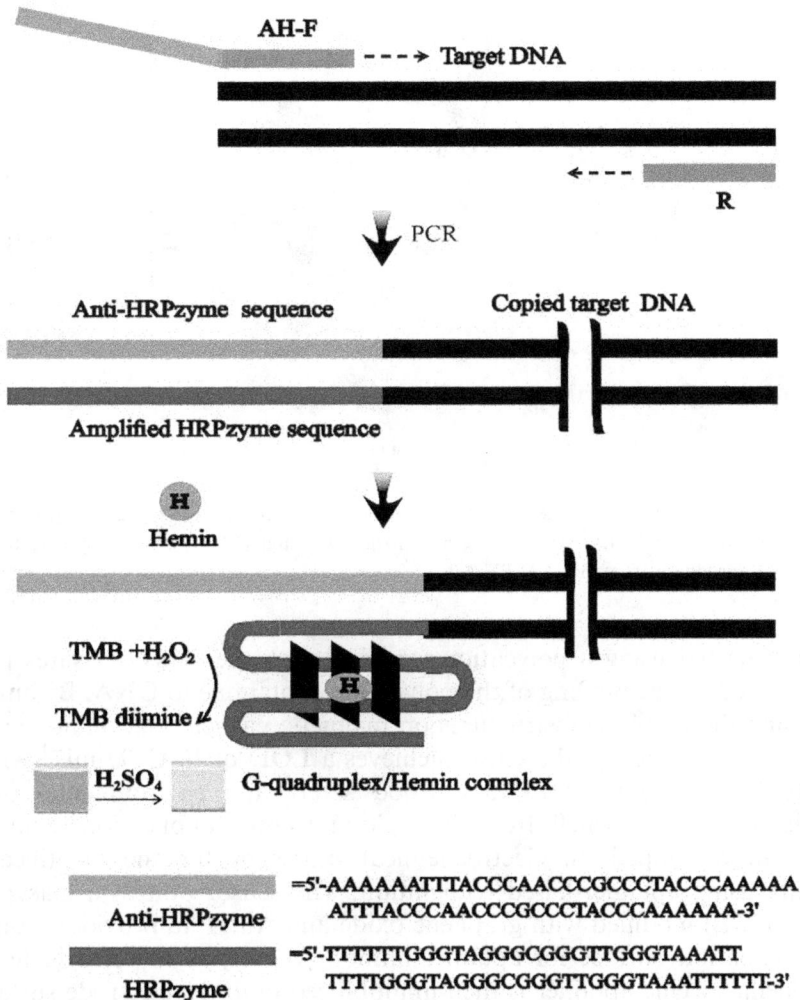

Figure 9.7. The colorimetric integrated PCR process schematic design for the instant identification of *V. parahaemolyticus* (forward primer: AH-F; reverse primer: R). Reproduced from [52]. CC BY 4.0.

an electrode or transducer within the biosensor. These biosensors offer high specificity, sensitivity, and potential for rapid detection, making them valuable tools for ensuring seafood safety [53].

Zhao *et al* present a new potentiometric aptasensor for detecting the food-borne bacterium *Vibrio alginolyticus*, shown in figure 9.8. The sensor uses DNA nano-structures containing aptamers as the recognition element, attached to magnetic beads for separation and concentration. Specifically, the DNA nanostructures are formed on magnetic microparticles modified with capture DNA strands. The aptamer DNA in the nanostructure specifically binds *V. alginolyticus* cells. This causes disassembly of the nanostructures, altering the DNA concentration on the beads. The beads are magnetically separated, and then the DNA concentration

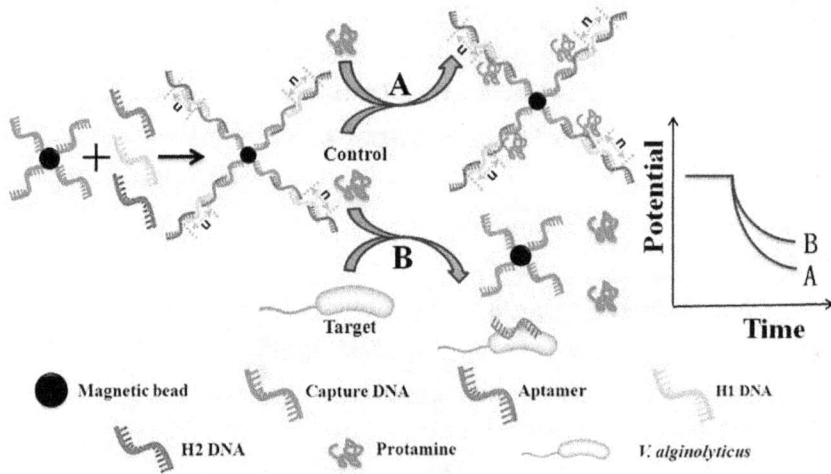

Figure 9.8. Schematic depiction of potentiometric aptasensor for *V. alginolyticus* using DNA nanostructure-altered magnetic beads, with (A) representing a control assay and (B) depicting a specific target assay configuration. Reproduced from [54]. CC BY 4.0.

change is detected using a polycation-sensitive electrode. This measures potential changes caused by the binding of the polycation protamine to DNA. By amplifying signals using the DNA nanostructures on magnetic particles and removing matrix interferences magnetically, the sensor achieves a LOD of 10 CFU ml^{-1} with good selectivity for *V. alginolyticus* [54]. Food safety and environmental pathogen monitoring could both benefit from the rapid and simple application of this sensor.

Ma *et al* developed an electrochemical biosensor for the identification of Salmonella using aptamer-based recognition. The sensor utilizes a glassy carbon electrode (GCE) modified with graphene oxide and AuNP to provide a conductive and biocompatible surface for aptamer immobilization, as depicted in figure 9.9. A Salmonella-specific aptamer is then immobilized onto the electrode surface. The binding of Salmonella cells to the aptamer causes changes in the electrochemical parameters between the electrode and the electrolyte solution. These changes are measured by electrochemical impedance spectroscopy (EIS), allowing quantification of the Salmonella concentration. They found that as more Salmonella binds to the aptamer on the electrode surface, there is an increment in the charge transfer resistance (impedance also denoted as R_{et}). This provides the basis for detection and quantification. In the region of 2.4–2400 CFU ml^{-1}, a strong linear association between impedance and Salmonella concentration was found. The biosensor demonstrated excellent sensitivity, with a LOD down to ~3 CFU ml^{-1}. It also showed good specificity for Salmonella compared to other bacteria [55].

Roig *et al* developed an impedimetric biosensor specifically designed for the detection of the marine pathogen *V. vulnificus* [56]. The commercial microelectrodes functionalized with species-specific single-stranded DNA fragments targeting the vvha gene. The biosensor's specificity and sensitivity were tested using both synthetic

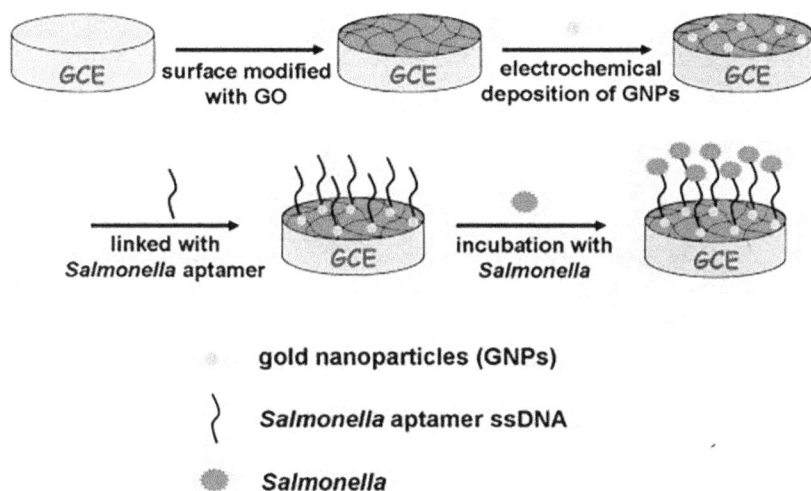

Figure 9.9. Schematic depiction of aptamer-mediated electrochemical salmonella detection. Reprinted from [55], Copyright (2014), with permission from Elseiver.

and natural DNA samples. The study demonstrated that the biosensor could accurately detect *V. vulnificus* in water samples, shown in figure 9.10(a).

Synthetic DNA samples in buffer were incubated to evaluate the biosensing response of the device before testing real samples. A notable decrease in impedance was observed after incubating with 1 pM, shown in figure 9.10(b). The frequency scan revealed that maximum sensitivity occurred at the lowest frequencies, which corresponds to the proximity of the DNA immobilized on the electrode surface. The impedance values consistently declined with higher DNA concentrations, reaching saturation before nanomolar levels, with 1 nM corresponding to 6 pg μl^{-1} of the target sequence. Given the significant signal observed with 1 pM, it is anticipated that even lower concentrations could be detected.

9.12 Piezoelectric biosensor

Piezoelectric-based biosensors function by detecting changes in mass that occur due to biomolecular interactions, such as those between an antibody and an antigen. These interactions cause a shift in the frequency of a piezoelectric crystal, which the sensor then measures. This change in frequency directly correlates with the mass change, allowing for the detection and quantification of the biomolecular event. The mechanism of detection entails the binding of a piezoelectric sensor with certain antibodies that resemble ligands, and then immersing the sensor into a sample solution that contains the target pathogen [57]. The binding of the target pathogen to the immobilized antibodies on the surface of the piezoelectric sensor causes a change in the mass of the sensor, which results in alteration in the frequency of the crystal. This alteration in frequency is then measured and used to determine the presence and concentration of the target pathogen. The principle of the piezoelectric immunosensors is depicted in figure 9.11.

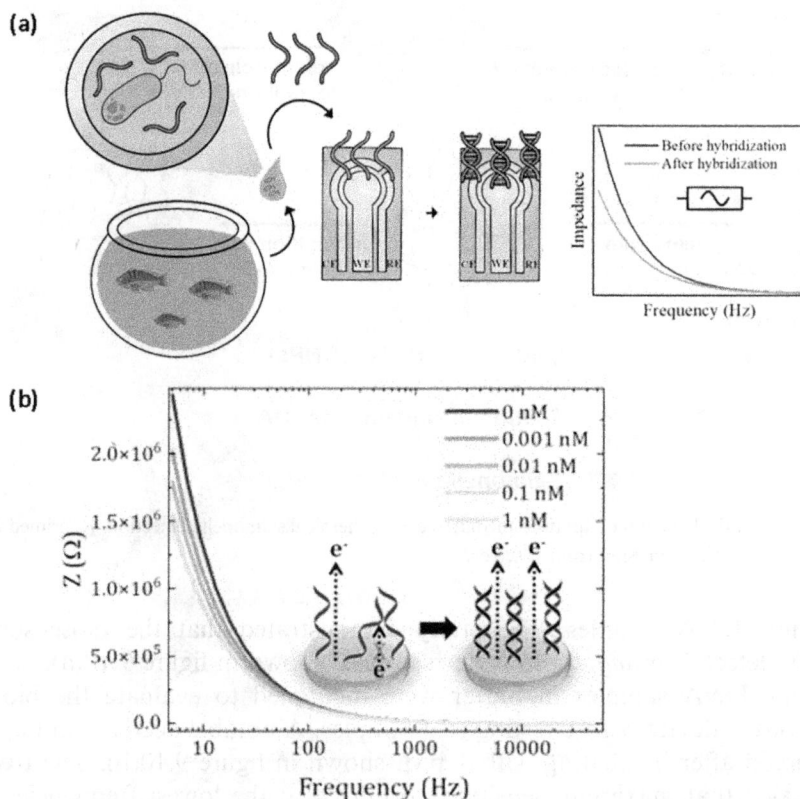

Figure 9.10. (a) Gold interdigitated ring array microelectrodes for detecting *V. vulnificus* by measuring changes in impedance. (b) Bode plot of raw impedance measurements with complementary DNA samples across concentrations ranging from 1 pM to 1 nM. Reprinted from [56], Copyright (2024), with permission from Elseiver.

Shi *et al* report a new method for detecting *P. aeruginosa* bacteria using a combination of aptamer recognition and piezoelectric sensing. The sensor complex consists of a magnetic bead functionalized with a *P. aeruginosa* aptamer and a partially complementary polyadenylated DNA strand. In the presence of target bacteria, the aptamer preferentially binds to *P. aeruginosa*, releasing the polyadenylated DNA into the solution. The released DNA can then readily adsorb onto a gold interdigital electrode owing to the strong affinity between adenine bases and gold. The binding of the polyadenylated DNA causes a sensitive frequency shift in a multichannel piezoelectric quartz crystal system connected to the electrode. This frequency change provides the analytical signal to detect the bacteria. Using this approach, the researchers achieved detection limits down to ∼9 CFU ml^{-1} in buffer solution [58]. Furthermore, the method is rapid, cost-effective, and label-free. By exploiting aptamer recognition and polyA-gold affinity, this novel biosensor enables selective detection of medically relevant bacteria.

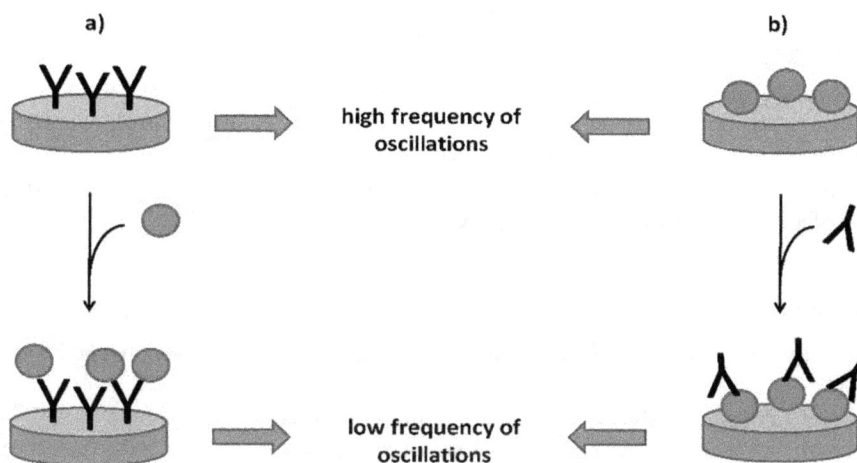

Figure 9.11. Piezoelectric immunosensors for detecting either an antigen (a) or an antibody (b). The piezoelectric crystal is represented as a blue disk, while the antibodies are represented as Y-shaped structures resembling common immunoglobulins. The antigen is illustrated as a red ball. Reproduced from [57]. CC BY 4.0.

Lian *et al* report the fabrication of a novel biosensor for specifically detecting the bacterium *S. aureus* [59]. The sensor utilizes *S. aureus* aptamer as the recognition element, immobilized on a graphene film chemically linked to gold interdigital electrodes. This electrode assembly connects to a piezoelectric quartz crystal system to transduce binding events into sensitive frequency change measurements. The aptamer self-assembles onto the conductive graphene layer through π–π stacking interactions. When target *S. aureus* cells are introduced, they bind to the aptamer, releasing them from the graphene surface. The electrode-graphene contact experiences a change in electrical characteristics as a result, which the piezoelectric sensor detects as a frequency shift response. A key innovation is the use of a diazonium-based cross-linked to strongly attach graphene onto the gold electrodes, enabling stable aptamer immobilization and sensor performance. The sensor can selectively detect *S. aureus* in 60 min over a linear range of 4.1×10^1 to 4.1×10^5 CFU ml^{-1}, down to a limit of 41 CFU ml^{-1}. They demonstrated its application in spiked milk samples, showing comparable results to conventional plating methods.

9.13 Conclusion and prospects

Beyond the limitations of traditional approaches, nucleic acid-based biosensors provide a potential alternative for the sensitive and quick detection of infectious microorganisms in marine food items. By leveraging the molecular recognition of nucleic acids coupled with various transduction mechanisms like optical, electrochemical, piezoelectric, and colorimetric techniques, these biosensors enable specific identification and quantification of seafood pathogens. While challenges remain in improving sensitivity, selectivity, and robustness for real-world applications, the prospects are exciting. These include integration of nanotechnology, multiplexed pathogen detection, development of portable and microfluidic devices, incorporation of machine learning and big data analytics, engineering of high-affinity aptamers, and efforts towards regulatory acceptance and standardization. As research advances, nucleic acid-based biosensors are poised to revolutionize seafood safety monitoring, protecting public health and contributing to the sustainable growth of the marine products industry.

References

[1] Bavisetty S C B, Vu H T K, Benjakul S and Vongkamjan K 2018 *Curr. Opin. Food Sci.* **20** 92–9

[2] Sarno E, Pezzutto D, Rossi M, Liébana E and Rizzi V 2021 *J. Food Prot.* **84** 2059–70

[3] Elbashir S, Parveen S, Schwarz J L, Rippen T E, Jahncke M L and DePaola A 2018 *Food Microbiol.* **70** 85–93

[4] Afreen M and Ucak I 2021 *EJAR* **5** 44–58

[5] Richiardi L, Pignata C, Fea E, Bonetta S and Carraro E 2023 *Water* **15** 2964

[6] Emmett R, Akkersdyk S, Yeatman H and Meyer B J 2013 *Nutrients* **5** 1098–109

[7] Jamioł-Milc D, Biernawska J, Liput M, Stachowska L and Domiszewski Z 2021 *Nutrients* **13** 1422

[8] Novoslavskij A, Terentjeva M, Eizenberga I, Valciņa O, Bartkevičs V and Bērziņš A 2015 *Ann. Microbiol.* **66** 1–15

[9] Ma B, Li J, Chen K, Yu X, Sun C and Zhang M 2020 *Foods* **9** 278

[10] Parlapani F F 2021 *Curr. Opin. Food Sci.* **37** 45–51

[11] Li P, Feng X, Chen B, Wang X, Liang Z and Wang L 2022 *Foods* **11** 3909

[12] Desdouits M, Reynaud Y, Philippe C and Guyader F S L 2023 *Microorganisms* **11** 2218

[13] Gyawali P, Fletcher G C, McCoubrey D-J and Hewitt J 2019 *Food Control* **99** 171–9

[14] Campos C and Lees D N 2014 *Appl. Environ. Microbiol.* **80** 3552–61

[15] Bonnin-Jusserand M, Copin S, Bris C L, Brauge T, Gay M, Brisabois A, Grard T and Midelet G 2017 *Crit. Rev. Food Sci. Nutr.* **59** 597–610

[16] Ndraha N, Huang L, Wu V C H and Hsiao H 2022 *Curr. Opin. Food Sci.* **48** 100927

[17] Baker-Austin C and Oliver J D 2017 *Environ. Microbiol.* **20** 423–30

[18] D'Souza C, Prithvisagar K S, Deekshit V K, Karunasagar I and Kumar B K 2020 *Microorganisms* **8** 999

[19] Iwamoto M, Ayers T, Mahon B E and Swerdlow D L 2010 *Clin. Microbiol. Rev.* **23** 399–411

[20] Kumar R, Datta T K and Lalitha K V 2015 *BMC Microbiol.* **15** 254

[21] Hennechart-Collette C, Dehan O, Fraisse A, Martin-Latil S and Pérelle S 2023 *Microorganisms* **11** 624

[22] Tsai Y-H *et al* 2022 *Microbiol. Spectr.* **10** 1–10

[23] Jami M, Ghanbari M, Zunabovic M, Domig K J and Kneifel W 2014 *Compr. Rev. Food Sci. Food Saf.* **13** 798–813

[24] Obaidat M M and Salman A E B 2017 *J. Food Prot.* **80** 414–9

[25] Ahmadi H, Anany H, Walkling-Ribeiro M and Griffiths M W 2015 *Food Bioprocess Technol.* **8** 1160–7

[26] Hamad G M, Ombarak R, Eskander M, Mehany T, Anees F, Elfayoumy R A, Omar S A, Roohinejad S and Abou-Alella S A-E 2022 *LWT—Food Sci. Technol.* **163** 113603

[27] Hamad G M, Hafez E E, Sobhy S, Mehany T, Elfayoumy R A, Elghazaly E M, Eskander M, Tawfik R, Hussein S M and Pereira L 2023 *Foods* **12** 1466

[28] Sivaraman G K, Gupta S S, Visnuvinayagam S, Muthulakshmi T, Elangovan R, Vivekanandan P, Balasubramanium G, Lodha T and Yadav A K 2022 *BMC Microbiol.* **22** 233

[29] Nnachi R C, Sui N, Ke B, Luo Z, Bhalla N, He D and Yang Z *Environ. Int.* **166** 107357

[30] Sarkar D J *et al Biosens. Bioelectron.* **219** 114771

[31] Peng X, Jiang L, Gong Y, Hu X, Peng L and Feng Y 2015 *Talanta* **132** 118–25

[32] Dasenaki M E and Thomaidis N S 2010 *Anal. Chim. Acta* **672** 93–102

[33] Bostan H B, Danesh N M, Karimi G, Ramezani M, Shaegh S A M, Youssefi K, Charbgoo F, Abnous K and Taghdisi S M 2017 *Biosens. Bioelectron.* **98** 168–79

[34] Ndraha N, Lin H-Y, Wang C-Y, Hsiao H and Lin H 2023 *Food Chem.* **7** 100183

[35] Naresh V and Lee J Y 2021 *Sensors* **21** 1109

[36] Du Y 2016 *Anal. Chem.* **89** 189–215

[37] Kim J and Oh S 2021 *Food Control* **121** 107575

[38] Deshmukh R, Joshi K, Bhand S and Roy U 2016 *Microbiol. Open* **5** 901–22

[39] Zhang Z, Yu Y, Wang M, Li J, Zhang Z, Liu J, Wu X, Lu A, Zhang G and Zhang B-T 2017 *Int. J. Mol. Sci.* **18** 2142

[40] Ning Y, Hu J and Lu F 2020 *Biomed. Pharmacother.* **132** 110902

[41] Kadam U S and Hong J C 2022 *Trends Environ. Anal. Chem.* **36** e00184

[42] Jia N, Xiong Y, Wang Y, Lu S, Zhang R, Kang Y and Du Y 2021 *J. Raman Spectrosc.* **53** 211–21

[43] Duan N, Chang B, Zhang H, Wang Z and Wu S 2016 *Int. J. Food Microbiol.* **218** 38–43

[44] Duan N, Yan Y, Wu S and Wang Z 2016 *Food Control* **63** 122–7

[45] Kakkar S, Gupta P, Kumar N and Kant K 2023 *Biosensors* **13** 249

[46] Nishi K, Isobe S-I, Zhu Y and Kiyama R 2015 *Sensors* **15** 25831–67

[47] Fan S, Ma C, Tian X, Ma X, Qin M, Wu H, Tian X, Lu J, Lyu M and Wang S 2021 *Front. Microbiol.* **12** 655845

[48] Ren Y, Cao L, Zhang X, Jiao R, Ou D, Wang Y, Zhang D, Shen Y, Ling N and Ye Y 2023 *Food Control* **145** 109412

[49] Ali A A, Altemimi A B, Alhelfi N and Ibrahim S A 2020 *Biosensors* **10** 58

[50] Kim D-M and Yoo S-M 2022 *Biosensors* **12** 532

[51] Wang K, Du L, Zhang L, Liu X, Yang T and Zeng H 2022 *Sens. Actuators* B **372** 132695

[52] Cheng K, Pan D, Teng J, Yao L, Ye Y, Xue F, Xia F and Chen W 2016 *Sensors* **16** 1600

[53] Thapa K, Liu W and Wang R 2021 *WIREs Nanomed. Nanobiotechnol.* **14** e1765

[54] Zhao G, Ding J, Han Y, Yin T and Qin W 2016 *Sensors* **16** 2052

[55] Ma X, Jiang Y, Jia F, Yu Y, Chen J and Wang Z 2014 *J. Microbiol. Methods* **98** 94–8

[56] Roig A P, Ibarlucea B, Amaro C and Cuniberti G 2024 *Biosens. Bioelectron.* X **100454**

[57] Pohanka M 2018 *Materials* **11** 448

[58] Shi X, Zhang J and He F 2019 *Biosens. Bioelectron.* **132** 224–9

[59] Lian Y, He F, Wang H and Tong F 2015 *Biosens. Bioelectron.* **65** 314–9

[60] Sun D, Fan T, Liu F, Wang F, Gao D and Lin J-M 2022 *Biosens. Bioelectron.* **212** 114390

IOP Publishing

Sensors for Marine Biosciences
Next-generation sensing approaches
Shyam S Pandey, Rout George Kerry and Kshitij RB Singh

Chapter 10

Proteomics and genomics-based innovation in biosensors for marine biology

Amit K Yadav, Damini Verma, Sumit K Yadav and Dhiraj Bhatia

Over the years, the marine ecosystem has been increasingly affected by the disputes caused by industrial activity and population growth along the shore. Activities that affect the water quality in the coastal zone include garbage disposal, port development, extraction, and dredging procedures. The maritime environment as well as the ecosystem it maintains are under more stress as a result of all of these and additional climate-related effects. While physical parameters are typically determined by sensors in marine studies, real-time biological and chemical characteristic data is becoming more and more in demand. Biosensors can provide a quicker, simpler, and more affordable option for testing, which makes it essential to gather the enormous amount of information needed for proteomic and genomic sensing. Combining these methods with biophysical sensors will eventually lead to breakthroughs in proteomics and genomics. The integration of proteomics and genomics into biosensor development holds great promise for advancing our understanding of marine biology, monitoring marine ecosystems, and addressing various environmental challenges. These innovations can enhance our ability to conserve marine life, manage marine resources sustainably, and respond to emerging threats in the world's oceans. Although amperometric sensors are still in use, new approaches and data-gathering technologies have greatly increased the sensitivity along with dependability of these sensors. To make critical decisions and safeguard this essential resource, this book chapter gives a quick summary and overview of recent biosensor innovations and their uses in marine biology's proteomics and genomics to monitor the ocean in real time.

10.1 Introduction

The ocean contains vast volumes of seawater, and takes part in many ecological processes *via* the general saltwater circulation [1], like climate systems, carbon cycles, and global heat cycles [2, 3]. Human activities including trash disposal, harbor construction in coastal areas, and mineral exploration can have a substantial impact on marine habitats [4]. Long-term ocean management and protection depend on an understanding of all the variables and mechanisms affecting the wellness of the marine ecosystem. Various organisms as well as chemical components are necessary to understand the marine resources and its environment, like methane (CH_4) [5], radon (Rn) [6, 7], ferrous ion (Fe^{2+}) [8], carbon dioxide (CO_2) [9], microbes [10], nutrients [11], as well as other seafood [12]. The sustainable growth of human communities and the use of marine resources depend on the measurement and exploration of the chemical materials as well as marine species in the sea. Conventional techniques for quantifying marine analytes require the direct collection of samples of seawater, biota, or sediment, at the shore or *via* ship, followed by analysis in central laboratories. Yet there are not enough sampling spots, limited sampling, and samples quickly degrade over time [13]. As a result, the distribution of analytes cannot be tracked in real time or space using conventional approaches. But in order to comprehend important processes in the sea more effectively, a broad variety of temporal and spatial investigations of some analytes such as Rn, CO_2, CH_4, hydrogen sulfide (H_2S), and helium (He),—are needed. The *in situ* technique has become more and more popular in the last few decades for emergency, mapping, and monitoring situations. Various chemical sensors as well as biosensors are developed for detecting marine analytes [14, 15]. The operational and mechanical stability, low weight, resistance to corrosion (comprising attack by biota and seawater corrosion), compactness, low power consumption, intelligent data gathering, and full automation have made *in situ* sensor platforms highly appealing [16, 17]. *In situ* systems were created to accomplish automated physicochemical and hydrological investigations directly in the field [18, 19].

The vast and diverse world of marine biology has always been a subject of intrigue and importance. The marine ecosystem, with its rich biodiversity, plays a crucial role in maintaining the planet's health and climate. However, studying this complex system poses numerous challenges, primarily due to its inaccessibility and the delicate nature of marine organisms. This is where the role of biosensors becomes pivotal. Biosensors were fabricated for a variety of analytes for studying fields like the medicinal [20, 21], food [22, 23], bioterrorism/defense, as well as environmental uses [24, 25]. The development and application of novel biosensors are covered in hundreds of new research articles published annually by many publications. Several distinct sensor set-ups were made possible, and some of the fabricated devices were commercialized and are used in everyday analysis. Nevertheless, only a small number of these advancements have been intended for use in marine environments [26]. The harsh working conditions and dearth of specialized facilities in the marine applications sector in contrast to, say, pharmaceutical or medical analysis, are the main causes of this industry falling behind. However, is this not a squandered chance? The justifications for taking measures at sea, the ways by which such measurements have changed, and the possible use of biosensors in marine surveillance initiatives in the future are all covered in the sections that follow.

In recent years, the fields of proteomics and genomics have made significant strides, providing deeper insights into the molecular mechanisms of organisms. In marine biology, proteomics and genomics significantly advance biosensor technology by enhancing the monitoring and understanding of marine ecosystems [27]. Genomics allows for the identification of genetic markers specific to various marine species, facilitating the development of biosensors that can detect and monitor the presence and abundance of these organisms in marine environments. This is crucial for tracking biodiversity, identifying invasive species, and assessing the health of marine populations. Moreover, genomic tools such as environmental DNA (eDNA) analysis enable the detection of species from water samples without the need for direct observation or capture [28]. Proteomics complements this by identifying protein biomarkers that indicate physiological states or stress responses in marine organisms. Proteomic data can be used to create biosensors that detect specific proteins associated with pollution, disease, or environmental changes, providing real-time insights into the health and condition of marine ecosystems [29]. The integration of proteomics and genomics with biosensor technology has opened up new avenues in marine biology research. These advanced biosensors can detect changes in the protein expression or genetic makeup of marine organisms in response to environmental changes. This provides a more accurate and comprehensive understanding of how these organisms interact with their environment. For instance, proteomics and genomics-based biosensors can be used to monitor water quality by detecting harmful substances that cause changes in the protein expression or genetic makeup of indicator species. They can also be used for investigating the climate change effects on marine organisms at a molecular level, providing valuable data that can inform conservation efforts. Moreover, these innovative biosensors have the potential to revolutionize the field of marine biotechnology. By understanding the unique proteins and genes of marine organisms, we can harness them for various applications, such as developing new pharmaceuticals, biofuels, and other valuable products.

However, the development and application of proteomics and genomics-based biosensors in marine biology are still in their nascent stages. There are various issues to overcome, including the complexity of marine samples, the need for high sensitivity and specificity, and the harsh conditions of the marine environment. Despite these challenges, the potential benefits of these advanced biosensors are immense, making this an exciting area of research. In this chapter, we will delve deeper into the world of proteomics and genomics-based biosensors in marine biology, exploring their potential applications, current challenges, and future directions. We hope that this chapter will provide a comprehensive overview of this innovative field and inspire further research and innovation.

10.2 Various strategies for marine measurements

The motives for measuring are equally diverse as the various tactics and instruments used to obtain marine measurements. The majority of samples were previously taken from the ships or shore, especially research vessels, that were either analyzed while

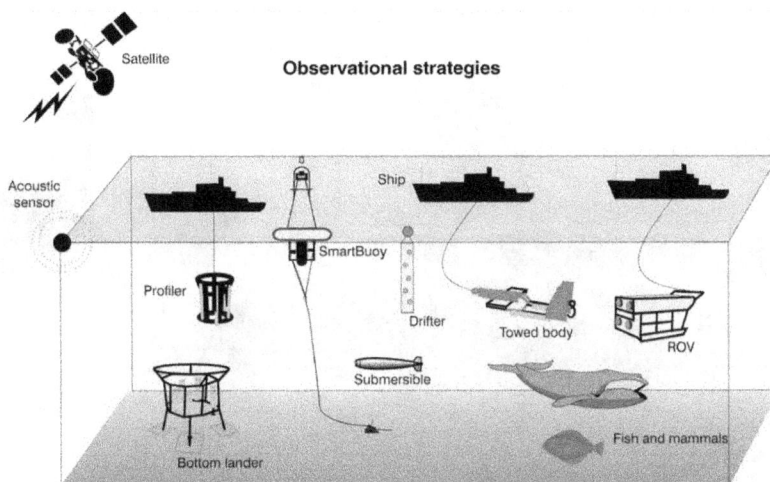

Figure 10.1. Schematic overview of marine observational strategies. Reproduced from [14], copyright (2005), with permission from Elsevier.

returning or after to the laboratory. However, an increasing number of instruments and platforms are being created to do automated measurements in the real world. Figure 10.1 provides a schematic illustration of the methods that have been utilized.

A crucial element in fostering synergy across diverse measurement modalities is achieving suitable temporal and spatial information resolution. A recent development in the field is marine animals tagging, including whales, seals, and fish, to study their actions and use them as subjects for 'researchers' to obtain data on hard-to-reach or distant locations. In the past, oceanographic factors including turbidity, depth, conductivity, and temperature, have been measured the most frequently using this method; however, an increasing number of biological and chemical variables are being investigated along with it. For instance, the use of optrodes or electrodes to measure oxygen, the automated wet chemistry analysis of nutrients (silicate, phosphate, and nitrate), and the determination of chlorophyll levels (as a substitute for algal biomass) using fluorometry. With CEFAS SmartBuoys, all of these parameters are acquired regularly (figure 10.2). Furthermore, it is possible to see how all of these assessment platforms might gain from the inclusion of new biosensors in order to expand our understanding of biological and chemical variables following the largely gathered physical evidence at this time. In addition to the remote measurement techniques, traditional laboratory analysis could benefit from the use of biosensors, for example as pre-screening tools, and field studies involving measurements in estuarine or coastal sites [14].

10.2.1 Key factors influencing measurements

It should come as no surprise that there are many different motivations for maritime study and surveillance given the significance of the oceans for transportation, waste

Figure 10.2. CEFAS 'SmartBuoy' with a payload of different sensors and measurement instruments ready to be deployed. Reproduced from [14], copyright (2005), with permission from Elsevier.

product disposal, food supply, raw material extraction, and leisure activities. Governments across the globe are compelled by several programs and legal requirements, along with scientific curiosity in ecosystem interactions and functions, to look into marine procedures, monitor supplies and temporal patterns, and assess the destiny and effects of releases later on (figure 10.3). For example:

(1) *Eutrophication:* Marine eutrophication remains a global topic that has been given the greatest importance for intervention at the European (EC) and Regional Seas (OSPAR) level *via* plans and directives aimed at managing the unfavorable effects of the enrichment of nutrients.

(2) *Ecotoxicology:* In the scientific field of ecotoxicology, the effects of chemical contact on biota are examined. Endocrine disruption, genotoxicity, muta- genicity, carcinogenicity, or cytotoxicity are examples of specific forms of activity. A pollutant's bioavailability is frequently more important to consider when determining the possible harm it can trigger than its exact level. The relationship between an individual's exposure, response, and outcomes at the ecological or even population scale is particularly significant.

(3) *Food safety:* A major factor in the examination of marine items, including shellfish, is food safety. If a suitable monitoring mechanism for contami- nants in seafood is not in place, the development of harmful algae and its resulting accumulation in filter feeders may have disastrous effects on human health. Further reasons to be concerned about toxic algae are their potential to kill huge numbers of fish, create localized depletion of oxygen, or just detract from the aesthetic appeal of coastal regions by forming scum or foam.

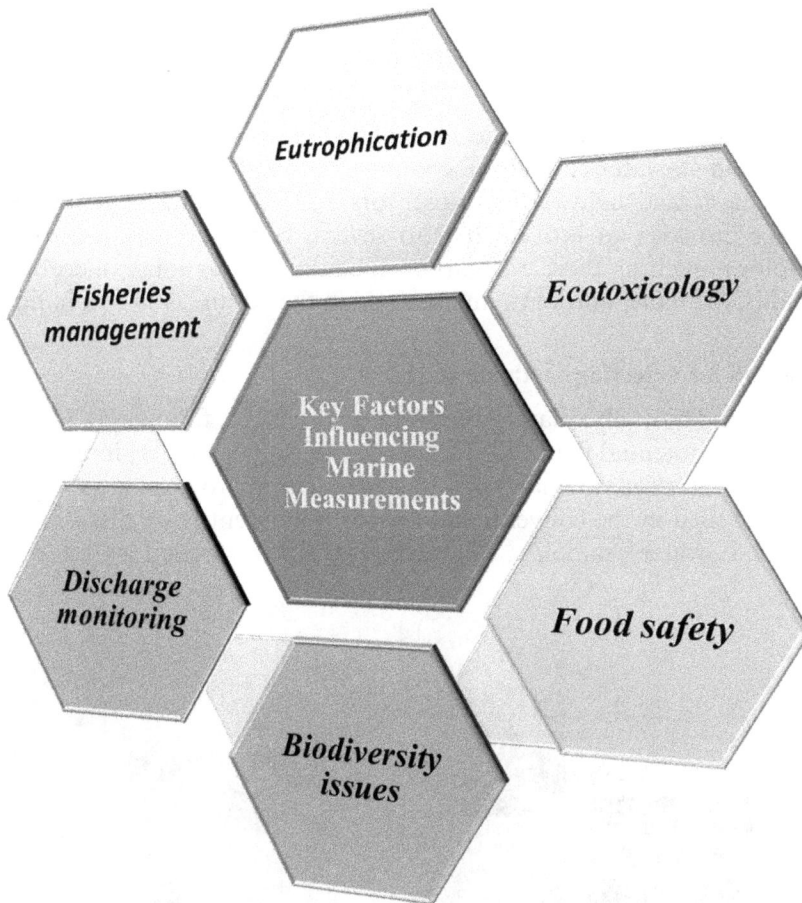

Figure 10.3. Schematic illustration showing the critical factors influencing marine measurements.

(4) *Biodiversity issues:* It is evident that problems with habitat and biodiversity protection are linked to human-caused effects and other types of waste in addition to fertilizer inputs.

(5) *Discharge monitoring:* An essential component of compliance control is the surveillance of recognized discharge sources, including oil platforms and industrial locations along the coast. Waste materials from operations like marina dredging must be analyzed for dangerous substances before being allowed to be disposed of in the sea.

(6) *Fisheries management:* Another area that gives rise to several measuring demands, including animal behavior research, is the management of fisheries and the evaluation of the corresponding fish stocks. Fisheries affect not just the species being targeted but additionally bycatch and the surrounding environment, all of which need to be evaluated. Research of such effects is necessary since fishing trawls have the potential to significantly alter the composition of the seabed and its related functions.

10.3 Role of biosensors in marine biology

Many chemical sensors and biosensors are designed for marine analytes [15]. On account of their operational and mechanical stability, corrosion resistance (such as biota attack and seawater corrosion), complete automation, intelligent data gathering, compactness, and lightweight, biosensors and sensors have piqued attention. This chapter provides an introduction to several biosensors and sensors used in marine biology, such as those for carbon dioxide, ferrous ions, microbes, radon, contaminants, methane, nutrients, and seafood in the ocean, as shown in figure 10.4.

10.3.1 Sensors for detecting methane (CH_4)

CH_4 is the second most abundant greenhouse gas on earth, following CO_2, and has a greater warming potential than CO_2 [30]. The main source of CH_4 in the water is the decomposition of organic materials [5]. Gas chromatography is the primary analytical tool used in the conventional approach of identifying dissolved methane from isolated water samples. Nevertheless, the sample is susceptible to

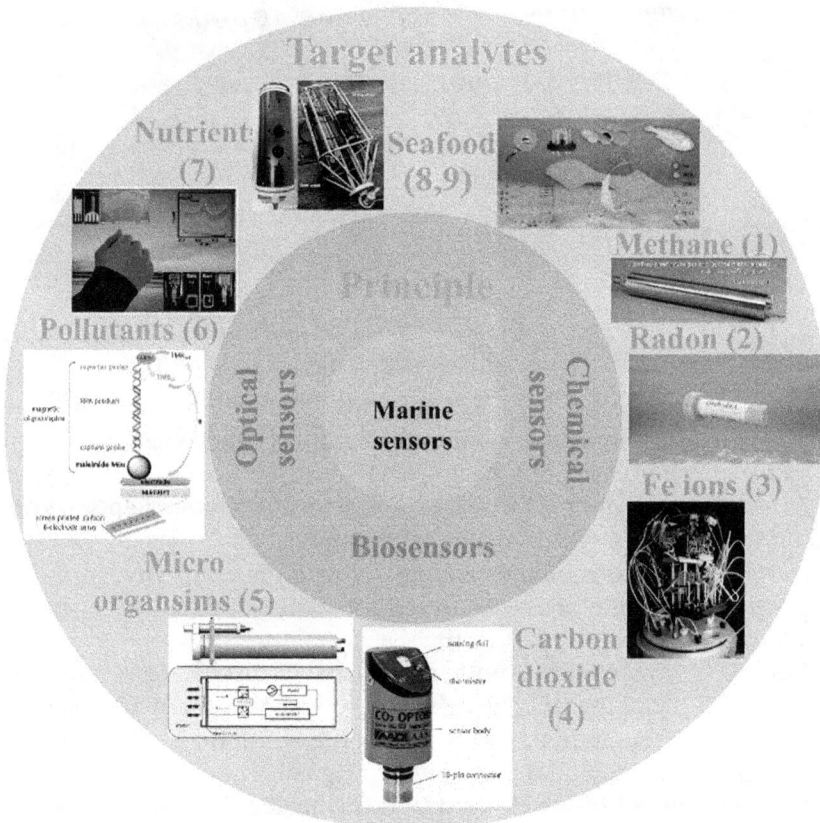

Figure 10.4. Schematic illustrations of marine sensors. Reproduced from [13] CC BY 4.0.

contamination, and methane may be released during the process of collecting and retaining the sample, leading to potential inaccuracies in the test results [31]. The German business Capsum invented the first *in situ* methane sensor based on SnO_2 for commercial use in 1999. To effectively separate the dissolved gases in seawater, a polydimethylsiloxane (PDMS) membrane is utilized. The linearity range vary from 10 to 4000 nM, and 1–30 min was the response time [32]. Du and co-workers developed an approach based on Raman spectroscopy for measuring dissolved methane in water [33]. Burton and co-authors developed a fiber refractometer for detecting methane in the deeper ocean [34]. Also, a new suboceanic detector was reported by Grilli *et al* that depends on absorption expertise boosted by an optical feedback hole for *in situ* determination of dissolved methane [35]. Hemond *et al* fabricated a stand-alone submerged mass spectrometer system, termed 'Nereus', for enabling *in situ* determination of dissolved vapours and gases of water at several ppb, like biological gases, atmospheric gases, and hydrocarbons [36]. Further, Gentz *et al* also developed a low-power underwater cryotrap-membrane-inlet system-coupled mass spectrometer [37].

10.3.2 Ferrous ion (Fe^{2+}) sensors

The two main iron chemical valence states in saltwater are ferric and ferrous ions, having ferrous ions being the most unstable and readily reduced to Fe^{3+}. Iron is a crucial redox-sensitive component of sea sediment, and it exerts a significant impact on various biogeochemical processes and events. These include the escape and accessibility of phosphorus, the cycling of sulfur, and the breakdown of organic matter [8]. A submersible voltammetric sensor that can monitor ferrous ions in the sediment–water junction in real time was reported by Tercier-Waeber *et al* [75] Milani *et al* developed a microfluidic *in situ* analyzer for measuring dissolved iron in vertical profiles and aquatic environments [38]. A spectrophotometric sensor along with microfluidic techniques are utilized in this system, and a unique in-cell diffusion process serves as the foundation for the mixed sample process as well as reagents. A polyazomethine/ascorbic acid-based fluorescent chemical sensor for Al^{3+} and ferrous ion measurement was investigated by Kamaci *et al* [76]. A unique planar optical sensor for displaying the two-dimensional pattern of dissolved ferrous ions was described by Zhu *et al* [39]. Guo *et al* described a unique device that combines collection, improvement, and quantitative analysis for *in situ* dissolved Fe(II) detection sediment pore water. To measure absorbance, the sensor was mounted on the clear poly(vinyl alcohol) membrane and ferrous ion indicator of ferrozine. The linearity for ferrous ions ranges from 0 to 200 μM having limit of detection (LOD) of 4.5 μM, and 10–30 min response time [40].

10.3.3 Sensors for detecting marine microorganisms

The ability to quickly identify aquatic microorganisms is important for comprehending the dynamics of marine environments and coastal fields. In the past decade, there has been fast development of biosensors that can identify planktonic organisms, particularly hazardous algae, and diseases [41]. Algal blooms are the primary sources of toxins in the ocean, posing an important risk to many regions worldwide [42]. For instance, the powerful neurotoxins produced by *A. ostenfeldii* and *Alexandrium tamarense* algae can be screened by shellfish of water for sustenance, accumulating in the shellfish, which subsequently becomes poisonous to humans along with other animals who consume the shellfish. A multi-biosensor chip for the *in situ* identification of harmful algae was reported by Diercks *et al* [77]. Further, McCoy *et al* developed a multiplex surface plasmon resonance (SPR) biosensor for detecting deadly blossom of *Alexandrium minutum* algae on the shore [43]. Also, an electrochemical DNA probe for *Ostreopsis cf* ovata detection was proposed by Toldra *et al* [78]. In addition to harmful algae, marine pathogens such as protozoa, viruses, and bacteria can contaminate seafood and ultimately infect humans. Liu *et al* reported a fast, portable, and simple fish-based microfluidic platform [44]. The system has the ability to jointly identify and assess the risk of infection from pathogens transmitted by water. Liu *et al* developed a lateral-flow DNA-aptasensor for Singapore iridovirus (SGIV) detection [45].

10.3.4 Sensors for carbon dioxide (CO_2)

The levels of carbon dioxide (CO_2) in seawater are rising in tandem with rising CO_2 emissions. The absorption of human originated CO_2 by the ocean can result in a reduction in the acidity of seawater and cause significant modifications to the sea's carbonate system [46]. Over the past two decades, a multitude of innovative sensors for carbon dioxide measurement have been documented. Lu *et al* developed a chemical pCO_2 sensor using optical fiber to measure surface seawater in real time [47]. Atamanchuk and co-workers developed a small, low-power consumption, and long-life optode for detecting pCO_2 [48]. The sensor has an operational life of over seven months when used underwater, and the highest accuracy of ± 2 µatm in the 200–1000 µatm pCO_2 range was achieved. Zhu *et al* reported a ratiometric planar optode fluorosensor for 2D imaging of pCO_2 distribution in overlying water and sediments [49]. A compact, inexpensive GasPro probe was used to measure pCO_2 *in situ* continuously along the shore of Panarea Island, Italy, as described by Graziani *et al* [79].

10.3.5 Sensors for seafood

A significant portion of the human diet is made up of fish and other seafood-related items. They are majorly comprised of various nutrients, like protein, minerals, unsaturated fatty acids, and vitamins [50]. Furthermore, in maritime environments, seafood may be polluted with algae toxins, raising serious concerns about food safety. In order to evaluate the freshness and safety of seafood, numerous innovative chemical and analytical biosensors, optical spectroscopic sensors, electronic tongues, eyes, noses, and nuclear magnetic resonance spectroscopic sensors have recently been fabricated. For example Mustafa and co-workers developed a portable nanoenzyme-based biosensor for measuring released hypoxanthine which is a nucleotide degradation product found in fish and meat, to screen the fish freshness [51]. A biosensor to measure the hypoxanthine content uses xanthine oxidase and cerium nanoparticles. It is possible to achieve an 89 μM LOD and a linear detection range till 597 μM. Chen *et al* fabricated a semiconductor comprising a metal oxide gas sensor that depends on mesoporous Au–ZnO nanospheres [52]. When seafood was allowed to degrade at 250 °C, the sensor demonstrated a 10-ppm sensitiveness of trimethylamine. A flow injection amperometric sensor for detecting formalin in fresh fish was described by Torrarit *et al*. The sensor was developed on a glassy carbon electrode(GCE) modified with carbon microspheres and palladium particles whose accuracy lies from 96% ± 1% and 105% ± 3% [53].

10.3.6 Sensors for detection of marine pollutants

In addition to the unsustainable use of biological assets, human interference has been responsible for the introduction of numerous toxins into the ocean in recent decades. These pollutants typically come into contact with ocean water from land-based origins through riverine supplies and urban or industrial activity. To a great extent, the pollutants found in ocean water comprise radionuclides, organic compounds like pharmaceuticals, veterinary medicines, microplastics, and pesticides, hydrocarbon pollutants (like fuel/oil), and stable trace elements (e.g., tin, cadmium, mercury, and lead) [54, 55]. The source and outflow of contaminants into the maritime environment are depicted in figure 10.5.

Recently, some novel chemical biosensors and sensors were developed to detect pollutants. Malzahn *et al*, for instance, described a wearable screen-printed biosensor containing neoprene undersea apparel made of synthetic rubber [57]. In marine samples, the biosensor could be employed to identify the presence of phenolic pollutants, nitroaromatic explosives, and detectable heavy metal

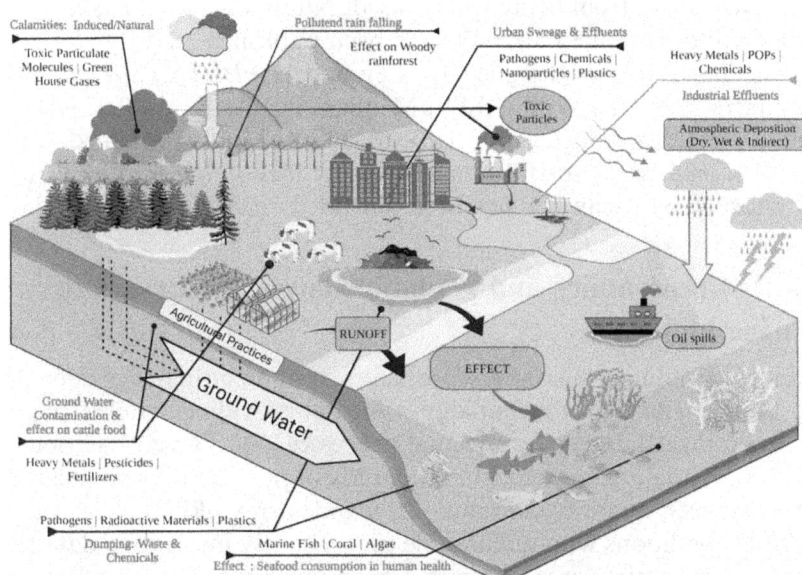

Figure 10.5. A schematic illustration depicting the origin and release of the mixed pollutants in the ecosystem. Reproduced from [56], copyright (2022), with permission from Elsevier.

contamination. Using an actuation component and a digital microfluidic diluter chip, Han *et al* constructed a completely automated whole-algae biosensor [58]. The biosensor has detection limits of $5.22\,\mu M$, $2.85\,mM$, $1.90\,\mu M$, and $0.65\,\mu M$, for nonylphenol, phenol, lead, and copper, respectively, in seawater. A microfluidic device combined with an oceanic phytoplankton motility sensor was described by Zheng *et al* [59] as a means of assessing the toxicity of pollutants. Pfannkuche *et al* utilized a surface-enhanced Raman scattering sensor for detecting polycyclic aromatic hydrocarbons in the Baltic Sea [50]. Additionally, Kolomijeca and colleagues described a mobile on-board surface-enhanced Raman spectroscopy (SERS) sensor [80] device with a $10\,s$ reaction time that is capable of detecting polycyclic aromatic hydrocarbons with a $0.3\,nM$ detection limit, in contrast to fluoranthene and anthracene.

10.3.7 Sensors for detecting marine nutrients

Numerous nutrients have been released into ocean ecosystems *via* agricultural waste, industrial effluent, and home sewage that result from increased urbanization, industrialization, as well as agricultural expansion [60]. One of the most difficult oceanic environmental issues is eutrophication, that results in red tides and toxic marine blooms [61].

A number of electrochemical sensors *in situ* have become commercially available. These include the CYCL phosphate (WV, Zatto Lane Danville, Wetlabs, USA), EcoLAB2 Wetlabs, Micro-Lab sensors, SUNAV2 (Nova Scotia, Halifax, Satlantic, Canada); and WIZ (SYSTEA, Anagni, Italy) sensors [62]. However, these sensors' low repeatability, short duration, limited detecting concentration range, and low accuracy prevent them from being widely used. Numerous novel *in situ* devices and sensors for detecting marine nutrients have been published recently. An automated electrochemical sensor was designed by Legrand *et al* for *in situ* monitoring of silicate in marine settings. With a $1.08\,\mu M$ quantification limit, $0.32\,\mu M$ detection limit, and a detection range from 1.63 to $132.8\,\mu M$ was attained. Further, a commercial automated sensor for simultaneously identifying nitrite, nitrate, silicic acid, and phosphate in seawater has been amended by Altahan *et al* [63].

10.4 Role of proteomics and genomics-based biosensors in marine biology

Proteomics and genomics-based biosensors represent a cutting-edge convergence of molecular biology and sensor technology, aimed at providing detailed and real-time insights into biological and environmental processes [64]. These biosensors leverage the comprehensive data generated from the study of proteins (proteomics) and genes (genomics) to detect, quantify, and monitor a wide array of biological markers and environmental conditions with remarkable precision. By integrating data from both proteomics and genomics, scientists gain a comprehensive understanding of marine organisms at the molecular level [65]. This integrative approach helps in deciphering the complex interactions between genetic information and protein function, ultimately leading to a deeper knowledge of marine biology and the strategies development for conservation and sustainable management of marine resources.

10.4.1 Genomics based biosensors

Human activity has affected every region of our oceans, resulting in significant reductions in the health and quantity of several marine species and their ecosystems [66]. Though it is currently gaining pace, the use of genomic information for marine conservation is still lagging behind other kinds of information [67]. Several fields of biomedical and biological research are fast changing due to the growing genomics field, which has made it possible to move from sequential investigations of individual genes to greater ecological approaches. Additionally, it entails the concurrent study of numerous elements and how they interact with the surroundings, ranging from routes to cell tissues to entire communities and organisms [68, 69]. It is the investigation of the whole set of DNA (the genome) of an organism. It involves sequencing, mapping, and analyzing the genetic material to understand the structure, function, evolution, and interaction of genes. Genomics helps identify genetic variations and how these variations influence biological processes and traits [70]. Genomics-based biosensors utilize genetic material (DNA and RNA) to identify specific sequences associated with various organisms or conditions. These biosensors can detect the presence of particular species, monitor genetic variations, and identify pathogens. Figure 10.6 shows the application of genomics in various areas of marine biology.

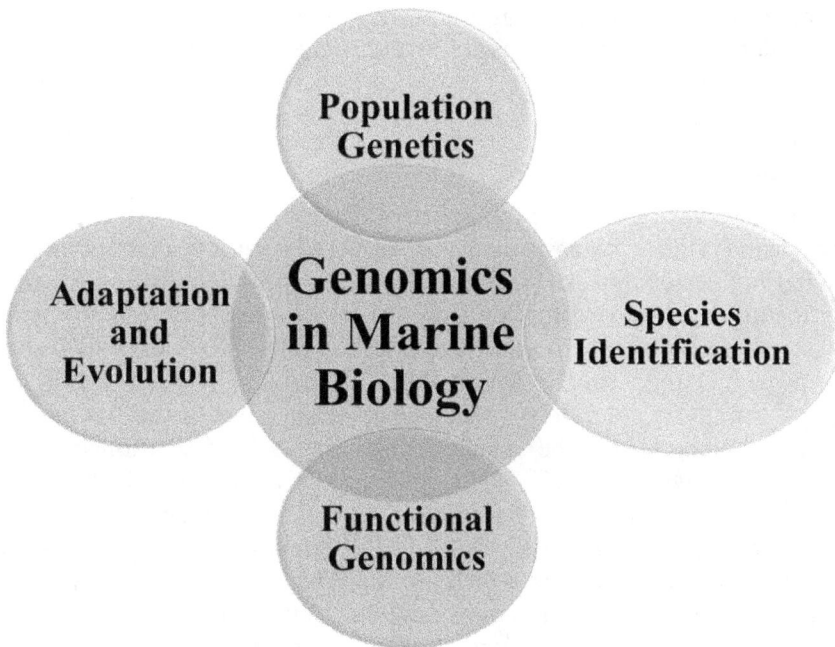

Figure 10.6. Schematic illustration showing the application of genomics in various areas of marine biology.

Genomics in marine biology:
  (i) **Species identification:** Genomic techniques such as DNA barcoding and environmental DNA (eDNA) analysis allow for the precise identification and cataloging of marine species, including cryptic or microscopic organisms that are otherwise difficult to detect.
 (ii) **Population genetics:** Genomic studies help in understanding the genetic diversity, population structure, and gene flow within and between marine species populations, aiding in conservation and management efforts.
(iii) **Adaptation and evolution:** Genomics provides insights into the genetic basis of adaptations to different marine environments, such as extreme temperatures, salinity levels, and pressures, revealing how marine organisms evolve and survive under varying conditions.
 (iv) **Functional genomics:** By studying gene expression and regulation, functional genomics helps in understanding how marine organisms respond to environmental changes, such as pollution, climate change, and habitat destruction.

Techniques used in genomics:
  (i) **DNA/RNA probes:** These are sequences of nucleic acids that hybridize with the target genetic material, allowing for specific detection.
 (ii) **CRISPR-Cas systems:** Adapted from bacterial immune systems, CRISPR-Cas technology can identify and cut specific DNA sequences, enabling precise detection of genetic mutations or the presence of specific organisms.
(iii) **Environmental DNA (eDNA):** This technique involves collecting and analyzing genetic material from environmental samples (such as water) to detect and monitor species without the need for physical capture or observation.

10.4.2 Proteomics based biosensors

Proteomics is a crucial approach to analyze functional genomic studies of microbes from the ocean [71]. The extensive study of proteins is known as proteomics, which studies vital molecules that perform a vast array of functions within organisms. It involves identifying and quantifying proteins and understanding their structures, functions, interactions, and modifications. Proteomics provides insights into the dynamic protein changes and activity within cells, tissues, or organisms in response to various stimuli or conditions [72]. The method has enhanced knowledge of physiological defenses and the regulatory structures that guarantee survival in potentially fatal environmental circumstances. The physiological examination of an uncultured bacterium endosymbiont from an underwater tubular worm has been effectively conducted using proteomics [73]. Proteomics-based biosensors focus on detecting and analyzing proteins, which are the functional molecules in cells. These biosensors can monitor the expression, modification, and interaction of proteins, providing insights into cellular processes and organism health. Figure 10.7 shows the application of proteomics in various areas of marine biology.

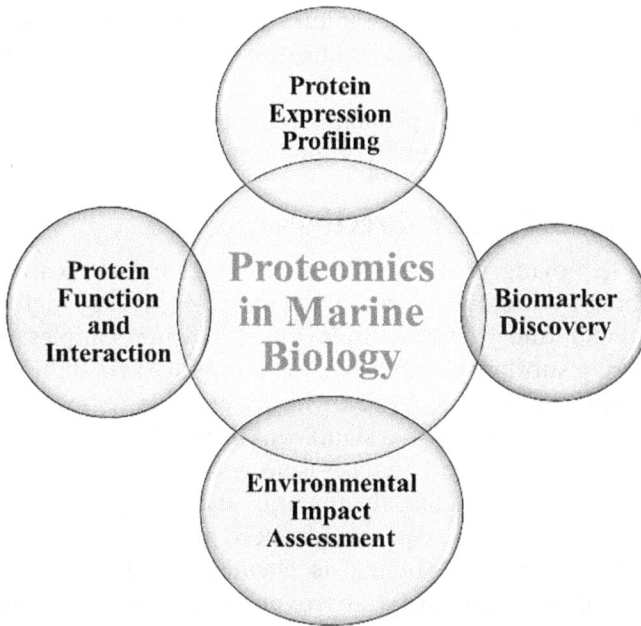

Figure 10.7. Schematic illustration showing the application of proteomics in various areas of marine biology.

Proteomics in marine biology:
(i) **Protein expression profiling:** Proteomics allows for the identification and quantification of proteins expressed by marine organisms, providing insights into their physiological and metabolic states under different environmental conditions [74].
(ii) **Biomarker discovery:** By identifying specific proteins associated with particular physiological states or stress responses, proteomics helps in discovering biomarkers that indicate health, disease, or exposure to pollutants in marine organisms.
(iii) **Protein function and interaction:** Understanding protein functions and their interactions is crucial for elucidating biological pathways and networks in marine organisms, helping to reveal the molecular mechanisms underlying various biological processes.
(iv) **Environmental impact assessment:** Proteomic analysis can assess the impact of environmental stressors, such as temperature changes, acidification, and toxins, on marine organisms by observing changes in protein expression and modifications.

Techniques used in proteomics:
(i) **Antibody-based detection:** Antibodies specific to target proteins are used to capture and quantify proteins of interest.

(ii) **Mass spectrometry:** This technique allows for the detailed analysis of protein composition and modifications, enhancing the specificity and sensitivity of biosensors.

(iii) **Protein chips:** Arrays of protein-specific probes can simultaneously detect multiple proteins, providing comprehensive profiles of protein expression.

10.5 Challenges and future directions

Developing and deploying proteomics and genomics-based biosensors to revolutionize marine biology presents several challenges. Achieving high sensitivity and specificity is difficult due to the complexity of marine samples, which contain numerous interfering substances. Ensuring the durability and stability of biosensors in harsh marine environments, characterized by extreme temperatures, high pressure, salinity, and biofouling, is also a significant hurdle. Standardizing protocols for sample collection, processing, and analysis is necessary to ensure reproducibility and reliability across different studies. Managing and analyzing the vast datasets generated by these biosensors requires advanced bioinformatics tools and interdisciplinary collaboration among biologists, chemists, engineers, and data scientists. Additionally, navigating regulatory approvals and addressing ethical concerns about the potential impacts on marine life and ecosystems add further complexity to the deployment of these advanced technologies.

Looking forward, advancements in nanotechnology, wearable biosensors, lab-on-a-chip systems, and AI-driven data analysis are expected to enhance biosensor capabilities. Integrating nanotechnology can enhance sensitivity and specificity, while wearable and implantable biosensors for marine organisms will enable continuous, real-time monitoring. Lab-on-a-chip systems can automate sample processing, making biosensors more efficient for field applications, and AI-driven data analysis will improve the extraction of meaningful patterns from complex datasets. Interdisciplinary research involving marine biologists, bioengineers, and data scientists, along with engaging citizen scientists, will drive innovation and practical deployment. International guidelines and standards will facilitate broader adoption, and increased funding will support these initiatives. Expanding applications to include more comprehensive conservation efforts and pollution monitoring will further enhance the impact of these advanced biosensors on marine ecosystem management.

10.6 Conclusion

The potential of proteomics and genomics-based biosensors to revolutionize marine biology is immense. By providing highly sensitive, specific, and real-time monitoring capabilities, these advanced biosensors offer unprecedented insights into the molecular underpinnings of marine organisms and ecosystems. They enable early detection of environmental changes, invasive species, and pollution, facilitating timely intervention and conservation efforts. Additionally, the integration of proteomic and genomic data enhances our understanding of complex biological processes, from gene expression to protein function, driving more informed and effective management of marine resources. As these technologies continue to evolve, they hold the promise of transforming marine biology into a more precise and predictive science, crucial for preserving the health and biodiversity of our oceans in the face of global environmental challenges.

In conclusion, this chapter highlights the transformative impact of integrating proteomics and genomics into biosensor technology for marine research. These innovative biosensors, rooted in the detailed molecular understanding provided by proteomics and genomics, offer unparalleled precision and sensitivity in monitoring marine environments. They enable the detection of specific species, assessment of ecosystem health, and identification of environmental stressors in real time. The advancements discussed underscore the potential for these technologies to revolutionize marine biology, offering new tools for conservation, pollution control, and sustainable resource management. As these biosensors become more sophisticated and accessible, they promise to deepen our understanding of marine ecosystems, paving the way for more proactive and effective stewardship of the oceans.

Acknowledgments

A.K.Y. expresses gratitude to the Indian Institute of Technology Gandhinagar (IITGN), India, for the financial support through the Early Career Fellowship. Both A.K.Y. and D.B. acknowledge IITGN for their financial support and the facilities provided.

References

[1] Buonocore E, Grande U, Franzese P P and Russo G F 2021 Trends and evolution in the concept of marine ecosystem services: an overview *Water* **13** 2060

[2] Fukuba T and Fujii T 2021 Lab-on-a-chip technology for *in situ* combined observations in oceanography *Lab Chip* **21** 55–74

[3] Ingeman K E, Samhouri J F and Stier A C 2019 Ocean recoveries for tomorrow's earth: hitting a moving target *Science (80–)* **363** eaav1004

[4] Neal C, Leeks G J L, Millward G E, Harris J R W, Huthnance J M and Rees J G 2003 Land ocean interaction: processes, functioning, and environmental management: a UK perspective *Sci. Total Environ.* **314** 801–19

[5] Niu M, Liang W and Wang F 2018 Methane biotransformation in the ocean and its effects on climate change: a review *Sci. China Earth Sci.* **61** 1697–713

[6] Eleftheriou G, Pappa F K, Maragos N and Tsabaris C 2020 Continuous monitoring of multiple submarine springs by means of gamma-ray spectrometry *J. Environ. Radioact.* **216** 106180

[7] Zhao S, Li M, Burnett W C, Cheng K, Li C, Guo J, Yu S, Liu W, Yang T and Dimova N T 2022 In-situ radon-in-water detection for high resolution submarine groundwater discharge assessment *Front. Mar. Sci.* **9** 1001554

[8] Gächter R and Müller B 2003 Why the phosphorus retention of lakes does not necessarily depend on the oxygen supply to their sediment surface *Limnol. Oceanogr.* **48** 929–33

[9] Le Quéré C, Raupach M R, Canadell J G, Marland G, Bopp L, Ciais P, Conway T J, Doney S C, Feely R A and Foster P 2009 Trends in the sources and sinks of carbon dioxide *Nat. Geosci.* **2** 831–6

[10] LaGier M J, Scholin C A, Fell J W, Wang J and Goodwin K D 2005 An electrochemical RNA hybridization assay for detection of the fecal indicator bacterium *Escherichia Coli Mar. Pollut. Bull.* **50** 1251–61

[11] Cao L, Xiang H, Xu J, Gao Y, Lin C, Li K, Li Z, Guo N, David P and He C 2022 Nutrient detection sensors in seawater based on ISI web of science database *J. Sensors* **2022** 5754751

[12] Pyz-Łukasik R, Chałabis-Mazurek A and Gondek M 2020 Basic and functional nutrients in the muscles of fish: a review *Int. J. Food Prop.* **23** 1941–50

[13] Liu Y, Lu H and Cui Y 2023 A review of marine *in situ* sensors and biosensors *J. Mar. Sci. Eng.* **11** 1469

[14] Kröger S and Law R J 2005 Biosensors for marine applications: we all need the sea, but does the sea need biosensors? *Biosens. Bioelectron.* **20** 1903–13

[15] Kumari C R U, Samiappan D, Kumar R and Sudhakar T 2019 Fiber optic sensors in ocean observation: a comprehensive review *Optik (Stuttg.)* **179** 351–60

[16] Yadav A K, Verma D, Sajwan R K, Poddar M, Yadav S K, Verma A K and Solanki P R 2022 Nanomaterial-based electrochemical nanodiagnostics for human and gut metabolites diagnostics: recent advances and challenges *Biosensors* **12** 733

[17] Yadav A K, Verma D, Kumar A, Kumar P and Solanki P R 2021 The perspectives of biomarkers based electrochemical immunosensors, artificial intelligence and the internet of medical things towards COVID-19 diagnosis and management *Mater. Today Chem.* **20** 100443

[18] Wang F, Zhu J, Chen L, Zuo Y, Hu X and Yang Y 2020 Autonomous and *in situ* ocean environmental monitoring on optofluidic platform *Micromachines* **11** 69

[19] Briciu-Burghina C, Power S, Delgado A and Regan F 2023 Sensors for coastal and ocean monitoring *Annu. Rev. Anal. Chem.* **16** 451–69

[20] Yadav A K, Verma D and Solanki P R 2023 Enhanced electrochemical biosensing of the Sp17 cancer biomarker in serum samples via engineered two-dimensional MoS_2 nanosheets on the reduced graphene oxide interface *ACS Appl. Bio Mater.* **6** 4250–68

[21] Yadav A K, Gulati P, Sharma R, Thakkar A and Solanki P R 2022 Fabrication of alkoxysilane substituted polymer-modified disposable biosensing platform: toward sperm protein 17 sensing as a new cancer biomarker *Talanta* **243** 123376

[22] Yadav A K, Verma D, Lakshmi G B V S, Eremin S and Solanki P R 2021 Fabrication of label-free and ultrasensitive electrochemical immunosensor based on molybdenum disulfide nanoparticles modified disposable ITO: an analytical platform for antibiotic detection in food samples *Food Chem.* **363** 130245

[23] Yadav A K, Verma D and Solanki P R 2021 Electrophoretically deposited L-cysteine functionalized MoS_2@MWCNT nanocomposite platform: a smart approach toward highly sensitive and label-free detection of gentamicin *Mater. Today Chem.* **22** 100567

[24] Verma D, Chauhan D, Mukherjee M, Das , Ranjan K R, Yadav A K and Solanki P R 2021 Development of MWCNT decorated with green synthesized AgNps-based electrochemical sensor for highly sensitive detection of BPA *J. Appl. Electrochem.* **51** 447–62

[25] Chaudhary N, Yadav A K, Sharma J G and Solanki P R 2021 Designing and characterization of a highly sensitive and selective biosensing platform for ciprofloxacin detection utilizing lanthanum oxide nanoparticles *J. Environ. Chem. Eng.* **9** 106771

[26] Kröger S, Piletsky S and Turner A P F 2002 Biosensors for marine pollution research, monitoring and control *Mar. Pollut. Bull.* **45** 24–34

[27] Quezada H, Guzmán-Ortiz A L, Díaz-Sánchez H, Valle-Rios R and Aguirre-Hernández J 2017 Omics-based biomarkers: current status and potential use in the clinic *Boletín Médico Del Hosp. Infant. México (English Ed.)* **74** 219–26

[28] Sun Y V and Hu Y-J 2016 Integrative analysis of multi-omics data for discovery and functional studies of complex human diseases *Adv. Genet* **93** 147–90

[29] Ponomarenko E A, Poverennaya E V, Ilgisonis E V, Pyatnitskiy M A, Kopylov A T, Zgoda V G, Lisitsa A V and Archakov A I 2016 The size of the human proteome: the width and depth *Int. J. Anal. Chem.* **2016** 7436849

[30] Zhang J, Wang Y, Wang Y, Bai Y, Feng X, Zhu J, Lu X, Mu L, Ming T and De Richter R 2022 Solar driven gas phase advanced oxidation processes for methane removal-challenges and perspectives *Chem. Eur. J.* **28** e202201984

[31] Liu L, Ryu B, Sun Z, Wu N, Cao H, Geng W, Zhang X, Jia Y, Xu C and Guo L 2019 Monitoring and research on environmental impacts related to marine natural gas hydrates: review and future perspective *J. Nat. Gas Sci. Eng.* **65** 82–107

[32] Aleksanyan M S 2010 Methane sensor based on $SnO_2/In_2O_3/TiO_2$ nanostructure *J. Contemp. Phys. (Armenian Acad. Sci.* **45** 77–80

[33] Du Z, Chen J, Ye W, Guo J, Zhang X and Zheng R 2015 Investigation of two novel approaches for detection of sulfate ion and methane dissolved in sediment pore water using raman spectroscopy *Sensors* **15** 12377–88

[34] Burton G, Melo L, Warwick S, Jun M, Bao B, Sinton D and Wild P 2014 Fiber refractometer to detect and distinguish carbon dioxide and methane leakage in the deep ocean *Int. J. Greenh. Gas Control* **31** 41–7

[35] Grilli R, Triest J, Chappellaz J, Calzas M, Desbois T, Jansson P, Guillerm C, Ferré B, Lechevallier L and Ledoux V 2018 SUB-OCEAN: subsea dissolved methane measurements using an embedded laser spectrometer technology *Environ. Sci. Technol.* **52** 10543–51

[36] Hemond H and Camilli R 2002 NEREUS: engineering concept for an underwater mass spectrometer *TrAC, Trends Anal. Chem.* **21** 526–33

[37] Gentz T and Schlüter M 2012 Underwater cryotrap-membrane inlet system (CT-MIS) for improved *in situ* analysis of gases *Limnol. Oceanogr. Methods* **10** 317–28

[38] Milani A, Statham P J, Mowlem M C and Connelly D P 2015 Development and application of a microfluidic in-situ analyzer for dissolved Fe and Mn in natural waters *Talanta* **136** 15–22

[39] Zhu Q and Aller R C 2012 Two-dimensional dissolved ferrous iron distributions in marine sediments as revealed by a novel planar optical sensor *Mar. Chem.* **136** 14–23

[40] Guo C, Ma M, Yuan D, Huang Y, Lin K and Feng S 2019 *In situ* measurement of dissolved Fe(II) in sediment pore water with a novel sensor based on c18-ferrozine concentration and optical imaging detection *Anal. Methods* **11** 133–41

[41] Bergamasco A, Nguyen H Q, Caruso G, Xing Q and Carol E 2021 Advances in water quality monitoring and assessment in marine and coastal regions *Water* **13** 1926

[42] Tang T, Effiong K, Hu J, Li C and Xiao X 2021 Chemical prevention and control of the green tide and fouling organism ulva: key chemicals, mechanisms, and applications *Front. Mar. Sci.* **8** 618950

[43] McCoy G R, McNamee S, Campbell K, Elliott C T, Fleming G T A and Raine R 2014 Monitoring a toxic bloom of alexandrium minutum using novel microarray and multiplex surface plasmon resonance biosensor technology *Harmful Algae* **32** 40–8

[44] Liu Y S, Deng Y, Chen C K, Khoo B L and Chua S L 2022 Rapid detection of microorganisms in a fish infection microfluidics platform *J. Hazard. Mater.* **431** 128572

[45] Liu J, Zhang X, Zheng J, Yu Y, Huang X, Wei J, Mukama O, Wang S and Qin Q 2021 A lateral flow biosensor for rapid detection of Singapore grouper iridovirus (SGIV) *Aquaculture* **541** 736756

[46] Iversen M H 2023 Carbon export in the ocean: a biologist's perspective *Ann. Rev. Mar. Sci.* **15** 357–81

[47] Lu Z, Dai M, Xu K, Chen J and Liao Y 2008 A high precision, fast response, and low power consumption *in situ* optical fiber chemical PCO2 sensor *Talanta* **76** 353–9

[48] Atamanchuk D, Tengberg A, Thomas P J, Hovdenes J, Apostolidis A, Huber C and Hall P O J 2014 Performance of a lifetime-based optode for measuring partial pressure of carbon dioxide in natural waters *Limnol. Oceanogr. Methods* **12** 63–73

[49] Zhu Q, Aller R C and Fan Y 2006 A new ratiometric, planar fluorosensor for measuring high resolution, two-dimensional PCO2 distributions in marine sediments *Mar. Chem.* **101** 40–53

[50] Pfannkuche J, Lubecki L, Schmidt H, Kowalewska G and Kronfeldt H-D 2012 The Use of surface-enhanced raman scattering (SERS) for detection of PAHs in the gulf of Gdańsk (Baltic Sea) *Mar. Pollut. Bull.* **64** 614–26

[51] Mustafa F, Othman A and Andreescu S 2021 Cerium oxide-based hypoxanthine biosensor for fish spoilage monitoring *Sens. Actuators* B **332** 129435

[52] Chen Y, Li Y, Feng B, Wu Y, Zhu Y and Wei J 2022 Self-templated synthesis of mesoporous Au–ZnO nanospheres for seafood freshness detection *Sens. Actuators* B **360** 131662

[53] Torrarit K, Kongkaew S, Samoson K, Kanatharana P, Thavarungkul P, Chang K H, Abdullah A F L and Limbut W 2022 Flow injection amperometric measurement of formalin in seafood *ACS Omega* **7** 17679–91

[54] Kuznetsova O V and Timerbaev A R 2022 Marine sediment analysis—a review of advanced approaches and practices focused on contaminants *Anal. Chim. Acta* **1209** 339640

[55] Ojemaye C Y and Petrik L 2019 Pharmaceuticals in the marine environment: a review *Environ. Rev.* **27** 151–65

[56] Saravanakumar K, SivaSantosh S, Sathiyaseelan A, Naveen K V, AfaanAhamed M A, Zhang X, Priya V V, MubarakAli D and Wang M-H 2022 Unraveling the hazardous impact of diverse contaminants in the marine environment: detection and remedial approach through nanomaterials and nano-biosensors *J. Hazard. Mater.* **433** 128720

[57] Malzahn K, Windmiller J R, Valdés-Ramírez G, Schöning M J and Wang J 2011 Wearable electrochemical sensors for *in situ* analysis in marine environments *Analyst* **136** 2912–7

[58] Han S, Zhang Q, Zhang X, Liu X, Lu L, Wei J, Li Y, Wang Y and Zheng G 2019 A digital microfluidic diluter-based microalgal motion biosensor for marine pollution monitoring *Biosens. Bioelectron.* **143** 111597

[59] Zheng G, Li Y, Qi L, Liu X, Wang H, Yu S and Wang Y 2014 Marine phytoplankton motility sensor integrated into a microfluidic chip for high-throughput pollutant toxicity assessment *Mar. Pollut. Bull.* **84** 147–54

[60] Wang H, Wang G and Gu W 2020 Macroalgal blooms caused by marine nutrient changes resulting from human activities *J. Appl. Ecol.* **57** 766–76

[61] Hu W, Li C, Ye C, Wang J, Wei W and Deng Y 2019 Research progress on ecological models in the field of water eutrophication: citespace analysis based on data from the isi web of science database *Ecol. Modell.* **410** 108779

[62] Aguilar D, Barus C, Giraud W, Calas E, Vanhove E, Laborde A, Launay J, Temple-Boyer P, Striebig N and Armengaud M 2015 Silicon-based electrochemical microdevices for silicate detection in seawater *Sens. Actuators* B **211** 116–24

[63] Altahan M F, Esposito M and Achterberg E P 2022 Improvement of on-site sensor for simultaneous determination of phosphate, silicic acid, nitrate plus nitrite in seawater *Sensors* **22** 3479

[64] Seib K L, Dougan G and Rappuoli R 2009 The key role of genomics in modern vaccine and drug design for emerging infectious diseases *PLoS Genet.* **5** e1000612

[65] Seib K L, Zhao X and Rappuoli R 2012 Developing vaccines in the era of genomics: a decade of reverse vaccinology *Clin. Microbiol. Infect.* **18** 109–16

[66] Halpern B S, Frazier M, Afflerbach J, Lowndes J S, Micheli F, O'Hara C, Scarborough C and Selkoe K A 2019 Recent pace of change in human impact on the world's ocean *Sci. Rep.* **9** 11609

[67] Taylor H R, Dussex N and van Heezik Y 2017 Bridging the conservation genetics gap by identifying barriers to implementation for conservation practitioners *Glob. Ecol. Conserv.* **10** 231–42

[68] Hollywood K, Brison D R and Goodacre R 2006 Metabolomics: current technologies and future trends *Proteomics* **6** 4716–23

[69] Joyce A R and Palsson B Ø 2006 The model organism as a system: integrating 'omics' data sets *Nat. Rev. Mol. Cell Biol.* **7** 198–210

[70] Van Oppen M J H and Coleman M A 2022 Advancing the protection of marine life through genomics *PLoS Biol.* **20** e3001801

[71] Schweder T, Markert S and Hecker M 2008 Proteomics of marine bacteria *Electrophoresis* **29** 2603–16

[72] Gajahin Gamage N T, Miyashita R, Takahashi K, Asakawa S and Senevirathna J D M 2022 Proteomic applications in aquatic environment studies *Proteomes* **10** 32

[73] Markert S, Arndt C, Felbeck H, Becher D, Sievert S M, Hügler M, Albrecht D, Robidart J, Bench S and Feldman R A 2007 Physiological proteomics of the uncultured endosymbiont of riftia pachyptila *Science (80–)* **315** 247–50

[74] Piñeiro C, Barros-Velázquez J, Vázquez , Figueras A and Gallardo J M 2003 Proteomics as a tool for the investigation of seafood and other marine products *J. Proteome Res.* **2** 127–35

[75] Tercier-Waeber M L, Buffle J, Confalonieri F, Riccardi G, Sina A, Graziottin F, Fiaccabrino G C and Koudelka-Hep M 1999 Submersible voltammetric probes for *in situ* real-time trace element measurements in surface water, groundwater and sediment-water interface *Meas. Sci. Technol.* **10** 1202–13

[76] Kamaci U D, Kamaci M and Peksel A 2021 A dual responsive colorimetric sensor based on polyazomethine and ascorbic acid for the detection of Al (III) and Fe (II) ions *Spectrochim. Acta Part A Mol. Biomol. Spectrosc.* **254** 119650

[77] Diercks S, Metfies K and Medlin L K 2008 Development and adaptation of a multiprobe biosensor for the use in a semi-automated device for the detection of toxic algae *Biosens. Bioelectron.* **23** 1527–33

[78] Toldra A, Alcaraz C, Diogene J, O'Sullivan C K and Campas M 2019 Detection of *Ostreopsis cf. ovata* in environmental samples using an electrochemical DNA-based biosensor *Sci. Total Environ.* **689** 655–61

[79] Graziani S, Beaubien S E, Bigi S and Lombardi S 2014 Spatial and temporal pCO$_2$ marine monitoring near Panarea Island (Italy) using multiple low-cost gas *Pro Sensors Environ. Sci. Technol.* **48** 12126–33

[80] Kolomijeca A, Kronfeldt H D and Kwon Y H 2013 A portable surface enhanced Raman spectroscopy (SERS) sensor system applied for seawater and sediment investigations on an Arctic sea-trial *Int. J. Offshore Polar* **23** 161–5

IOP Publishing

Sensors for Marine Biosciences
Next-generation sensing approaches
Shyam S Pandey, Rout George Kerry and Kshitij RB Singh

Chapter 11

Commercial aspect of sensors in marine biosciences and its future prospects

Olugbemi T Olaniyan, Young N Wike, Charles O Adetunji, Gloria E Okotie and Chioma A Ohanenye

Over the past 35 years biological detectors have been developed, and during the last 15 years, studies in this area have grown significantly in popularity. The biological sensor works as follows: the transducer transforms the way the body reacts to an identifiable message that is capable of being determined electrochemically, visually, through sound, manually, by thermal analysis, or via the internet. Biological detectors remain the subject of broad research and use in a wide range of applications, from national defense and food hygiene to public health and surveillance of the environment. Biological detectors have been made using a variety of naturally occurring detection components, such as cofactors, enzymes, antibodies, microbes, organelle-like structures, tissues, and cells derived from other living things. This chapter is aimed at enumerating the commercial aspect of sensors derived from marine biosciences and their future prospects.

11.1 Introduction

Over the past 35 years, biological detectors have been developed, and during the last 15 years, studies in this area have grown significantly in popularity. Despite being the most established type of biological detector, just a single type—glucose—has seen prevalent commercial achievement at the point of sale thanks to electrochemical biological detectors [1]. A detector appeared to be a probe of some kind in the beginning phases (the 1960s and 1970s), maybe because of a vision that was inextricably connected to pH, ion-selective, or oxygen electrodes. Study of biological sensors, also known as bioelectrodes, enzyme electrodes, or biocatalytic membrane electrodes, can be found in previous works [2]. In recent times, the concept has been expanded to encompass detectors concealed within sizable automated instruments [3]. Certain individuals believe that electrophoresis, chromatography, or mass spectrometry are suitable detector components [4]. As a result, numerous detectors utilized in biological applications—such as those measuring temperature, pressure, electrocardiograms, pH, Ca^{2+}, catecholamines, and the like—are not biological detectors. On the other hand, one could legitimately classify surface plasmon resonance (SPR) devices as employing biological sensors [3, 5]. Effective detectors may involve labeled micron-sized particles inserted into the cytosol of specific cells, which transmit data visually.

In order to produce a signal that can be measured proportionately to the level of the evaluations, a biological sensor is a device for analysis that combines a component that recognizes living things with a physical transducer [6–12]. The biological sensor works as follows: the transducer transforms the way the body reacts to an identifiable message that is capable of being determined electrochemically, visually, through sound, manually, by thermal analysis, or via the internet. The measurement result is subsequently associated with the amount of substance under analysis [7, 8, 12, 13]. Biological detectors remain the subject of broad research and use in a wide range of applications, from national defense and food hygiene to public health and surveillance of the environment, since Clark and Lyon created the first biological detector for detecting sugar in 1962 [1, 11, 14–16].

Biological detectors have been made using a variety of naturally occurring detection components, such as cofactors, enzymes, antibodies, microbes, organelle-like structures, tissues, and cells derived from other living things [9]. Because of their exceptional accuracy and precision, enzymes are an extremely frequently utilized biological component for detection [17]. On the other hand, the extraction of enzymes is expensive and laborious. Furthermore, there is a chance that the *in vitro* setting will cause enzyme activity to decline [13]. Microorganisms (such as yeast, bacteria, and algae) present a potential substitute for the production of biological detectors due to the fact that they are capable of being generated in large quantities via cell culture. Additionally, cells from microbial species possess more effective sustainability and equilibrium *in vitro* and are somewhat less difficult to work with than cells from larger creatures like plants, animals, and humans [13]. These characteristics may significantly streamline the manufacturing method and improve the effectiveness of biological detectors.

Small algae are typically suggested as biological receptors in detectors to assess contaminants in water, but they are difficult to work with and must first be properly collected, reduced, and trapped on alternating membranes. As a community-forming microalga that forms macroscopic mats, the kind of algae used in this investigation doesn't involve an encapsulation process, which facilitates and expedites the setup of each evaluation. The suggested biological sensor demonstrated intermediate responsiveness to carbonyl and high responsiveness to atrazine, but it took some time to identify the harmful effects of heavy metals. This may have been caused by chemical variables like the existence of organic finalizing agents and the acidic nature of the testing medium, or biological variables like the adsorption properties of the cell walls and particular routes for confinement, metabolization, and discharge of the utilized algae organisms.

Massive volumes of salt water are found in the marine environment, and the ocean's overall distribution of salt water contributes to many ecological systems guidelines [18], including those pertaining to worldwide temperature phases, carbon phases, as well as weather structures [19, 20]. Human-induced processes like disposing of trash and marina setup, along with mineral resource research, can have a major effect on underwater ecosystems [21, 22]. For prolonged management as well as safeguarding the ocean, it is crucial to comprehend all the elements and mechanisms that affect its well-being. Knowledge of the marine ecosystem and its wealth of assets requires knowledge of a wide range of chemical substances along with organisms, including methane [23, 24], radon [25–31], ferrous ions [31, 32–34], carbon dioxide [35, 36], microbes [37, 38], nutrients [39, 40], and seafood [41, 42]. Exploiting marine resources and ensuring the long-term growth of humanity require the discovery and measurement of chemicals and creatures in the marine environment.

Conventional techniques for quantifying marine analytical substances require the direct collection of specimens of water from the sea, debris, or organisms at the coast, either by ship or directly, followed by analysis in central testing facilities. Nonetheless, specimens swiftly deteriorate as time passes, places for sampling remain constrained, and the rate of sampling is small. As a result, the spread of analytical substances cannot be tracked in real time or space using traditional techniques. But in order to comprehend key events in the sea more effectively, a broad spectrum of time- and space-dependent investigations of certain analytical substances—like gases of radon (Rn), methane (CH_4), carbon dioxide (CO_2), hydrogen sulfide (H_2S), and helium (He)—are important. The *in situ* method has become more popular in recent years for tracking, visualization, and times of emergency. Marine analytical substances are the focus of numerous chemical-based and biosensor designs [43–45]. The physical and functional equilibrium, total computerization, minimal energy utilization, knowledgeable gathering of information, ability to withstand rust (such as ocean water rust as well as organism penetration), reduction in size, and compact size of *in situ* sites and detectors have piqued curiosity greatly. To accomplish computerized hydrological, chemical, and physical examinations outdoors, *in situ* structures were established [19, 21, 46, 47].

11.2 Classification of biosensors

Current reports of microorganism biological detectors can be divided into two primary categories according to the detection method: visual and electrochemical biological detectors. The method most frequently employed in electrochemical microorganism biological detectors is amperometry. Uses in the environment have made significant utilization of amperometric microorganism biological detectors [17, 48, 49]. Nevertheless, this approach takes a lot of time and isn't appropriate for remote surveillance [50]. The conductometric microorganism biological detector has a responsive and rapid reaction to evaluations, which makes it enticing. To this end, a conductometric biological detector was built to identify pesticides and heavy metal ions in water samples utilizing the biological receptor of *Chlorella vulgaris* micro-algaeas [51].

By gauging the voltage variance within the electrode that is being used and the electrode serving as the reference, which is distinguished by a particular membrane, potentiometric microorganism biological detectors are able to determine the quantity of evaluations. The latest invention for the specific and quick identification of the cephalosporin class of antimicrobial agents is a potentiometric biological detector that utilizes the pH electrode tweaked by porous *Pseudomonas aeruginosa* [52]. Protons were produced close to the pH electrode as a result of the microorganism layer's enzyme reaction causing the breakdown of cephalosporin. The functioning electrode's and the reference electrode's changing electric potential disparities caused the reaction. There has also been an investigation into determining the presence of beta-lactam contaminants in milk using a different potentiometric biosensor [53]. The biological detectors are categorized as follows.

11.2.1 Luminous microorganism biological detectors

These can be classified as either *in vivo* or *in vitro*, depending on their method of identification. Utilizing mutated microbes with a transcriptional combination involving a stimulated booster and a messenger DNA encoding luminescent amino acids, *in vivo* luminescent microbe biosensors work. Because of its appealing equilibrium and responsiveness, the protein known as green fluorescent protein (GFP), which is encoded by the GFP genetic material, is one of the most widely used devices. Contemporary imaging devices have the ability to identify the luminescence that GFP emits with minimal or no damage to the host system [54].

11.2.2 Biologically luminescent microbiosensors

A biologically luminescent microbiosensor gauges the variations in illumination that microbes that exist release. In actuality, the lux gene-coded luciferase is what changes illumination and reacts in a dose-related way to the subject of analysis. Microbes can have essential or stimulated regulation of the activity of the lux gene. The lux genetic material is present essentially in the detecting microorganism, and the introduction of chemical substances of interest will cause an immediate shift in biological luminescence. A biologically luminescent microorganism biological detector inspired by *Vibrio fischeri* was created for the quick assessment of the poisoning of certain prevalent contaminants in an ongoing flow structure, relying on the observation that the amount of light generated by the microorganisms might be decreased in the presence of poisonous substances [55]. For the quick and precise tetracycline leftovers test in chicken muscle tissue, a tetracycline luminous whole-cell biological detector was created using membrane-permeabilizing intermediary poly-myxin and the band-sensitizing agent EDTA [56].

11.2.3 Sodium channel-based biological detectors

A tissue biological detector was developed by utilizing the impact of paralytic shellfish poison toxins (saxitoxin, gonyautoxin, and tetrodotoxin) as sodium channel inhibitors [57, 58]. The frog bladder membrane, which is abundant in sodium channels, was used by the researchers to protect a sodium electrode and incorporate it into a circulation cell. By examining the movement of sodium ions, the researchers were able to identify the existence of a poisonous substance. In the instance of tetrodotoxin, the biological detector had the ability to identify quantities that were over an order size less than the biological experiment's threshold of identification. The hazardous quantities of the contaminants are linked with the values found using the mouse in a biological experiment. One example is the neuronal system biological detector that Kulagina *et al* [61] created, which takes advantage of the external action potentials that are induced by saxitoxin and brevetoxin-3. Growing cultivated human neural networks from developing mouse spinal cord tissue over a 64-site microelectrode array allowed for the construction of the biological detector. Both of these contaminants reduced the average surge rate of spinal cord neural circuits, regardless of having different effects on the nervous system (brevetoxin-3 increases the firing of sodium channels, whereas saxitoxin blocks the development of action potentials).

The saxitoxin and brevetoxin-3 thresholds for identification were 12 and 296 pg l^{-1} in buffer and 28 and 430 pg l^{-1} in 25-fold diluted ocean water, respectively. The tremendous responsiveness of the neural connection accounts for these extraordinarily low readings. Furthermore, the collection reacted to the presence of algae that produced toxins but did not react to isolates of the same algal genera that did not produce toxins. Even though this general method is unable to precisely recognize or measure specific toxins, its use as a means of detection is unquestionably warranted.

11.2.4 Biological detectors designed around enzyme inhibition

Currently, Hamada-Sato *et al* have created a detector for biological purposes that integrates pyruvate oxidase's absorption of phosphate ions and protein phosphatase 2 A inhibition into a flow infusion evaluation structure [62]. But only the second enzyme becomes trapped during the immobilization phase, which is carried out in a microtube. However, they used a biological detector that is fifty times more responsive than an ELISA to determine optical absorbance with an admissible limit of 0.1 ng ml^{-1}.

11.2.5 Methane (CH$_4$) monitoring devices

Methane (CH$_4$) has a greater warming impact compared to carbon dioxide, making it the second-most significant greenhouse gas in the world [63, 64]. The decomposition of biological material is the main source of methane in the marine environment [39]. Methane can be taken up, disintegrated, gaseous in nature, or fully hydrated. Methane moves through the ocean by condensation or dissemination. Methane leaves underwater from below methane stores on the mainland and ocean gradients in certain sea regions, where it is primarily produced by the naturally occurring gas moisturize sheet in the seafloor debris [65, 66]. The water-soluble form of natural gas is highly susceptible to variations in either climate or tension. Natural gas moisture in the sediment on the seafloor has become less stable over the past decade due to tidal pressure variations and the ongoing warming of the ocean. As a result, the degree of disintegrated methane in ocean water is unusual, and the methane hydrate continues to break down or disintegrate, releasing an enormous quantity of methane gas into the marine environment [67].

Microbes in salt water break down and consume a portion of the gas called methane present in the water. The procedure of breakdown in marine water necessitates an enormous quantity of air and generates copious amounts of CO_2, exacerbating coral bleaching and impacting the overall carbon content of the worldwide marine environment and carbon cycle structure [68]. By means of air–sea swapping, a different amount of methane gas escapes towards the surrounding environment, which has a major effect on climate change [67, 68].

However, in recent decades, natural gas hydrate—a source of methane energy—has been regarded as a possible renewable energy source with the highest effectiveness and most clean burning [68]. Even though the carbon content of methane is more than two times that of all petroleum products, the carbon dioxide generated by the burning of methane is substantially lower compared with that of petroleum oil and coal [69]. Finding stores of natural gas hydrates depends critically on the identification of anomalous variations in the amount of methane in the ocean. In earlier times, passive or separate specimen evaluations had been the basis for the majority of methane metrics [70, 71]. Gas chromatography is the primary analytical tool used in the conventional approach to identifying disintegrated methane from separate water specimens. Nevertheless, methane may diffuse during specimen gathering and preservation, making it effortless for the specimen to become polluted. This could lead to a discrepancy in the evaluation outcome [72, 73].

Methane flow patterns were determined recently through the development of hydroacoustic imaging methods and bubble catcher evaluations [74, 75]. Weber *et al* for instance, investigated a technique to calculate the methane gas movement collected from the ocean floor. It uses split-beam and multibeam ultrasound devices in combination with acoustic mapping methodologies as its foundation [76]. A stereo-camera underwater detector bubble container was used to monitor ocean gas discharge locations, according to Jordt *et al* For the acoustic polls, the structure may supply fluxes or bubble size distributions [77]. Nonetheless, *in situ* sensors—such as mass spectrometry detectors [78], optical detectors that utilize infrared absorbance spectroscopy [79], Raman spectra [80], and electrochemical conductivity detectors centered on semiconductor gas sensor components [81, 82]—can observe disintegrated methane instantaneously. A few of these are readily accessible detecting techniques for electrochemical conductivity as well as off-axis incorporated chamber production spectroscopy.

As soon as methane crosses the gas–liquid separation membrane and gets to the outermost layer of the semiconductor test (such as SnO_2), it electrochemically interacts with oxygen according to the heating power, changing the conductivity of SnO_2. This reaction is the basis for the operation of electrochemical conductivity detectors, which utilize semiconductor sensors to detect gaseous substances. The percentage of methane thereafter is determined. The German company Capsum developed the initial commercial SnO_2-based *in situ* methane detector in 1999 [83–85]. For the purpose of separating the dissolved gases in ocean water, a polydimethylsiloxane (PDMS) barrier is utilized. The sensing container contains sections such as temperature, humidity, and methane sensor probes. The methane-detecting probe detects the methane gas isolated by the PDMS membrane.

The methane-detecting probe's operational environmental conditions are monitored by the temperature as well as moisture detector probes, accordingly. The detector is capable of being operated typically once the moisture content within the detection probe reaches 100%. This device's highest performing level of water is 2000 m, its sensitivity range is 10–4000 nM, and its reactivity period is typically 1–30 min. Nevertheless, the accuracy, specificity, and sensitivity of the equipment may be impacted if additional gases that are traveling along the PDMS membrane break down on the semiconductor. The harmless, fast, and extremely accurate nature of optical detectors makes them useful. Additionally, there are a lot of commercial products with optical *in situ* methane detectors available.

11.2.6 Radon (RN) detection sensors

Given that exertion alters the degree of radon, an inert nonchemical reactive gas, comprehending fundamental procedures in the waters around us is crucial. Radon is useful as an indicator of certain geophysical processes that occur in marine as well as water-based settings, like determining the piston speed of air and sea gases and locating groundwater that moves toward water at the surface [86, 87]. Radon-222 is appropriate for measurement and application in investigations like dissemination from particles [88], seismic event research [89], naval groundwater emissions [42, 47], land-sea relationship [90], springs [91], carbon dioxide gathering and preservation obligations [92], air–sea communication [93], the discharge of greenhouse gases and global warming [94], and renewable energy sources (particularly in the deep sea) [95]. This is because the half-life of radon-222 (3.83 days) is significantly longer compared with that of radon-219 (3.96 s) and radon-220 (55.6 s) [96].

In laboratories, liquid oscillation counting is commonly used for radon measurement [97]. Alternatively, radon can be extracted from water samples using a radon removal line framework, transferred to Lucas alpha oscillation cells, and then counted using a counting device after a three-hour span of sealing to enable the radioactive balance of the radon-222 descendants (218Po, 214Po) to reach [98]. This method has been widely applied in a number of global investigation initiatives and offers the greatest evaluation effectiveness, which is nearly 300% [99, 100]. But this approach is laborious and fraught with problems with organization.

The RAD-7 radon-in-air tracking framework and gamma-ray spectrometer are just two examples of numerous innovative *in situ* detectors and detection mechanisms that have been documented in the last few decades to conduct constant radon metrics in marine life. A computerized radon-222 evaluating apparatus for determining the changing pattern of groundwater participation in the area near the coast has been described by Burnett *et al* [26]. A submersible pump-powered constant supply of water that travels through an air–water exchange system, which transfers radon from a circulating flow of water to a sealed air cycle, is used by the device to identify radon-222. To determine the level of radon-222, the air stream is channeled into an industrial RAD-7 radon-in-air tracker (Durridge Co., Inc., Bedford, MA, USA) that gathers and evaluates the α-emitting daughters, Po-214 and Po-218.

11.2.7 Sensors for ferrous ions (FE^{2+})

In the outermost layer of the earth, iron (Fe) is found in large quantities (about 5.04%), only surpassed by oxygen (O), silicon (Si), and aluminum (Al). Ferrous ion and ferric ion are the two primary chemical-based forms of the valence of iron in seawater; ferrous ion is particularly unstable and readily reduced to Fe^{3+} [101, 102]. Iron is a crucial redox-responsive substance in debris from the ocean and is involved in numerous biological, geological, and chemical procedures and techniques, including the discharge and accessibility of phosphorus, the periodic turnover of sulfur, and the recovery of minerals from biological material [38–41]. For instance, despite the fact that naturally accessible disintegrated iron concentrations in the seawater are inadequate, they regulate the development of phytoplankton and nitrogen fixation, which subsequently impacts ocean-atmosphere carbon dioxide exchange and global warming [103].

According to a few researchers, adding a particular quantity of iron to the ocean may encourage the development of phytoplankton as well as hasten carbon dioxide's contact from the atmosphere to the bottom of the sea. This will help lower the Earth's carbon dioxide concentration as well as lessen the impact of climate change [104, 105]. On the other hand, iron that is present in excess of what is needed for physical needs might prove harmful as well as function as a blocker of certain enzymes, leading to oxidative damage as well as irregularities in the chemical breakdown of iron in multiple biological reactions [106, 107]. For this reason, precise measurements of Fe ions within the seawater are crucial for biological, geological, and chemical research.

Within particles, there is a significant slope in the spatial arrangement of the amount of Fe ions, ranging from numerous micromolars across millimeters to centimeter-straight measures [108, 109]. Nevertheless, the amounts in ocean water range from 0.1 to 10 nM, making it extremely difficult to determine them accurately [110]. The atom absorption spectrometer (AAS), inductively coupled plasma mass spectrometer (ICP-MS), and spectrophotometer are the mainstays of traditional techniques for identifying disintegrated ferrous ions. These instruments have been utilized to analyze the examined water from pores [111, 112]. Nevertheless, ferrous ions rapidly oxidize in the atmosphere, and this could harm the typical arrangement of ferrous ions in particles, and the process of collection is difficult and takes a while [113, 114].

Furthermore, despite the fact that the ability to identify iron in ocean water is capable of being significantly improved by applying these techniques used in laboratories, their application is limited to *in situ* identification due to the costly and labor-intensive instruments as well as the intricate management procedures. A variety of approaches, such as diffusive gradients in thin film (DGT) or diffusive equilibrium in thin film (DET) techniques, optical procedures, and electrochemical procedures, were created for the purpose of detecting ferrous ions *in situ*. Of these, voltametric microelectrode-based detectors are already accessible to consumers, while additional techniques are still in the laboratory. Electrically charged tiny sensors, which evaluate upward as well as geographically inserted identities of particles using sub-millimeter image quality, provide an acceptable replacement to traditional techniques [115]. An underwater voltammetric test that may detect ferrous ions in the sediment-water link instantaneously *in situ* has been published by Tercier-Waeber *et al* [116]. The technique for determining the concentration uses squarewave cathodic stripping voltammetry, which yields a detection limit of 0.1 μM and a reaction period ranging from 5 to 10 min. For the purpose of measuring minor metals *in situ*, a method is now commercially viable [117, 118].

A submersible multiphysicochemical investigator framework was built just recently [119] for *in situ* tiny metal detection in various waterways and coastlines. It is made up of three distinct circulation units designed to gauge free ions of metallic substances and flexible metal organisms, and it has become an enhancement over the previous model. The detector isn't required to be updated for eight days of constant evaluation, thanks to the structure. Nevertheless, when substances that are organic are added to the particles, the outer layer of the electrode becomes quickly disconnected, and the precision of detecting them diminishes [120–122].

A different strategy for *in situ* small-molecule metals measurement depends on the diffusive variations in thin film (DGT) or diffusive balance in thin film approach [104, 123–125]. The measurement channel for diffusive balance in thin film is a hydrogel with a 95% moisture content, and its working concept is identical to the one used for kidney replacement therapy. The difference in the concentration of the intended analytical substance in the exterior aquatic environment and the hydrogel (a dispersion stage) propels DET diffusion. The level of the slope continually drops as the distribution procedure moves forward until it settles at a value of zero, at which point the procedure of sampling is finished and the dispersion balance is attained. Both the diffusive balance in thin film as well as the DGT methods depend on Fick's first law of diffusion, which makes them comparable.

By managing the procedure of ion exchange via the hydrogel, the DGT electronic devices achieve the statistical buildup and evaluation of the desired analytical substance while separating the ion exchange resin from the testing solution. The time frame for sampling for both DET and DGT is typically a few hours, and they are the same *in situ* inactive methods for collecting data [126–128]. To determine the spatial arrangement of ferrous ions in particles, it may be utilized in conjunction with colorimetric techniques [116, 129]. A colorimetric DET technique, for instance, was proposed by Pages *et al* [130] to perform a two-dimensional evaluation of co-distributed iron dioxide and sulfide in algae debris porewaters.

In order to obtain a two-dimensional representation of the iron that is dispersed and the phosphate in the particles of sediment pore water, Cesbron *et al* merged the DET and spectrophotometric methods [116]. This technique's exceptional spectral clarity (4.5 nm) and sub-millimeter identification (60 μm visualization acquisition per pixel) are its benefits. Furthermore, new optical techniques for identifying ferrous ions in particles and seawater have been reported recently [43, 129].

These techniques include straight optical detectors and microfluidic *in situ* analytical instruments. A microfluidic *in situ* analyzer was created to measure dispersed iron in ocean bodies and vertical profiles [43]. A spectrophotometric detector and microfluidic techniques are combined in this system, and a unique in-cell process of diffusion serves as the foundation for the incorporated method used for specimens and reagents.

11.2.8 Detectors for carbon dioxide (CO_2)

The amount of CO_2 in the water from the sea is rising in tandem with rising greenhouse gas emissions. It was previously shown that the ocean eventually received over 25% of carbon dioxide greenhouse gases. As a result, carbon dioxide primarily accumulates in the seawater [42, 43]. Human-caused carbon dioxide consumption by the seafloor may dramatically shift the sea carbonate structure and cause a drop in seawater pH [130–132]. The marine carbon dioxide and ocean carbonate structures are characterized by six variables: pH, bicarbonate ionic levels, total alkaline content, inorganic carbon dissolved in water, partial pressure of carbon dioxide, as well as carbonate ionic levels. Among them, pH and the partial pressure of carbon dioxide are frequently employed in measurements [133]. In laboratory settings, regular tests and samplings of seawater take a lot of time and effort. Additional exciting factors include prolonged, independent, *in situ* tests, which can yield improved temporary and geographical information [134, 135].

Gas analysis, electrochemical, wet-chemical, and brightly colored optode measurements are the four methods that underpin the majority of carbon dioxide detectors. Just the gas analysis detector has been made accessible for purchase out of all of them. The equilibrator is the essential component of the gas detector for carbon dioxide transfers, which includes bubbler, shower form, laminar circulation, and percolated-bed forms [135–138]. Gas sensors are capable of transferring ocean water CO_2 into a gaseous state for evaluation. Although the membrane-based equilibrator has the advantage of being portable and appropriate for *in situ* distribution, changes in pressure at varying sea depths can readily impact the membrane, which in turn can impact reaction times. The pH change caused by carbon dioxide is measured by electrochemical evaluations using pH detectors [139, 140].

Potentiometric microelectrodes are commonly utilized for pH measurement purposes. Nevertheless, these detectors' precision is limited to ±0.01 pH units [40]. In order to satisfy the precision demands, the optimal pH reading must be precise to within 0.001 units [139]. The spectrophotometric structure, which gauges the absorbing capacity of ocean water after adding colorimetric pH markers, is typically the foundation of wet-chemical approaches for gauging pH changes in seawater [140, 141]. The unstable nature of the illumination origin, the degradation of the gas separation membrane, and the ease of biofouling are some of the difficulties associated with the installation of *in situ* spectrophotometers. The analyte-sensitive signals in gas porous membranes serve as the primary basis for the optodes used for identifying disintegrated CO_2 in ocean water [131]. Optodes have been used for real-time measurements at sea with the benefit of requiring little electricity, no need for mechanical components or reagents, and zero waste generation.

Numerous innovative detectors for measuring carbon dioxide have been documented over the last 20 years [132, 142, 143]. For surface seawater, Lu *et al* stated the use of an optical fiber-based *in situ* chemical partial pressure carbon dioxide detector [141]. A ratiometric straight optode fluorosensor was suggested by Zhu *et al* [132] for two-dimensional visualization of the spatial distribution of the partial pressure of carbon dioxide in particles and surrounding water. By employing a small, inexpensive GasPro probe, Graziani *et al* documented the *in situ*, constant observation of the partial pressure of carbon dioxide at the coast of Panarea Island (Italy) [144]. Temperature cross-sensitivity is a problem for all of the aforementioned detectors utilized for *in situ* carbon dioxide evaluation, necessitating more readings of temperature [131]. A planar optical detector for the concurrent assessment of temperature, pH, carbon dioxide, and oxygen [145]. It brought together a pair of optically distinct, multifaceted detection systems with multi-layer components.

11.2.9 The detectors for the identification of marine microbes

The field of molecular science places significant emphasis on the recognition of organisms as well as the characterization of microorganisms. Comprehending the relationship between coastlines and marine environments is made possible by the swift detection of microbes that live in water [37, 38]. The past few decades have witnessed an unprecedented growth of biological sensors for facilitating the identification of infectious agents and planktonic microbes, particularly hazardous bacteria [146]. The primary source of toxins in the seawater is hazardous bacteria, which are putting numerous regions in global danger [147, 148]. For instance, the algae *Alexandrium tamarense* and *A. ostenfeldii* have the ability to produce strong neurotoxins that are subsequently removed from the water by shellfish and stored in the shellfish. These neurotoxins ultimately become harmful to people as well as creatures when shellfish that contain them are consumed [149].

Shellfish can become harmful to mankind with as little as one or two hundred cells per liter of some exceptionally tiny amounts of harmful organisms. These microorganisms have been effectively measured using whole-cell luminescence in live hybridization, sandwich combination experiments, and PCR-based tests; however, these techniques require the transportation of specimens to specifically designed research centers, as well as the use of expensive technology and skilled operators [150–152]. A multi-biosensor component for the *in situ* identification of harmful algae has been documented by Diercks *et al* [153]. With the help of a set of sixteen gold electrodes, the biological sensors were able to identify as many as fourteen species of interest at once. Numerous surface plasmon resonance biological sensors were described by McCoy *et al* [154] as a means of tracking the harmful *Alexandrium minutum* algae growth along the shorelines.

There was a clear correspondence between the identification outcomes and the light microscope outcomes. An electrochemical genetic material biological sensor has been suggested by Toldra *et al* [39] for the detection of *Ostreopsis* cf. *ovata*. The strain in question is detected on electrode sets using maleimide-coated magnetic beads using sandwich mixing and isothermal regenerated polymerase enhancement. In addition to harmful algae, marine organisms such as bacteria, viruses, and protozoa can contaminate seafood and ultimately contaminate humans [155, 156].

Certain infectious agents, like *Vibrio cholera* and the cholera bacteria, come from naturally occurring aquatic environments [157], whereas others, like streptococci from feces and *Escherichia coli*, come from the defecation, urine, or shedding of either human or animal hosts and are transported into coastal waterways by waterways or wastewater [158]. Because every infectious agent has a distinct molecular makeup, there is no standard approach to quantifying these infectious agents. Genetic information and biological sensors are the foundation of a particular intriguing technique [153, 159, 160]. Over the past few decades, a few innovative biological detection systems have been created. As an illustration, Liu *et al* described a fish-based microfluidic system that is quick, large-scale, lightweight, and simple to use [153]. The system in question has the ability to operate in tandem to identify and assess the risk of contamination from pathogens that are transmissible through water. A lateral-flow genetic material-aptamer-based biological sensor for the detection of Singapore grouper iridovirus (SGIV) was described by Liu *et al* [153]. For SGIV identification, two DNA aptamers with different functions were utilized: one of them for target separation and another for the DNA strand displacement enhancement response. Both aptamers were directed against SGIV-infected tissues. In contrast to PCR techniques, this approach requires no cutting-edge technology and can detect results in as little as 90 min.

11.2.10 The detectors for monitoring the contaminants in the sea

In addition to the inappropriate use of natural assets, human activity has been responsible for the introduction of numerous contaminants into the marine environment in recent years [161, 162]. These pollutants typically come into contact with ocean water from terrestrial origins by means of aquatic components, factories, or municipal operations. Particulate matter is often released into the oceans directly by petroleum and natural gas misuse and vessel operations [163, 164]. The primary contaminants found in water from the ocean are contaminants from hydrocarbons like gasoline and petroleum, chemical substances like microplastics, chemical fertilizers, animal care medications, and drugs, reliable minor components like cadmium, lead, mercury, and tin, and radionuclides [163, 165, 166]. The natural environment may be greatly impacted by the places of origin, quantity, dispersion, and persistent nature of such contaminants [167, 168]. The healthy functioning of the aquatic ecosystem depends on the specific and delicate tracking of such contaminants, which is an essential problem in marine scientific studies [169].

Sampling, observation, and measurement are the three primary methods used to evaluate the marine ecosystem. Sea sampling is done by vessels, which then send the specimens to the lab or within. The minimal levels of the aquatic pollutants make measurement challenging. As a result, such analytical substances ought to be detectable by the detectors at the ng l^{-1} or pg l^{-1} quantities [171–173]. Contaminant identification has made use of a few traditional analytical techniques [174]. For instance, liquid and gas chromatography are the primary methods used in the identification of organic pollutants in marine water specimens. Ecological testing is another common application for mass spectrometry [175, 176]. Furthermore, a number of channels, including wanderers, diving vessels, and orbiting objects, were built for computerized metrics.

Such approaches depend on observational techniques, satellite and aircraft pictures, and sampling directly [169]. The majority of computerized and distant gauges rely on tracking variations in seawater's characteristics, including depth, viscosity, insulation, and climate [47]. Furthermore, freshly developed chemical detectors and biological sensors have been created lately for tracking pollutants [177–179]. Malzahn *et al* for instance, described an embedded screen-printed biological sensor on neoprene submerged clothing made of rubber-like material [40]. In ocean water specimens, biological sensors are capable of being utilized to track the presence of phenolic pollutants, nitroaromatic weapons of mass destruction, as well as contaminants from heavy metals. Using a stimulating component and a computerized microfluidic diluter device, Han *et al* created an entirely automated whole-algae biological sensor [170]. The biological sensor has thresholds for detection of 0.65 μM, 1.90 μM, 2.85 mM, and 5.22 μM for the concentrations of lead, copper, phenol, and nonylphenol in ocean water, accordingly.

11.2.11 Aquatic nutrient detection sensors

The development of microbes depends on the vitamins and minerals found in the seawater, including phosphate, silicate, and nitrate [39, 40]. Numerous nutrients are flowing into the oceans by means of sewage from factories, crop residue, and household waste as a result of increased population growth, industrialization, and farming expansion [180, 181]. Among the more difficult issues affecting the oceans is eutrophication, and this can result in waves of red and toxic aquatic explosions [182, 183]. Tracking aquatic efficiency, forecasting the detrimental effects of eutrophication, and comprehending the changing patterns of aquatic ecosystems all benefit from gauging these elements [46, 184]. Conventional techniques for tracking nutrients involve taking samples of ocean water and bringing them to a lab for routine assessments like luminescence, the measurement of color, and spectroscopic measurement [185].

It is challenging to pinpoint short-term aquatic happenings and ascertain the movement and variability of nutrients using such techniques [186–188]. Furthermore, for certain saturated areas, the amount present is at the nM threshold, but the majority of traditional techniques are able to identify concentrations (nitrate, phosphate, as well as silicate) exceeding 0.1 μM [33, 189]. Thus, a viable substitute for catastrophic cautioning mechanisms in the aquatic setting is an *in situ* effortless detector along with an enhanced threshold for detection that is capable of being utilized to conduct lasting and constant measurements. The *in situ* detectors for aquatic nutrients are divided into three categories based on evaluation principles: colorimetric, visual, and electrochemical instruments [190].

There are certain *in situ* electrochemical measurement devices available for purchase. Nevertheless, such detectors' inadequate consistency, brief lifespan, limited identification quantity spectrum, and lack of precision prevent them from being widely used [191, 192]. The colorimetric detector uses a great deal of electricity and an enormous amount of chemicals to identify the substance being measured according to hue reactions at a given wavelength. The devices are hence unsuitable for purposes requiring continuous tracking [193]. In contrast, reagent-less UV sensor technologies that rely on direct absorption of ultraviolet (UV) are suitable for uses involving constant surveillance [194]. Numerous innovative *in situ* detectors and tools for tracking aquatic nutrients have been developed over the past decade [195, 196]. An independent electrochemical detector was created by Legrand *et al* for *in situ* silicate identification in aquatic settings [41].

11.2.12 The detectors for aquatic products

In human nutrition, aquatic creatures along with various marine life foods are essential. Numerous nutrients, including amino acids, unsaturated fats, vitamins, and minerals, are abounding in them [41, 42]. Aquatic products, on the other hand, are extremely susceptible to deterioration due to their neutral pH, significant water activity, minimal connective tissue content, and autolytic enzymes [196]. The spoiling procedure quickly degrades the animal's appearance, consistency, flavor, and aroma [197, 198]. Furthermore, in marine environments, aquatic products can become polluted with algal toxic substances, raising serious concerns about food safety. Thus, trustworthy techniques are required to assess the health and nutritional value of aquatic products. Numerous cutting-edge detection techniques and innovations, including those involving chemicals, physical, electrical, and biological sensors outlined below, were documented for use in evaluating the hygiene and nutritional value of aquatic products.

The precise proliferation of spoiling organisms, oxidation of lipids, peroxide amount, overall unstable essential nitrogen, polyunsaturated fatty acid composition, and additional unstable amino acids are a few examples of the physical as well as chemical variables that are capable of being utilized to determine the nutritional value and freshness of aquatic products [198–201]. Nevertheless, a variety of variables, like organisms, maturity, fishing region, time of year, and animal diet, affect these variables [202]. Conventional means of detection require specially trained workers and are laborious, costly, harmful, and labor-intensive [203]. In the past few decades, a few ecologically sound and safe methods have become popular. In order to evaluate the quality and safety of aquatic products, numerous innovative logical and chemical-based biological sensors, computerized noses, eyes, and tongues, visual and electromagnetic spectroscopic detectors, were recently developed [204–206].

For example, Mustafa *et al* observed a handheld nanoenzyme-based biological sensor to measure the shelf life of seafood by monitoring its discharge of hypoxanthine, a byproduct of nucleotide deterioration in meat and seafood [207]. To measure the hypoxanthine content, the biological sensor uses xanthine oxidase and cerium nanoparticles. A detection threshold of 89 μM is capable of being attained, and the linear identification span is capable of reaching 597 μM. Mesoporous Au–ZnO nanospheres served as the basis for a type of semiconductor metallic oxide gas detector that Chen *et al* investigated [208]. When seafood was allowed to deteriorate at 250 °C, the detector demonstrated an effective identification of 10 ppm trimethylamine. A pressure infusion amperometric detector was described by Torrarit *et al* [209] to identify formalin in seafood that is fresh.

An ultrasensitive permeable electrode-capped natural gas detector for the detection of unstable amino acid gas in seafood has been effectively created by Chang *et al* [210]. The detector can identify levels of ammonia that are as small as 100 parts per billion in just one moment. A bioelectronic nose, gas chromatography–mass spectroscopy (GC–MS), and sensory assessment were all incorporated into the fast-tracking device for fish nutritional value that Lee *et al* investigated [211]. A resistance detector constructed from WO_3 nanosheets with Au tiny particles, Zhao *et al* disclosed an ultraefficient trimethylamine gas detector for quick evaluation of fish and shellfish freshness [42]. Trimethylamine is detectable by the sensor at levels as tiny as 0.5 ppm.

11.2.13 Utilizing microbiological cells as biological detection components

When used as biological material detectors to create biological sensors, microorganisms offer several benefits. They are widely distributed and have the capacity to break down a large variety of chemical molecules. Microbes are highly adaptive to changing circumstances in the environment and can eventually break down novel compounds. Additionally, microorganisms can be genetically modified by genetic engineering or alteration, and they are inexpensive sources of enzymes found inside cells [33, 34]. Because the particular kinds of bacteria utilized in biological sensors have distinctive substrate wavelengths that might not correlate nicely with the wavelengths of compounds found in the sample, choosing a suitable environment is crucial. It is frequently preferable to modify bacteria for the purpose of inducing desired metabolic processes and absorption systems through cultivation in an environment that includes the right materials [35–37]. A number of biological processes are using microorganism biological detectors, which use the light discharge coming from luminous microbes as a delicate, quick, and simple assay [38, 39]. Biologically luminescent microbes can be identified in a variety of natural settings, from land (*Photorhabdus luminescens*) to aquatic (*V. fischeri*). Genetically modified microbes were additionally utilized to create biologically luminescent entire-cell biological detectors to track heavy metals, pesticides, and organic pollution. The microbes utilized in these biological detectors are usually grown using a DNA fragment that has been engineered to ensure that the transcription factor that detects the analysis of interest controls the luciferase-coding genetic material. The inheritable regulation system also activates the production of luciferase, which generates light that luminometers are able to identify when these tiny organisms' breakdown organic contaminants. [40].

11.2.14 Uses of microorganisms as biological detectors in the environment

Microorganism biological detectors are primarily used in the ecological area [41–50]. Microorganism biological detectors were created to measure biological oxygen demand, which is a measure of the total quantity of organic substances present in wastewater. Biological oxygen demand detectors evaluate the degree of decrease in oxygen by taking advantage of the rapid responses of microbes connected to electrodes. An oxygen microelectrode containing trapped bacterial cells in polyvinyl alcohol was combined to create a microbial biological detector that can determine the amount of accessible natural carbon present in harmful particles. The biological detector makes it possible to estimate accessible organic material that is dissolved at a tiny level in debris profiles [51]. Biological oxygen demand: biological detectors have been reported to utilize transducers that utilize optical fiber [52] and calorimetric [53] technology.

A few of the biggest categories of contaminants in the atmosphere are halogen-containing hydrocarbons, which are utilized as bubbling agents, flame retardants, chemicals, medicines, and intermediates in the synthesis of polymers. Microorganism biological analyses employ trapped *Rhodococcus* strain cells that included alkyl-halide hydrolase. From halogen-containing hydrocarbons, the enzyme that can be found in the cell releases halogen ions [54]. The creation of a microbial detector involved an extension of these investigations [55]. The detector has a one-week shelf life at 277 K when kept dry. Currently, a gram-positive actinomycete-like organism has demonstrated higher prospects and may hold promise in the development of a wide-ranging biological detector for halogen-containing hydrocarbons. It exhibits an array of characteristics for the dehalogenation of hydrocarbons with halogen [56].

11.3 Microbial biosensor applications in the food, fermentation and related fields

The need for fast and precise analytical equipment for dietary and oxidation evaluation is increasing and has continued to do so in the past few decades. A diverse range of scientific techniques is needed for the high-quality guarantee of dietary substances, as required by both the private sector and public medical care organizations. Keeping track of dietary variables, flavorings, food-borne illnesses, microorganism counts, durability evaluation, as well as manufacturing attributes such as aroma and stench all require evaluation. Numerous detectors, including smart noses [60, 61] and those that utilize proteins and immune systems [57–59], are being documented. Additionally, microorganism biological detectors have demonstrated promise in the assessment of diet [59].

Numerous studies on microorganism biological detectors for proteins, including glutamic acid (*Bacillus subtilis*), tryptophan (*Pseudomonas fluorescens*), and tyrosine (*Aeromonas phenologenes*), are obtainable [62, 63]. Phenylalanine identification is required for determining the presence of hyperphenylalaninemia in neonates as well as for the nutritional treatment of the condition. It is additionally necessary for the methodical monitoring of the phenylalanine microbial metabolism. There is a study on a microorganism biological detector that uses trapped *Proteus vulgaris* cells in calcium alginate on an amperometric air electrode [64]. The cell's phenylalanine deaminase converts phenylalanine to phenylacetic acid.

11.4 The significance of microbiological biological detectors in maritime settings

Emerging reports of seafood as well as crustaceans' intoxication caused by various phycotoxins have highlighted the critical need to rely on suitable sensor innovations. Numerous procedures for analysis, including immune testing methods, chromatographic approaches, enzyme inhibition-based tests, and biological tests, are being designed for this particular reason. Nonetheless, simple-to-use, quick, as well as affordable equipment that can handle intricate frameworks remain necessary. Biological detectors present themselves as viable biological tools for quick, easy, affordable, and trustworthy contaminants assessment, either as a supplement or a substitute to traditional statistical methods.

The aquatic algae *Spirulina subsalsa* is connected to a Clark-type oxygen electrode in a circulation framework as part of a whole-cell detector structure that is used to analyze the ocean. This method estimates the degree of contamination based on changes in transpiration [65]. Chemicals used in agriculture, surfactants, and chlorophenols were among the compounds examined. For atrazine, simazine, isoproturon, and diuron, an analogous method utilizing *C. vulgaris* and fiber optic sensor detection has been established with sub-ppb detection thresholds for photosystem 2 blockers [66]. Full-cell detectors have an edge in this situation because they may gauge bodily reactions and provide data regarding accessibility, both of which have significance for aquatic procedures. The drawbacks include the fact that the information gathered is typically more vague compared to that generated by enzymatic or affinity detectors and may require chemical examination to determine the connection that exists between pollutants.

The contaminants released into aquatic ecosystems by petroleum exploration systems' drilling fluids as well as generated water are an issue unique to nautical surveillance [67–69]. The chemicals used for generating water can range from hydrocarbons to biological agents, emulsifiers, surfactants, anti-corrosion agents, and antifoam agents. These substances may exhibit hazardous consequences that are either short-term or long-term. An estimate of the harmful effects or danger presented by utilizing a complicated and perhaps important system of life might serve as a beneficial instrument in analyzing the impact, considering the broad array of substances that could be involved. A number of microbes are now altered genetically to react to specific pressures or toxins, and these microbes are referred to as 'biological detectors for ecological pressures' in previous research [70–73].

The pervasive presence of repellent substances in water and debris, most notably in the form of tributyltin (TBT), is a further concern that is specific to the aquatic setting. Tributylin continues to exist in numerous particles even though the International Maritime Organization (IMO) has approved an outright restriction, and oceanographers and authorities will continue to struggle with its measurement and elimination for decades afterward [74]. The consequences of organotinson biota that have been best studied include endocrine interruption, direct contamination, oyster shell thickening, and a decrease in the recruitment of their young phases.

11.5 Prospects

The advancement of aquatic detection systems as well as techniques has significantly sped up over the past decade, in part due to growing awareness of the importance of the marine ecosystem. The foundation of conventional methods of measurement involves taking specimens coming from the ground or from a vessel and sending them to a lab. The entire process is lengthy and laborious, and it is challenging to determine the flow of analytical substances because these substances may alter immediately after departing from their initial state. To gain more insight into significant marine procedures, it is crucial to conduct *in situ* time- and space-dependent studies of certain substances, such as CO_2, radon, and methane gases. With enhanced automation as well as instantaneous evaluation, *in situ* detecting systems are capable of being tiny in dimension, reduced in power usage, and stable over a prolonged period. They can also be exceptionally picky and react rapidly to trace quantities of target analytical substances.

The common *in situ* aquatic sensors—such as ferrous ion, carbon dioxide, radon, methane, and detectors to identify microbes, contaminants, vitamins and minerals, and seafood—as well as their purposes have been outlined above. Finding petroleum moisturizer storage tanks depends on the identification of methane, which can be done using mass spectrometry, optical, and electrochemical conductance detectors. In order to investigate maritime subsurface water leaks, radon must be detected. Radon detectors consist of gamma-ray spectrometry, pulsed-ionization-chamber-based detecting structures, and radon-in-air detection devices. A lot of biogeochemical responses and procedures, including phosphorus discharge and accessibility, depend on the identification of ferrous ions. Ferrous ion detectors consist of electrochemical detectors, DET or DGT detectors, and optical detectors.

Human-caused carbon dioxide utilization in the marine environment has the potential to modify the sea carbonate structure. GasPro probes, optode-based detectors, and optical detectors are examples of carbon dioxide detectors. Knowing the changing patterns of coastlines as well as aquatic environments depends on the identification of microbes. Examples of microbe detectors are genetic material detectors for infectious agents and bacteria, RNA biological sensors for bacteria, and plasmon resonance (PR) biological sensors for seaweed. Pollution identification is crucial when considering the aquatic ecosystem. Some of the detectors used to identify toxins are whole-algae biological sensors, which identify the presence of lead, copper, phenol, and nonylphenol in ocean water; phytoplankton from the ocean motion detectors, which evaluate polluting substance poisoning; and surface-enhanced Raman scattering detector systems, which identify polycyclic aromatic hydrocarbons. Nutritional value identification is a critical component of evaluating aquatic environment enrichment, and available detectors for this purpose consist of colorimetric, optical, and electrochemical apparatuses.

The primary purpose of aquatic product identification is to evaluate the health and nutritional value of different types of marine life. Chemical biological detectors, electronic noses, eyes, and tongues, nuclear magnetic resonance spectroscopic detectors, and optical spectroscopic detectors are some of the detectors utilized to analyze the quality of marine life. Instead of concentrating on keeping track of particular data collection regions, subsequent studies on the invention and application of aquatic and biological detectors ought to focus on the creation of lightweight, intelligent, durable, inexpensive, unique, and diverse detecting systems for *in situ* tracking of these aquatic analytical substances. In the years to come, these *in situ* detectors can be incorporated into floating vessels (like gliders, vessels of chance, independent submerged automobiles, and distantly operated automobiles) and stable/fixed systems (like vessels and landers) in accordance with varying functioning depths. On these structures, these detectors can be effortlessly performed for constantly monitoring target substances without inspection.

One of the most important future requirements is the need to bring consumers detectors that can be readily incorporated into portable as well as attached aquatic systems and that, when applied to massive or huge amounts of data, offer enhanced statistics by means of machine learning and artificial intelligence techniques. Furthermore, these detector systems show promise for developing coastline monitoring systems to gather additional actual time and multifaceted, spatially resolved information, which will offer knowledge for managing aquatic environmental problems like greenhouse gases released into the atmosphere, groundwater discharges from beneath the surface, pollution of aquatic ecosystems, and methane and CO_2 source sinks more skillfully. Robust computerized evaluating structures, when paired with state-of-the-art *in situ* detectors on appropriate platforms, may generate extensive permanent observational information concerning aquatic matters.

Furthermore, a deeper comprehension of several important marine procedures, including eroding, being moved, and deposition, can be gained through the *in situ* temporal as well as spatial investigation of these analytical substances. Nevertheless, maintaining the precision, stability, and consistency of aquatic detectors is extremely difficult due to the highly volatile and variable nature of the oceanic setting, which frequently experiences significant daily, cyclical, and geographical shifts in salt content and temperature. It will take a lot of interdisciplinary verification work before those detectors can be used in real-world applications. To maximize the efficiency of the detectors, materials science, chemistry, and electronics must be integrated. One significant method for cutting down on detector dimensions and electrical consumption is microfluidics.

In order to construct these *in situ* detectors, anti-biofouling techniques are required because the buildup of plants, animals, and microbes on the surface of the detector can bio-foul the instrument and reduce its ability to function. The durability and reliability of aquatic permanent detectors should be enhanced by the investigation of novel antifouling substances and procedures. Furthermore, real-world ocean water situations with changes in temperature as well as other chemical and physical qualities ought to be used to assess sensor efficiency.

References

[1] Kissinger P T 2005 Biosensors—a perspective *Biosens. Bioelectron.* **20** 2512–6

[2] Arnold M A and Meyerhoff M E 1984 Ion-selective electrodes *Anal. Chem.* **56** 20R–48R

[3] Aldridge S 2004 Biosensors offer advantages for screening *Genet. Eng. News* **24** 25

[4] Huynh B H, Fogarty B A, Lunte S M and Martin R S in press On-line coupling of microdialysis sampling withmicrochip-based capillary electrophoresis *Anal. Chem.* **76**

[5] Hitt E 2004 Label-free methods are not problem free *Drug Discov. Devel.* **7** 34–42

[6] Belkin S 2003 Microbial whole-cell sensing systems of environmental pollutants *Curr. Opin. Microbiol.* **6** 206–12

[7] Cunningham A J 1998 *Introduction to Bioanalytical Sensors* (New York/Chichester: Wiley)

[8] Eggins B R 2002 *Chemical Sensors and Biosensors* (Chichester: Wiley)

[9] Lei Y, Chen W and Mulchandani A 2006a Microbial biosensors *Anal. Chim. Acta* **568** 200–10

[10] Sadana A 2001 *Engineering Biosensors: Kinetics and Design Applications* (San Diego, CA/ London: Academic)

[11] Wilson G S and Gifford R 2005 Biosensors for real-time in vivo measurements *Biosens. Bioelectron.* **20** 2388–403

[12] Wilson J S 2005 *Sensor Technology Handbook* (Amsterdam: Elsevier) Yagi K 2007 Applications of whole-cell bacterial sensors in biotechnology and environmental science *Appl. Microbiol. Biotechnol.* **73** 1251–8

[13] Byfield M P and Abuknesha R A 1994 Biochemical aspects of biosensors *Biosens. Bioelectron.* **9** 373–400

[14] Amine A, Mohammadi H, Bourais I and Palleschi G 2006 Enzyme inhibition-based biosensors for food safety and environmental monitoring *Biosens. Bioelectron.* **21** 1405–23

[15] Lazcka O, Del Campo F J and Munoz F X 2007 Pathogen detection: A perspective of traditional methods and biosensors *Biosens. Bioelectron.* **22** 1205–17

[16] Patolsky F, Zheng G F and Lieber C M 2006 Nanowire-based biosensors *Anal. Chem.* **78** 4260–9

[17] D'Souza S F 2001b Microbial biosensors *Biosens. Bioelectron.* **16** 337–53

[18] Buonocore E, Grande U, Franzese P P and Russo G F 2021 Trends and evolution in the concept of marine ecosystem services: an overview *Water* **13** 2060

[19] Fukuba T and Fujii T 2021 Lab-on-a-chip technology for *in situ* combined observations in oceanography *Lab Chip* **21** 55–74

[20] Ingeman K E, Samhouri J F and Stier A C 2019 Ocean recoveries for tomorrow's Earth: hitting a moving target *Science* **363** eaav1004

[21] Mills G and Fones G 2012 A review of *in situ* methods and sensors for monitoring the marine environment *Sens. Rev.* **32** 17–28

[22] Neal C, Leeks G J L, Millward G, Harris J R W, Huthnance J M, Rees J G, Millward E and G 2003 Land ocean interaction: processes, functioning and environmental management: a UK perspective *Sci. Total Environ.* **314–316** 801–19

[23] Niu M Y, Liang W Y and Wang F P 2018 Methane biotransformation in the ocean and its effects on climate change: a review *Sci. China Earth Sci.* **61** 1697–713

[24] Valentine D L 2011 Emerging topics in marine methane biogeochemistry *Annu. Rev. Mar. Sci.* **3** 147–71

[25] Zhao S B, Li M, Burnett W C, Cheng K, Li C Q, Guo J J, Yu S L, Liu W, Yang T, Dimova N T *et al* 2022 In-situ radon-in-water detection for high resolution submarine groundwater discharge assessment *Front. Mar. Sci.* **9** 1001554

[26] Burnett W C and Dulaiova H 2003 Estimating the dynamics of groundwater input into the coastal zone via continuous radon-222 measurements *J. Environ. Radioactiv.* **69** 21–35

[27] Eleftheriou G, Pappa F K, Maragos N and Tsabaris C 2020 Continuous monitoring of multiple submarine springs by means of gamma-ray spectrometry *J. Environ. Radioactiv.* **216** 106180

[28] Tsabaris C, Androulakaki E G, Prospathopoulos A, Alexakis S, Eleftheriou G, Patiris D L, Pappa F K, Sarantakos K, Kokkoris M and Vlastou R 2019 Development and optimization of an underwater *in situ* cerium bromide spectrometer for radioactivity measurements in the aquatic environment *J. Environ. Radioactiv.* **204** 12–20

[29] Tsabaris C, Patiris D L and Lykousis V K 2011 An *in situ* spectrometer for continuous monitoring of radon daughters in aquatic environment *Nucl. Instrum. Methods Phys. Res. Sect.* A **626** S142–4

[30] Tsabaris C, Androulakaki E G, Alexakis S and Patiris D L 2018 An *in situ* gamma-ray spectrometer for the deep ocean *Appl. Radiat. Isotopes* **142** 120–7

[31] Pavlidou A, Papadopoulos V P, Hatzianestis I, Simboura N, Patiris D and Tsabaris C 2014 Chemical inputs from a karstic submarine groundwater discharge (SGD) into an oligotrophic Mediterranean coastal area *Sci. Total Environ.* **488** 1–13

[32] Turekian K K 1977 Fate of metals in oceans *Geochim. Cosmochim. Acta* **41** 1139–44

[33] Gächter R and Müller B 2003 Why the phosphorus retention of lakes does not necessarily depend on the oxygen supply to their sediment surface *Limnol. Oceanogr.* **48** 929–33

[34] Wells M L, Price N M and Bruland K W 1995 Iron chemistry in seawater and its relationship to phytoplankton: a workshop report *Mar. Chem.* **48** 157–82

[35] Le Quéré C, Raupach M R, Canadell J G, Marland G, Bopp L, Ciais P, Conway T J, Doney S C, Feely R A, Foster P *et al* 2009 Trends in the sources and sinks of carbon dioxide *Nat. Geosci.* **2** 831–6

[36] Millero F J 2007 The marine inorganic carbon cycle *Chem. Rev.* **107** 308–41

[37] LaGier M J, Scholin C A, Fell J W, Wang J and Goodwin K D 2005 An electrochemical RNA hybridization assay for detection of the fecal indicator bacterium *Escherichia coli* *Mar. Pollut. Bull.* **50** 1251–61

[38] Haley S T, Cavender J F and Murray T E 1999 Detection of *Alexandrium tamarensis* by rapid PCR analysis *Biotechniques* **26** 88–91

[39] Li M, Xu K, Watanabe M and Chen Z 2007 Long-term variations in dissolved silicate, nitrogen, and phosphorus flux from the Yangtze River into the East China Sea and impacts on estuarine ecosystem *Estuar. Coast. Shelf Sci.* **71** 3–12

[40] Cao L, Xiang H, Xu J, Gao Y, Lin C, Li K, Li Z, Guo N, David P, He C *et al* 2022 Nutrient detection sensors in seawater based on ISI web of science database *J. Sens.* **2022** 5754751

[41] Legrand D C, Mas S, Jugeau B, David A and Barus C 2021 Silicate marine electrochemical sensor *Sens. Actuators* B **335** 129705

[42] Zhao C, Shen J B, Xu S S, Wei J, Liu H Q, Xie S Q, Pan Y J, Zhao Y and Zhu Y H 2022 Ultra-efficient trimethylamine gas sensor based on Au nanoparticles sensitized WO_3 nanosheets for rapid assessment of seafood freshness *Food Chem.* **392** 133318

[43] Pyz-Łukasik R, Chałabis-Mazurek A and Gondek M 2020 Basic and functional nutrients in the muscles of fish: a review *Int. J. Food Prop.* **23** 1941–50

[44] Moore T S, Mullaugh K M, Holyoke R R, Madison A S, Yucel M and Luther G W 2009 Marine chemical technology and sensors for marine waters: potentials and limits *Annu. Rev. Mar. Sci.* **1** 91–115

[45] Kroger S and Law R J 2005 Biosensors for marine applications—we all need the sea, but does the sea need biosensors? *Biosens. Bioelectron.* **20** 1903–13

[46] Kumari C R U, Samiappan D, Kumar R and Sudhakar T 2019 Fiber optic sensors in ocean observation: a comprehensive review *Optik* **179** 351–60

[47] Wang F, Zhu J M, Chen L F, Zuo Y F, Hu X J and Yang Y 2020 Autonomous and *in situ* ocean environmental monitoring on optofluidic platform *Micromachines* **11** 68

[48] Briciu-Burghina C, Power S, Delgado A and Regan F 2023 Sensors for coastal and ocean monitoring *Ann. Rev. Anal. Chem.* **16** 451–69

[49] Rogers K R 2006 Recent advances in biosensor techniques for environmental monitoring *Anal. Chim. Acta* **568** 222–31

[50] Yagi K 2007 Applications of whole-cell bacterial sensors in biotechnology and environmental science *Appl. Microbiol. Biotechnol.* **73** 1251–8

[51] Kara S, Keskinler B and Erhan E 2009 A novel microbial BOD biosensor developed by the immobilization of *P. Syringae* in micro-cellular polymers *J. Chem. Technol. Biotechnol.* **84** 511–8

[52] Chouteau C, Dzyadevych S, Durrieu C and Chovelon J M 2005 A bi-enzymatic whole cell conductometric biosensor for heavy metal ions and pesticides detection in water samples *Biosens. Bioelectron.* **21** 273–81

[53] Kumar S, Kundu S, Pakshirajan K and Dasu V V 2008 Cephalosporins determination with a novel microbial biosensor based on permeabilized *pseudomonas aeruginosa* whole cells *Appl. Biochem. Biotechnol.* **151** 653–64

[54] Ferrini A M, Mannoni V, Carpico G and Pellegrini G E 2008 Detection and identification of β-lactam residues in milk using a hybrid biosensor *J. Agric. Food Chem.* **56** 784–8

[55] Pickup J C, Hussain F, Evans N D, Rolinski O J and Birch D J S 2005 Fluorescence-based glucose sensors *Biosens. Bioelectron.* **20** 2555–65

[56] Stolper P, Fabel S, Weller M G, Knopp D and Niessner R 2008 Whole-cell luminescence-based flow-through biodetector for toxicity testing *Anal. Bioanal. Chem.* **390** 1181–7

[57] Virolainen N E, Pikkemaat M G, Elferink J W A and Karp M T 2008 Rapid detection of tetracyclines and their 4-epimer derivatives from poultry meat with bioluminescent biosensor bacteria *J. Agric. Food Chem.* **56** 11065–70

[58] Cheun B, Endo H, Hayashi T, Nagashima Y and Watanabe E 1996 Development of an ultra high sensitive tissue biosensor for determination of swellfish poisoning, tetrodotoxin *Biosens. Bioelectron.* **11** 1185

[59] Kulagina N V, Mikulski C M, Gray S, Ma W, Doucette G J, Ramsdell J S and Pancraziio J J 2006 Detection of marine toxins, Brevetoxin-3 and Saxitoxin, in seawater using neuronal networks *Environ. Sci. Technol.* **40** 578

[60] Hamada-Sato N, Minamitani N, Inaba Y, Nagashima Y, Kobayashi T, Imada C and Watanabe E 2004 Development of amperometric sensor system for measurement of diarrheic shellfish poisoning (DSP) toxin, okadaic acid *Sens. Mater.* **16** 99–107

[61] Ruppel C D and Kessler J D 2017 The interaction of climate change and methane hydrates *Rev. Geophys.* **55** 126–68

[62] Zhang J, Wang Y Y, Wang Y, Bai Y, Feng X, Zhu J H, Lu X H, Mu L W, Ming T Z, de Richter R *et al* 2022 Solar driven gas phase advanced oxidation processes for methane removal-challenges and perspectives *Chem. Eur. J.* **28** e202201984

[63] Judd A G 2004 Natural seabed gas seeps as sources of atmospheric methane *Environ. Geol.* **46** 988–96

[64] Liu H T, Zhan L S and Lu H L 2022 Mechanisms for upward migration of methane in marine sediments *Front. Mar. Sci.* **9** 1031096

[65] Kelley D S and Fruh-Green G L 1999 Abiogenic methane in deep-seated mid-ocean ridge environments: insights from stable isotope analyses *J. Geophys. Res. Solid Earth* **104** 10439–60

[66] Valentine D L, Kastner M, Wardlaw G D, Wang X C, Purdy A and Bartlett D H 2005 Biogeochemical investigations of marine methane seeps, Hydrate Ridge, Oregon *J. Geophys. Res. Biogeo.* **110** G02005

[67] Amouroux D, Roberts G, Rapsomanikis S and Andreae M O 2002 Biogenic gas (CH$_4$, N$_2$O, DMS) emission to the atmosphere from near-shore and shelf waters of the north-western Black Sea *Estuar. Coast. Shelf Sci.* **54** 575–87

[68] Rosentreter J A, Borges A V, Deemer B R, Holgerson M A, Liu S D, Song C L, Melack J, Raymond P A, Duarte C M, Allen G H *et al* 2021 Half of global methane emissions come from highly variable aquatic ecosystem sources *Nat. Geosci.* **14** 225–30

[69] Sloan E D 2003 Fundamental principles and applications of natural gas hydrates *Nature* **426** 353–9

[70] Chong Z R, Yang S H B, Babu P, Linga P and Li X S 2016 Review of natural gas hydrates as an energy resource: prospects and challenges *Appl. Energy* **162** 1633–52

[71] Gentz T, Damm E, von Deimling J S, Mau S, McGinnis D F and Schluter M 2014 A water column study of methane around gas flares located at the West Spitsbergen continental margin *Cont. Shelf Res.* **72** 107–18

[72] Westbrook G K *et al* 2009 Escape of methane gas from the seabed along the West Spitsbergen continental margin *Geophys. Res. Lett.* **36** L15608

[73] Boulart C, Mowlem M C, Connelly D P, Dutasta J P and German C R 2008 A novel, low-cost, high performance dissolved methane sensor for aqueous environments *Opt. Express* **16** 12607–17

[74] Liu L P *et al* 2019 Monitoring and research on environmental impacts related to marine natural gas hydrates: review and future perspective *J. Nat. Gas Sci. Eng.* **65** 82–107

[75] Greinert J, Artemov Y, Egorov V, De Batist M and McGinnis D 2006 1300-m-high rising bubbles from mud volcanoes at 2080 m in the Black Sea: hydroacoustic characteristics and temporal variability *Earth Planet. Sc. Lett.* **244** 1–15

[76] Sahling H *et al* 2014 Gas emissions at the continental margin west of Svalbard: mapping, sampling, and quantification *Biogeosciences* **11** 6029–46

[77] Weber T C, Mayer L, Jerram K, Beaudoin J, Rzhanov Y and Lovalvo D 2014 Acoustic estimates of methane gas flux from the seabed in a 6000 km^2 region in the Northern Gulf of Mexico *Geochem. Geophy. Geosy.* **15** 1911–25

[78] Jordt A, Zelenka C, von Deimling J S, Koch R and Koser K 2015 The bubble box: towards an automated visual sensor for 3D analysis and characterization of marine gas release sites *Sensors* **15** 30716–35

[79] Garcia-Betancourt M L and Masson M 2004 Environmental and geologic application of solid-state methane sensors *Environ. Geol.* **46** 1059–63

[80] Du Z F, Chen J, Ye W Q, Guo J J, Zhang X and Zheng R 2015 Investigation of two novel approaches for detection of sulfate ion and methane dissolved in sediment pore water using raman spectroscopy *Sensors* **15** 12377–88

[81] Hemond H and Camilli R 2002 NEREUS: engineering concept for an underwater mass spectrometer *Trac-Trends Anal. Chem.* **21** 526–33

[82] Tang Q J, Wang X Y, Zhang H, Liu Z and Guan X F 2022 Thin-film samarium nickelate as a potential material for methane sensing *J. Mater. Res.* **37** 3816–30

[83] Lechevallier L, Grilli R, Kerstel E, Romanini D and Chappellaz J 2019 Simultaneous detection of C_2H_6, CH_4, and delta C-13-CH_4 using optical feedback cavity-enhanced absorption spectroscopy in the mid-infrared region: towards application for dissolved gas measurements *Atmos. Meas. Tech.* **12** 3101–9

[84] Aleksanyan M S 2010 Methane sensor based on SnO_2/In_2O_3/TiO_2 nanostructure *J. Contemp. Phys. (Armen. Acad. Sci.)* **45** 77–80

[85] Fukasawa T, Hozumi S, Morita M, Oketani T and Masson M 2006 Dissolved methane sensor for methane leakage monitoring in methane hydrate production *Proc. the Oceans 2006 Conf. (Boston, MA)* 449–54

[86] Fukasawa T, Oketani T, Masson M, Groneman J, Hara Y and Hayashi M 2008 Optimized METS sensor for methane leakage monitoring *Proc. of the Int. Conf. OCEANS 2008 and MTS/IEEE Kobe Techno-Ocean (Kobe, Japan, 8–11 April)* pp 456–99

[87] Schubert M, Paschke A, Lieberman E and Burnett W C 2012 Air–water partitioning of Rn-222 and its dependence on water temperature and salinity *Environ. Sci. Technol.* **46** 3905–11

[88] Burnett W C *et al* 2006 Quantifying submarine groundwater discharge in the coastal zone via multiple methods *Sci. Total Environ.* **367** 498–543

[89] Li C Q *et al* 2022 Further refinements of a continuous radon monitor for surface ocean water measurements *Front. Mar. Sci.* **9** 1047126

[90] Corbett D R, Burnett W C, Cable P H and Clark S B 1998 A multiple approach to the determination of radon fluxes from sediments *J. Radioanal. Nucl. Chem.* **236** 247–52

[91] Chen Z, Li Y, Liu Z F, Wang J, Zhou X C and Du J G 2018 Radon emission from soil gases in the active fault zones in the Capital of China and its environmental effects *Sci Rep.* **8** 16772

[92] Quindos L S, Soto J and Villar E 1983 Determination of radon daughter products near land-sea discontinuities by gamma spectrometry *J. Aerosol Sci.* **14** 495–8

[93] Choubey V M, Bartarya S K and Ramola R C 2000 Radon in Himalayan springs: a geohydrological control *Environ. Geol.* **39** 523–30

[94] Elio J, Ortega M F, Nisi B, Mazadiego L F, Vaselli O, Caballero J, Quindos-Poncela L S, Sainz-Fernandez C and Pous J 2015 Evaluation of the applicability of four different radon measurement techniques for monitoring CO_2 storage sites *Int. J. Greenh. Gas Control* **41** 1–10

[95] Taguchi S, Tasaka S, Matsubara M, Osada K, Yokoi T and Yamanouchi T 2013 Air-sea gas transfer rate for the Southern Ocean inferred from Rn-222 concentrations in maritime air and a global atmospheric transport model *J. Geophys. Res. Atmos.* **118** 7606–16

[96] Lee K, Ha K, Lee S H, Yoon Y, Kim D H and Kim Y 2021 A continuous radon monitoring system for integration into the climate change observation network *J. Radioanal. Nucl. Chem.* **330** 547–54

[97] Fouladi-Fard R, Amraei A, Fahiminia M, Hosseini M R, Mahvi A H, Oskouei A O, Fiore M and Mohammadbeigi A 2020 Radon concentration and effective dose in

drinking groundwater and its relationship with soil type *J. Radioanal. Nucl. Chem.* **326** 1427–35

[98] Freyer K, Treutler H C, Dehnert J and Nestler W 1997 Sampling and measurement of radon-222 in water *J. Environ. Radioactiv.* **37** 327–37

[99] Stringer C E and Burnett W C 2004 Sample bottle design improvements for radon emanation analysis of natural waters *Health Phys.* **87** 642–6

[100] Key R M, Brewer R L, Stockwell J H, Guinasso N L and Schink D R 1979 Some improved techniques for measuring radon and radium in marine sediments and in seawater *Mar. Chem.* **7** 251–64

[101] Broecker W S, Li Y H and Cromwell J 1967 Radium-226 and Radon-222—concentration in Atlantic and Pacific Oceans *Science* **158** 1307–10

[102] Rose A L and Waite T D 2002 Kinetic model for Fe(II) oxidation in seawater in the absence and presence of natural organic matter *Environ. Sci. Technol.* **36** 433–44

[103] Fitzsimmons J N and Conway T M 2023 Novel insights into marine iron biogeochemistry from iron isotopes *Annu. Rev. Mar. Sci.* **15** 383–406

[104] Boye M, Nishioka J, Croot P, Laan P, Timmermans K R, Strass V H, Takeda S and de Baar H J W 2010 Significant portion of dissolved organic Fe complexes in fact is Fe colloids *Mar. Chem.* **122** 20–7

[105] Martin J H 1990 Glacial-interglacial CO_2 change: the iron hypothesis *Paleoceanography* **5** 1–13

[106] Barbeau K, Rue E L, Bruland K W and Butler A 2001 Photochemical cycling of iron in the surface ocean mediated by microbial iron(III)-binding ligands *Nature* **413** 409–13

[107] Frias-Espericueta M G, Voltolina D and Osuna-Lopez J I 2003 Acute toxicity of copper, zinc, iron, and manganese and of the mixtures copper–zinc and iron-manoanese to whiteleg shrimp *Litopenaeus vannamei* postlarvae *Bull. Environ. Contam. Toxicol.* **71** 68–74

[108] Dong Y Q, Song G L, Zhang J W, Gao Y H, Wang Z M and Zheng D J 2022 Biocorrosion induced by red-tide alga-bacterium symbiosis and the biofouling induced by dissolved iron for carbon steel in marine environment *J. Mater. Sci. Technol.* **128** 107–17

[109] Severmann S, McManus J, Berelson W M and Hammond D E 2010 The continental shelf benthic iron flux and its isotope composition *Geochim. Cosmochim. Acta* **74** 3984–4004

[110] Aller R C, Heilbrun C, Panzeca C, Zhu Z B and Baltzer F 2004 Coupling between sedimentary dynamics, early diagenetic processes, and biogeochemical cycling in the Amazon-Guianas mobile mud belt: coastal French Guiana *Mar. Geol.* **208** 331–60

[111] Millero F J, Yao W S and Aicher J 1995 The speciation of Fe(II) and Fe(III) in natural waters *Mar. Chem.* **50** 21–39

[112] Stookey L L 1970 Ferrozine—a new spectrophotometric reagent for iron *Anal. Chem.* **42** 779–81

[113] Stockdale A, Davison W and Zhang H 2009 Micro-scale biogeochemical heterogeneity in sediments: a review of available technology and observed evidence *Earth-Sci. Rev.* **92** 81–97

[114] Blomqvist S 1991 Quantitative sampling of soft-bottom sediments—problems and solutions *Mar. Ecol. Prog. Ser.* **72** 295–304

[115] Troup B N, Bricker O P and Bray J T 1974 Oxidation effect on the analysis of iron in the interstitial water of recent anoxic sediments *Nature* **249** 237–9

[116] Suarez E M, Lepkova K, Forsyth M, Tan M K Y, Kinsella B and Machuca L L 2022 *In situ* investigation of under-deposit microbial corrosion and its inhibition using a multi-electrode array system *Front. Bioeng. Biotech.* **9** 803610

[117] Tercier-Waeber M L, Buffle J, Confalonieri F, Riccardi G, Sina A, Graziottin F, Fiaccabrino G C and Koudelka-Hep M 1999 Submersible voltammetric probes for *in situ* real-time trace element measurements in surface water, groundwater and sediment-water interface *Meas. Sci. Technol.* **10** 1202–13

[118] Tercier M L, Buffle J and Graziottin F 1998 Novel voltammetric *in situ* profiling system for continuous real-time monitoring of trace elements in natural waters *Electroanal* **10** 355–63

[119] Howell K A, Achterberg E P, Braungardt C B, Tappin A D, Turner D R and Worsfold P J 2003 The determination of trace metals in estuarine and coastal waters using a voltammetric *in situ* profiling system *Analyst* **128** 734–41

[120] Lin M Y, Hu X P, Pan D W and Han H T 2018 Determination of iron in seawater: from the laboratory to *in situ* measurements *Talanta* **188** 135–44

[121] Tercier-Waeber M L, Confalonieri F, Riccardi G, Sina A, Noel S, Buffle J and Graziottin F 2005 Multi physical–chemical profiler for real-time *in situ* monitoring of trace metal speciation and master variables: development, validation and field applications *Mar. Chem.* **97** 216–35

[122] Ma S, Luther G W, Keller J, Madison A S, Metzger E, Emerson D and Megonigal J P 2008 Solid-state Au/Hg microelectrode for the investigation of Fe and mn cycling in a freshwater wetland: implications for methane production *Electroanal* **20** 233–9

[123] Anschutz P, Sundby B, Lefrancois L, Luther G W and Mucci A 2000 Interactions between metal oxides and species of nitrogen and iodine in bioturbated marine sediments *Geochim. Cosmochim. Acta* **64** 2751–63

[124] Brendel P J and Luther G W 1995 Development of a gold amalgam voltammetric microelectrode for the determination of dissolved Fe, Mn, O_2, and S(-II) in porewaters of marine and freshwater sediments *Environ. Sci. Technol.* **29** 751–61

[125] Amato E D, Simpson S L, Belzunce-Segarra M J, Jarolimek C V and Jolley D F 2015 Metal fluxes from porewaters and labile sediment phases for predicting metal exposure and bioaccumulation in benthic invertebrates *Environ. Sci. Technol.* **49** 14204–12

[126] Fones G R, Davison W and Hamilton-Taylor J 2004 The fine-scale remobilization of metals in the surface sediment of the North-East Atlantic *Cont. Shelf Res.* **24** 1485–504

[127] Guan D, Wei T, Yuan Z, Li G and Chen Z 2021 A review of researches on bioavailability and interfacial processes of arsenic based on passive sampling techniques: progress and prospect *Acta Pedol. Sin.* **58** 344–56

[128] Harper M P, Davison W and Tych W 1997 Temporal, spatial, and resolution constraints for *in situ* sampling devices using diffusional equilibrium: dialysis and DET *Environ. Sci. Technol.* **31** 3110–9

[129] Li C, Ding S M, Yang L Y, Wang Y, Ren M Y, Chen M S, Fan X F and Lichtfouse E 2019 Diffusive gradients in thin films: devices, materials and applications *Environ. Chem. Lett.* **17** 801–31

[130] Metzger E, Viollier E, Simonucci C, Prevot F, Langlet D and Jezequel D 2013 Millimeter-scale alkalinity measurement in marine sediment using DET probes and colorimetric determination *Water Res.* **47** 5575–83

[131] Pages A, Teasdale P R, Robertson D, Bennett W W, Schafer J and Welsh D T 2011 Representative measurement of two-dimensional reactive phosphate distributions and co-distributed iron(II) and sulfide in seagrass sediment porewaters *Chemosphere* **85** 1256–61

[132] Zhu Q Z and Aller R C 2012 Two-dimensional dissolved ferrous iron distributions in marine sediments as revealed by a novel planar optical sensor *Mar. Chem.* **136** 14–23

[133] Woosley R J, Millero F J and Wanninkhof R 2016 Rapid anthropogenic changes in CO_2 and pH in the Atlantic Ocean: 2003–2014 *Glob. Biogeochem. Cycles* **30** 70–90

[134] Thomas H, Prowe A E F, van Heuven S, Bozec Y, de Baar H J W, Schiettecatte L S, Suykens K, Kone M, Borges A V, Lima I D *et al* 2007 Rapid decline of the CO_2 buffering capacity in the North Sea and implications for the North Atlantic Ocean *Glob. Biogeochem. Cycles* **21** GB4001

[135] Iversen M H 2023 Carbon export in the ocean: a biologist's perspective *Annu. Rev. Mar. Sci.* **15** 357–81

[136] Marion G M, Millero F J, Camões M F, Spitzer P, Feistel R and Chen C T A 2011 PH of seawater *Mar. Chem.* **126** 89–96

[137] Clarke J S, Achterberg E P, Connelly D P, Schuster U and Mowlem M 2017 Developments in marine pCO_2 measurement technology; towards sustained *in situ* observations *Trac-Trends Anal. Chem.* **88** 53–61

[138] Graziani S, Beaubien S E, Bigi S and Lombardi S 2014 Spatial and temporal pCO(2) marine monitoring near Panarea Island (Italy) using multiple low-cost gaspro sensors *Environ. Sci. Technol.* **48** 12126–33

[139] Borisov S M, Seifner R and Klimant I 2011 A novel planar optical sensor for simultaneous monitoring of oxygen, carbon dioxide, pH and temperature *Anal. Bioanal. Chem.* **400** 2463–74

[140] Schrumpf A, Lengerer A, Schmid N and Kazda M 2021 Portable measurement system for *in situ* estimation of oxygen and carbon fluxes of submerged plants *Front. Plant Sci.* **12** 765089

[141] Zhang Z H, Li M, Yang T, Zang Z X, Li N, Zheng R E and Guo J J 2023 Seconds-scale response sensor for *in situ* oceanic carbon dioxide detection *Anal. Chem.* **95** 3577–86

[142] Li M, Du B L, Guo J J, Zhang Z H, Lu Z Y and Zheng R E 2022 A low-cost *in situ* CO_2 sensor based on a membrane and NDIR for long-term measurement in seawater *J. Oceanol. Limnol.* **40** 986–98

[143] Bergamasco A, Nguyen H Q, Caruso G, Xing Q G and Carol E 2021 Advances in water quality monitoring and assessment in marine and coastal regions *Water* **13** 1926

[144] Lee J, Choi J, Fatka M, Swanner E, Ikuma K, Liang X W, Leung T and Howe A 2020 Improved detection of mcyA genes and their phylogenetic origins in harmful algal blooms *Water Res.* **176** 115730

[145] Tang T, Effiong K, Hu J, Li C and Xiao X 2021 Chemical prevention and control of the green tide and fouling organism ulva: key chemicals, mechanisms, and applications *Front. Mar. Sci.* **8** 618950

[146] Penna A and Magnani M 1999 Identification of *Alexandrium* (Dinophyceae) species using PCR and rDNA-targeted probes *J. Phycol.* **35** 615–21

[147] Metfies K, Huljic S, Lange M and Medlin L K 2005 Electrochemical detection of the toxic dinoflagellate *Alexandrium ostenfeldii* with a DNA-biosensor *Biosens. Bioelectron.* **20** 1349–57

[148] Guillou L, Nezan E, Cueff V, Denn E E L, Cambon-Bonavita M A, Gentien P and Barbier G 2002 Genetic diversity and molecular detection of three toxic dinoflagellate genera (*Alexandrium*, *Dinophysis*, and *Karenia*) from French coasts *Protist* **153** 223–38

[149] Kim C J and Sako Y 2005 Molecular identification of toxic *Alexandrium tamiyavanichii* (Dinophyceae) using two DNA probes *Harmful Algae* **4** 984–91

[150] Diercks S, Metfies K and Medlin L K 2008 Development and adaptation of a multiprobe biosensor for the use in a semi-automated device for the detection of toxic algae *Biosens. Bioelectron.* **23** 1527–33

[151] McCoy G R, McNamee S, Campbell K, Elliott C T, Fleming G T A and Raine R 2014 Monitoring a toxic bloom of *Alexandrium minutum* using novel microarray and multiplex surface plasmon resonance biosensor technology *Harmful Algae* **32** 40–8

[152] Liu Y S, Deng Y L, Chen C K, Khoo B L and Chua S L 2022 Rapid detection of microorganisms in a fish infection microfluidics platform *J. Hazard. Mater.* **431** 128572

[153] Liu J X, Zhang X Y, Zheng J Y, Yu Y P, Huang X H, Wei J G, Mukama O, Wang S W and Qin Q W 2021 A lateral flow biosensor for rapid detection of Singapore grouper iridovirus (SGIV) *Aquaculture* **541** 736756

[154] Zielinski O, Busch J A, Cembella A D, Daly K L, Engelbrektsson J, Hannides A K and Schmidt H 2009 Detecting marine hazardous substances and organisms: sensors for pollutants, toxins, and pathogens *Ocean Sci.* **5** 329–49

[155] Doni L, Martinez-Urtaza J and Vezzulli L 2023 Searching pathogenic bacteria in the rare biosphere of the ocean *Curr. Opin. Biotechnol.* **80** 102894

[156] Dutta D, Kaushik A, Kumar D and Bag S 2021 Foodborne pathogenic vibrios: antimicrobial resistance *Front. Microbiol.* **12** 638331

[157] Kay D, Bartram J, Pruss A, Ashbolt N, Wyer M D, Fleisher J M, Fewtrell L, Rogers A and Rees G 2004 Derivation of numerical values for the World Health Organization guidelines for recreational waters *Water Res.* **38** 1296–304

[158] Straub T M and Chandler D P 2003 Towards a unified system for detecting waterborne pathogens *J. Microbiol. Meth.* **53** 185–97

[159] Hong S R, Jeong H D and Hong S 2010 QCM DNA biosensor for the diagnosis of a fish pathogenic virus VHSV *Talanta* **82** 899–903

[160] Toldra A, Furones M D, O'Sullivan C K and Campas M 2020 Detection of isothermally amplified ostreid herpesvirus 1 DNA in Pacific oyster (*Crassostrea gigas*) using a miniaturised electrochemical biosensor *Talanta* **207** 120308

[161] Kuznetsova O V and Timerbaev A R 2022 Marine sediment analysis—a review of advanced approaches and practices focused on contaminants *Anal. Chim. Acta* **1209** 339640

[162] Saravanakumar K, SivaSantosh S, Sathiyaseelan A, Naveen K V, AfaanAhamed M A, Zhang X, Priya V V, MubarakAli D and Wang M H 2022 Unraveling the hazardous impact of diverse contaminants in the marine environment: detection and remedial approach through nanomaterials and nano-biosensors *J. Hazard. Mater.* **433** 128720

[163] Kroger S, Piletsky S and Turner A P F 2002 Biosensors for marine pollution research, monitoring and control *Mar. Pollut. Bull.* **45** 24–34

[164] Camara J S, Montesdeoca-Esponda S, Freitas J, Guedes-Alonso R, Sosa-Ferrera Z and Perestrelo R 2021 Emerging contaminants in seafront zones. environmental impact and analytical approaches *Separations* **8** 95

[165] Justino C I L, Freitas A C, Duarte A C and Santos T 2015 Sensors and biosensors for monitoring marine contaminants *Trends Environ. Anal.* **6** 21–30

[166] Han S, Zhang Q, Zhang X C, Liu X M, Lu L, Wei J F, Li Y C, Wang Y H and Zheng G X 2019 A digital microfluidic diluter-based microalgal motion biosensor for marine pollution monitoring *Biosens. Bioelectron.* **143** 111597

[167] Pfannkuche J, Lubecki L, Schmidt H, Kowalewska G and Kronfeldt H D 2012 The use of surface-enhanced Raman scattering (SERS) for detection of PAHs in the Gulf of Gdansk (Baltic Sea) *Mar. Pollut. Bull.* **64** 614–26

[168] Ronan J M and McHugh B 2013 A sensitive liquid chromatography/tandem mass spectrometry method for the determination of natural and synthetic steroid estrogens in seawater and marine biota, with a focus on proposed water framework directive environmental quality standards *Rapid Commun. Mass Spectrom.* **27** 738–46

[169] Sanchez-Avila J, Quintana J, Ventura F, Tauler R, Duarte C M and Lacorte S 2010 Stir bar sorptive extraction-thermal desorption-gas chromatography–mass spectrometry: an effective tool for determining persistent organic pollutants and nonylphenol in coastal waters in compliance with existing directives *Mar. Pollut. Bull.* **60** 103–12

[170] Lisboa N S, Fahning C S, Cotrim G, dos Anjos J P, de Andrade J B, Hatje V and da Rocha G O 2013 A simple and sensitive UFLC-fluorescence method for endocrine disrupters determination in marine waters *Talanta* **117** 168–75

[171] Kotke D, Gandrass J, Xie Z Y and Ebinghaus R 2019 Prioritised pharmaceuticals in German estuaries and coastal waters: occurrence and environmental risk assessment *Environ. Pollut.* **255** 113161

[172] Humbert K, Debret M, Morin C, Cosme J and Portet-Koltalo F 2022 Direct thermal desorption-gas chromatography-tandem mass spectrometry versus microwave assisted extraction and GC–MS for the simultaneous analysis of polyaromatic hydrocarbons (PAHs, PCBs) from sediments *Talanta* **250** 123735

[173] Monaghan J, Xin Q, Aplin R, Jaeger A, Heshka N E, Hounjet L J, Gill C G and Krogh E T 2022 Aqueous naphthenic acids and polycyclic aromatic hydrocarbons in a meso-scale spill tank affected by diluted bitumen analyzed directly by membrane introduction mass spectrometry *J. Hazard. Mater.* **440** 129798

[174] Schnitzler L, Zarzycki J, Gerhard M, Konde S, Rexer K-H, Erb T J, Maier U G, Koch M, Hofmann M R and Moog D 2021 Lensless digital holographic microscopy as an efficient method to monitor enzymatic plastic degradation *Mar. Pollut. Bull.* **163** 111950

[175] Peron O, Rinnert E, Lehaitre M, Crassous P and Compere C 2009 Detection of polycyclic aromatic hydrocarbon (PAH) compounds in artificial sea-water using surface-enhanced Raman scattering (SERS) *Talanta* **79** 199–204

[176] Kolomijeca A, Kronfeldt H D and Kwon Y H 2013 A portable surface enhanced Raman spectroscopy (SERS) sensor system applied for seawater and sediment investigations on an arctic sea-trial *Int. J. Offshore Polar* **23** ISOPE–13-23-3-161

[177] Wang H, Wang G C and Gu W H 2020 Macroalgal blooms caused by marine nutrient changes resulting from human activities *J. Appl. Ecol.* **57** 766–76

[178] Hong H S, Chen N W and Wang D L 2015 River-estuary-coast continuum: biogeochemistry and ecological response to increasing human and climatic changes—editorial overview *Estuar. Coast. Shelf Sci.* **166** 144–5

[179] Hu W, Li C H, Ye C, Wang J, Wei W W and Deng Y 2019 Research progress on ecological models in the field of water eutrophication: CiteSpace analysis based on data from the ISI web of science database *Ecol. Model.* **410** 108779

[180] Nehir M, Esposito M, Begler C, Franke C, Zielinski O and Achterberg E P 2021 Improved calibration and data processing procedures of OPUS optical sensor for high-resolution *in situ* monitoring of nitrate in seawater *Front. Mar. Sci.* **8** 663800

[181] Chen X Y, Zhou G H, Mao S and Chen J H 2018 Rapid detection of nutrients with electronic sensors: a review *Environ. Sci. Nano* **5** 837–62

[182] Nightingale A M, Beaton A D and Mowlem M C 2015 Trends in microfluidic systems for *in situ* chemical analysis of natural waters *Sens. Actuator* B **221** 1398–405

[183] Prien R D 2007 The future of chemical *in situ* sensors *Mar. Chem.* **107** 422–32

[184] Daly K L, Byrne R H, Dickson A G, Gallager S M, Perry M J and Tivey M K 2004 Chemical and biological sensors for time-series research: current status and new directions *Mar. Technol. Soc. J.* **38** 121–43

[185] Daniel A *et al* 2020 Toward a harmonization for using *in situ* nutrient sensors in the marine environment *Front. Mar. Sci.* **6** 773

[186] Aguilar D *et al* 2015 Silicon-based electrochemical microdevices for silicate detection in seawater *Sens. Actuators* B **211** 116–24

[187] Jonca J, Fernandez V L, Thouron D, Paulmier A, Graco M and Garcon V 2011 Phosphate determination in seawater: toward an autonomous electrochemical method *Talanta* **87** 161–7

[188] Johnson K S, Needoba J A, Riser S C and Showers W J 2007 Chemical sensor networks for the aquatic environment *Chem. Rev.* **107** 623–40

[189] Ye Z P, Yang J Q, Zhong N, Tu X, Jia J N and Wang J D 2020 Tackling environmental challenges in pollution controls using artificial intelligence: a review *Sci. Total Environ.* **699** 134279

[190] Grand M M *et al* 2019 Developing autonomous observing systems for micronutrient trace metals *Front. Mar. Sci.* **6** 35

[191] Yang Z M, Li C, Chen F, Liu C, Cai Z F, Cao W X and Li Z H 2022 An *in situ* analyzer for long-term monitoring of nitrite in seawater with versatile liquid waveguide capillary cells: development, optimization and application *Mar. Chem.* **245** 104149

[192] Altahan M F, Esposito M and Achterberg E P 2022 Improvement of on-site sensor for simultaneous determination of phosphate, silicic acid, nitrate plus nitrite in seawater *Sensors* **22** 3479

[193] Cuartero M, Crespo G, Cherubini T, Pankratova N, Confalonieri F, Massa F, Tercier-Waeber M L, Abdou M, Schafer J and Bakker E 2018 *In situ* detection of macronutrients and chloride in seawater by submersible electrochemical sensors *Anal. Chem.* **90** 4702–10

[194] Barus C, Romanytsia I, Striebig N and Garcon V 2016 Toward an *in situ* phosphate sensor in seawater using square wave voltammetry *Talanta* **160** 417–24

[195] Machado M C, Vimbela G V and Tripathi A 2021 Creation of a low cost, low light bioluminescence sensor for real time biological nitrate sensing in marine environments *Environ. Technol.* **43** 4002–9

[196] Cozzolino D and Murray I 2012 A review on the application of infrared technologies to determine and monitor composition and other quality characteristics in raw fish, fish products, and seafood *Appl. Spectrosc. Rev.* **47** 207–18

[197] Li J R, Lu H X, Zhu J L, Wang Y B and Li X P 2009 Aquatic products processing industry in China: challenges and outlook *Trends Food Sci. Tech.* **20** 73–7

[198] Parlapani F F, Mallouchos A, Haroutounian S A and Boziaris I S 2014 Microbiological spoilage and investigation of volatile profile during storage of sea bream fillets under various conditions *Int. J. Food Microbiol.* **189** 153–63

[199] Li X P, Chen Y, Cai L Y, Xu Y X, Yi S M, Zhu W H, Mi H B, Li J R and Lin H 2017 Freshness assessment of turbot (Scophthalmus maximus) by Quality Index Method (QIM), biochemical, and proteomic methods *LWT Food Sci. Technol.* **78** 172–80

[200] Yao L, Luo Y K, Sun Y Y and Shen H X 2011 Establishment of kinetic models based on electrical conductivity and freshness indictors for the forecasting of crucian carp (*Carassius carassius*) freshness *J. Food Eng.* **107** 147–51

[201] Onal A 2007 A review: current analytical methods for the determination of biogenic amines in foods *Food Chem.* **103** 1475–86

[202] Cheng J H, Sun D W, Zeng X A and Liu D 2015 Recent advances in methods and techniques for freshness quality determination and evaluation of fish and fish fillets: a review *Crit. Rev. Food Sci.* **55** 1012–25

[203] Cheng J H, Sun D W, Han Z and Zeng X A 2014 Texture and structure measurements and analyses for evaluation of fish and fillet freshness quality: a review *Compr. Rev. Food Sci. Food Saf.* **13** 52–61

[204] Franceschelli L, Berardinelli A, Dabbou S, Ragni L and Tartagni M 2021 Sensing technology for fish freshness and safety: a review *Sensors* **21** 1373

[205] Hassoun A and Karoui R 2017 Quality evaluation of fish and other seafood by traditional and nondestructive instrumental methods: advantages and limitations *Crit. Rev. Food Sci.* **57** 1976–98

[206] Karunathilaka S R, Ellsworth Z and Yakes B J 2021 Detection of decomposition in mahi-mahi, croaker, red snapper, and weakfish using an electronic-nose sensor and chemometric modeling *J. Food Sci.* **86** 4148–58

[207] Mustafa F, Othman A and Andreescu S 2021 Cerium oxide-based hypoxanthine biosensor for fish spoilage monitoring *Sensors Actuator* B **332** 129435

[208] Chen Y Q, Li Y X, Feng B X, Wu Y, Zhu Y H and Wei J 2022 Self-templated synthesis of mesoporous Au-ZnO nanospheres for seafood freshness detection *Sens. Actuators* B **360** 131662

[209] Torrarit K, Kongkaew S, Samoson K, Kanatharana P, Thavarungkul P, Chang K H, Abdullah A F L and Limbut W 2022 Flow injection amperometric measurement of formalin in seafood *ACS Omega* **7** 17679–91

[210] Chang L Y, Chuan M Y, Zan H W, Meng H F, Lu C J, Yeh P H and Chen J N 2017 One-minute fish freshness evaluation by testing the volatile amine gas with an ultrasensitive porous-electrode-capped organic gas sensor system *ACS Sens.* **2** 531–9

[211] Lee K M, Son M, Kang J H, Kim D, Hong S, Park T H, Chun H S and Choi S S 2018 A triangle study of human, instrument and bioelectronic nose for non-destructive sensing of seafood freshness *Sci Rep.* **8** 547